T4-AHR-524

Department of Economic
and Social Affairs

Statistics Division   Series V No. 31

World
Statistics
Pocketbook

Containing data available
as of December 2006

United Nations   New York 2007

The **Department of Economic and Social Affairs** of the United Nations Secretariat is a vital interface between global policies in the economic, social and environmental spheres and national action. The Department works in three main interlinked areas: (i) it compiles, generates and analyses a wide range of economic, social and environmental data and information on which States Members of the United Nations draw to review common problems and to take stock of policy options; (ii) it facilitates the negotiations of Member States in many intergovernmental bodies on joint courses of action to address ongoing or emerging global challenges; and (iii) it advises interested Governments on the ways and means of translating policy frameworks developed in United Nations conferences and summits into programmes at the country level and, through technical assistance, helps build national capacities.

Note

The designations employed and the presentation of material in this publication do not imply the expression of any opinion whatsoever on the part of the Secretariat of the United Nations concerning the legal status of any country, territory, city or area or of its authorities, or concerning the delimitation of its frontiers or boundaries.

The term "country" as used in this publication also refers, as appropriate, to territories or areas.

The designations "developed" and "developing" regions are intended for statistical convenience and do not necessarily express a judgement about the stage reached by a particular country or area in the development process.

Visit the United Nations World Wide Web site on the Internet:
For the Department of Economic and Social Affairs,
http://www.un.org/esa/desa/
For statistics and statistical publications,
http://unstats.un.org/unsd/
For UN publications, https://unp.un.org/

ST/ESA/STAT/SER.V/31
United Nations Publication
Sales No. E.07.XVII.4
ISBN-13: 978-92-1-161499-2
Inquiries should be directed to:
United Nations Publications
New York, NY 10017

Preface

This World Statistics Pocketbook is the twenty-sixth compilation of basic economic and social indicators for countries and areas of the world, prepared by the United Nations Statistics Division, Department of Economic and Social Affairs. It responds to General Assembly resolution 2626 (XXV), in which the Secretary-General is requested to supply basic national data that will increase international public awareness of countries' development efforts.

The indicators shown are selected from the wealth of international statistical information compiled regularly by the Statistics Division and Population Division of the United Nations, and the statistical services of the United Nations specialized agencies and of other international organizations and institutions.

This issue of the World Statistics Pocketbook generally covers the years 2000 and 2006. The statistics included for each year shown are those most recently compiled and made available by the international statistical services from official national sources, supplemented by international estimates in some fields. Statistical sources and methods are described in the section "Technical notes", in the "Data dictionary" and in footnotes. Statistics presented are in general the latest available to the United Nations Statistics Division as of December 2006.

Readers wishing to consult more detailed statistics and descriptions of technical methods used in their collection and compilation are referred to the more specialized publications listed in the Introduction and in the reference lists at the end of this publication.

Introduction

Considerable progress has been made in the last two decades towards standardization of statistical definitions worldwide, for example in the wide scope of topics covered by the 1968 and 1993 versions of the System of National Accounts[1], the United Nations Principles and Recommendations for Population and Housing Censuses[2] and the recommendations on statistics of the International Labour Organization among many. The internationally recommended definitions used in the present publication are given in the "Data dictionary", beginning on p. 227, with citations to the original sources.

In addition, the section "Technical notes", beginning on p. 217, contains short descriptions of the sources for the indicators presented here and the methodology used in their compilation at national and international levels, and describes some of their limitations and differences from international standards.

Readers interested in more detailed time-series and data should consult the following major publications:

United Nations
  Statistical Yearbook
  United Nations publication, Series S [16]*
  Demographic Yearbook
  United Nations publication, Series R [10]
  National Accounts Statistics
  United Nations publication, Series X [14], [15]
  International Trade Statistics Yearbook
  United Nations publication, Series G [12]
  Energy Statistics Yearbook
  United Nations publication, Series J [11]
  Monthly Bulletin of Statistics
  United Nations publication, Series Q [13]
  World Population Prospects (biennial)
  United Nations publication [19]
  World Urbanization Prospects (biennial)
  United Nations publication [18]

Food and Agriculture Organization of the
United Nations (Rome)
  FAO Yearbook: Production [2]
International Labour Office (Geneva)
  Key Indicators of the Labour Market [4]
  Yearbook of Labour Statistics [7]
International Monetary Fund (Washington DC)
  *International Financial Statistics (monthly
  and annual)* [6]
United Nations Educational, Scientific and Cultural
Organization (UNESCO) Institute for Statistics, Montreal,
  the *UNESCO Institute Statistics Database* [20]
World Tourism Organization (Madrid)
  Yearbook of Tourism Statistics [25]

The *World Statistics Pocketbook* is prepared by the
Statistical Services Branch of the Statistics Division,
Department of Economic and Social Affairs of the United
Nations Secretariat. The editor is Virgilio Castillo.
Salomon Cameo is the software developer. They are
assisted by Paul Narain.

* Numbers in brackets refer to numbered entries listed in "Statistical
sources" at the end of this publication.
1 United Nations, Commission of the European Communities, International
Monetary Fund, Organisation for Economic Co-operation and Development
and World Bank (1994), *System of National Accounts 1993* (*SNA 1993*)
(joint publication, United Nations publication Sales No. E.94.XVII.4).
2 United Nations (1998), *Principles and Recommendations for Population
and Housing Censuses Revision 1*, Statistical Office, Series M, No. 67,
Rev.1 (United Nations publication, Sales No. E.98.XVII.8).

Contents

viii

Explanatory notes and abbreviations

...	Data not available
—	Magnitude zero
<	Magnitude not zero, but less than half of the unit employed
—<	Magnitude not zero, but negative and less than half of the unit employed
p.a.	Per annum
Km^2	Square kilometre
000 Mt	Thousand metric tons
	Decimal figures are always preceded by a period (.).

Conversion coefficients and factors

The metric system of weights and measures has been employed in *World Statistics Pocketbook*. The following table shows the equivalents of the basic metric, British imperial and United States units of measurement:

Area	1 square kilometre	= 0.386102 square mile
Weight or mass	1 ton	= 1.102311 short tons, or
		= 0.987207 long ton
	1 kilogram	= 35.273962 avdp. ounces
		= 2.204623 avdp. pounds
Distance	1 kilometre	= 0.621371 mile

Country and area tables

Afghanistan

Region	South-central Asia
Largest urban agglom. (pop., 000s)	Kabul (2994)
Currency	afghani
Population in 2006 (proj., 000s)	31082
Surface area (square kms)	652090
Population density (per square km)	48
United Nations membership date	19 November 1946

Economic indicators	2000	2005
Exchange rate (national currency per US$) [ab]	3000.00	50.10[c]
Tourist arrivals (000s)	4[d]	...
GDP (million current US$)	2963	6504
GDP (per capita current US$)	125	218
Gross fixed capital formation (% of GDP)	15.0	16.0
Primary energy production (000s Mt oil equiv.)	136	75[e]
Motor vehicles (per 1,000 inhabitants)	0.6	0.5[f]
Telephone lines (per 100 inhabitants)	0.1	0.3
Internet users, estimated (000s)	1.0	30.0

Total trade		Major trading partners			2005
	(million US$)	(% of exports)		(% of imports)	
Exports	365.0	Pakistan	82	Japan	19
Imports	2218.0	India	6	Pakistan	18
		Russian Fed.	4	China	14

Social indicators	2000-2006
Population growth rate 2000-2005 (% per annum)	4.6
Population aged 0-14 years (%)	47.0
Population aged 60+ years (women and men, % of total)	5.0/4.0
Sex ratio (women per 100 men)	94
Life expectancy at birth 2000-2005 (women and men, years)	46/46
Infant mortality rate 2000-2005 (per 1,000 births)	149
Total fertility rate 2000-2005 (births per woman)	7.5
Contraceptive use (% of currently married women)	5[g]
Urban population (%)	23
Urban population growth rate 2000-2005 (% per annum)	6.1
Rural population growth rate 2000-2005 (% per annum)	4.2
Refugees and others of concern to UNHCR[h]	911679
Government education expenditure (% of GNP)	1.6[i]
Primary-secondary gross enrolment ratio (w and m per 100)	33/82
Third-level students (women and men, % of total)	20/80
Television receivers (per 1,000 inhabitants)	67
Parliamentary seats (women and men, % of total)	26/74

Environment	2000-2006
Threatened species	38
Forested area (% of land area)	2
CO2 emissions (000s Mt of carbon dioxide/per capita)	704/0.0
Energy consumption per capita (kilograms oil equiv.)	12
Precipitation (mm)	312[j]
Average minimum and maximum temperatures (centigrade)[j]	5.5/19.6

a Afghanistan redenominated its currency, Afghani in 2002. The old 1000 Afghani is 1 Afghani. b Principal rate. c September 2006. d 1997. e 2004. f 2002. g The data refer only to the Eastern Region, South-eastern Region, and two provinces of the Central Region of Afghanistan. h Provisional. i 1981. j Kabul.

Albania

Region	Southern Europe
Largest urban agglom. (pop., 000s)	Tirana (388)
Currency	lek
Population in 2006 (proj., 000s)	3147
Surface area (square kms)	28748
Population density (per square km)	109
United Nations membership date	14 December 1955

Economic indicators	2000	2005
Exchange rate (national currency per US$)	142.64	95.59[a]
Consumer price index (2000=100)	100	119[b]
Unemployment (percentage of labour force)	16.8	14.4[c]
Balance of payments, current account (million US$)	−156	−571
Tourist arrivals (000s)	32	42[c]
GDP (million current US$)	3709	8538
GDP (per capita current US$)	1211	2728
Gross fixed capital formation (% of GDP)	39.0	46.0
Labour force participation, adult female pop. (%)	49.7[d]	49.3[e]
Labour force participation, adult male pop. (%)	74.8[d]	71.2[e]
Employment in industrial sector (%)	6.6	13.5
Employment in agricultural sector (%)	71.8	58.4
Agricultural production index (1999-2001=100)	100	104[c]
Food production index (1999-2001=100)	100	105[c]
Primary energy production (000s Mt oil equiv.)	729	931[c]
Motor vehicles (per 1,000 inhabitants)	51.4	85.2[f]
Telephone lines (per 100 inhabitants)	4.9	8.6
Internet users, estimated (000s)	3.5	188.0

Total trade		Major trading partners			2005
	(million US$)	(% of exports)		(% of imports)	
Exports	662.2	Italy	72	Italy	29
Imports	2634.6	Greece	10	Greece	16
		Germany	3	Turkey	7

Social indicators	2000-2006
Population growth rate 2000-2005 (% per annum)	0.4
Population aged 0-14 years (%)	27.0
Population aged 60+ years (women and men, % of total)	13.0/11.0
Sex ratio (women per 100 men)	102
Life expectancy at birth 2000-2005 (women and men, years)	77/71
Infant mortality rate 2000-2005 (per 1,000 births)	25
Total fertility rate 2000-2005 (births per woman)	2.3
Contraceptive use (% of currently married women)	75[g]
Urban population (%)	45
Urban population growth rate 2000-2005 (% per annum)	2.1
Rural population growth rate 2000-2005 (% per annum)	−0.8
Refugees and others of concern to UNHCR[h]	91
Government education expenditure (% of GNP)	2.8
Primary-secondary gross enrolment ratio (w and m per 100)	86/88
Third-level students (women and men, % of total)	62/38
Television receivers (per 1,000 inhabitants)	318
Intentional homicides (per 100,000 inhabitants)	7
Parliamentary seats (women and men, % of total)	7/93

Environment	2000-2006
Threatened species	43
Forested area (% of land area)	36
CO2 emissions (000s Mt of carbon dioxide/per capita)	3045/1.0
Energy consumption per capita (kilograms oil equiv.)	551
Average minimum and maximum temperatures (centigrade)	1.8/30.7

a August 2006. b June 2006. c 2004. d 2001. e 2002. f 2003. g For women aged 15-44 in union or married. h Provisional.

Algeria

Region	Northern Africa
Largest urban agglom. (pop., 000s)	Algiers (3200)
Currency	dinar
Population in 2006 (proj., 000s)	33354
Surface area (square kms)	2381741
Population density (per square km)	14
United Nations membership date	8 October 1962

Economic indicators	2000	2005
Exchange rate (national currency per US$)[a]	75.34	72.08[b]
Consumer price index (2000=100)	100	120[c]
Industrial production index (1995=100)	97	101
Unemployment (percentage of labour force)[d]	27.3[e]	17.7[f]
Tourist arrivals (000s)[g]	866	1234[f]
GDP (million current US$)	54790	102257
GDP (per capita current US$)	1799	3112
Gross fixed capital formation (% of GDP)	21.0	25.0
Labour force participation, adult female pop. (%)	11.8[h]	...
Labour force participation, adult male pop. (%)	77.5[h]	...
Employment in industrial sector (%)	24.3[e]	26.0[f]
Employment in agricultural sector (%)	21.1[e]	20.7[f]
Agricultural production index (1999-2001=100)	96	117[f]
Food production index (1999-2001=100)	95	117[f]
Primary energy production (000s Mt oil equiv.)	175104	188399[f]
Motor vehicles (per 1,000 inhabitants)	88.2	90.5
Telephone lines (per 100 inhabitants)	5.8	7.8
Internet users, estimated (000s)	150.0	1920.0

Total trade		Major trading partners			2005
	(million US$)[f]	(% of exports)[f]		(% of imports)[f]	
Exports	32082.6	USA	24	France	23
Imports	18307.7	Italy	16	Italy	8
		France	11	Germany	7

Social indicators	2000-2006
Population growth rate 2000-2005 (% per annum)	1.5
Population aged 0-14 years (%)	30.0
Population aged 60+ years (women and men, % of total)	7.0/6.0
Sex ratio (women per 100 men)	98
Life expectancy at birth 2000-2005 (women and men, years)	72/70
Infant mortality rate 2000-2005 (per 1,000 births)	37
Total fertility rate 2000-2005 (births per woman)	2.5
Contraceptive use (% of currently married women)	64
Urban population (%)	63
Urban population growth rate 2000-2005 (% per annum)	2.7
Rural population growth rate 2000-2005 (% per annum)	−0.3
Foreign born (%)	0.8[i]
Refugees and others of concern to UNHCR[j]	94408
Government education expenditure (% of GNP)	4.5
Primary-secondary gross enrolment ratio (w and m per 100)	95/96
Third-level students (women and men, % of total)	51/49
Television receivers (per 1,000 inhabitants)	189
Parliamentary seats (women and men, % of total)	5/95

Environment	2000-2006
Threatened species	73
Forested area (% of land area)	1
CO2 emissions (000s Mt of carbon dioxide/per capita)	163946/5.1
Energy consumption per capita (kilograms oil equiv.)	1240
Precipitation (mm)	687[k]
Average minimum and maximum temperatures (centigrade)[k]	11.7/23.1

a Official rate. b October 2006. c May 2006. d Persons aged 15 years and over. e 2001. f 2004. g Including nationals residing abroad. h 1996. i Estimated data. j Provisional. k Dar-El-Beida.

American Samoa

Region	Oceania-Polynesia
Largest urban agglom. (pop., 000s)	Pago Pago (55)
Currency	US dollar
Population in 2006 (proj., 000s)	65[a]
Surface area (square kms)	199
Population density (per square km)	288

Economic indicators	2000	2005
Consumer price index (2000=100)[b]	100	126[c]
Tourist arrivals (000s)	44	36[d]
Agricultural production index (1999-2001=100)	100	100[e]
Food production index (1999-2001=100)	100	100[e]
Motor vehicles (per 1,000 inhabitants)	121.7[f]	123.3
Telephone lines (per 100 inhabitants)	...	25.2[d]

Social indicators	2000-2006
Population growth rate 2000-2005 (% per annum)	2.3
Population aged 0-14 years (%)	39.0
Population aged 60+ years (women and men, % of total)[g]	6.0/5.0[h]
Sex ratio (women per 100 men)	96[ij k]
Life expectancy at birth 2000-2005 (women and men, years)	76/69[k l]
Infant mortality rate 2000-2005 (per 1,000 births)	6
Total fertility rate 2000-2005 (births per woman)	4.0
Urban population (%)	91
Urban population growth rate 2000-2005 (% per annum)	2.9
Rural population growth rate 2000-2005 (% per annum)	−2.8
Government education expenditure (% of GNP)	8.1[m]
Third-level students (women and men, % of total)	54/46[n]
Television receivers (per 1,000 inhabitants)	325

Environment	2000-2006
Threatened species	27
Forested area (% of land area)	60
CO2 emissions (000s Mt of carbon dioxide/per capita)	292/4.7
Energy consumption per capita (kilograms oil equiv.)	1596[o]

a 2005. b Excluding rent. c 1st quarter 2006. d 2001. e 2003. f 1999. g De jure population but including armed forces stationed in the area. h 1990. i Including armed forces stationed in the area. j Refers to de jure population count. k 2000. l Published by the Secretariat of the Pacific Community. m 1981. n 1995. o 2002.

Andorra

Region	Southern Europe
Largest urban agglom. (pop., 000s)	Andorra-la-Vella (22)
Currency	euro[a]
Population in 2006 (proj., 000s)	70
Surface area (square kms)	468
Population density (per square km)	150
United Nations membership date	28 July 1993

Economic indicators	2000	2005
Exchange rate (national currency per US$)[b]	192.36[c]	0.79[d]
Tourist arrivals (000s)	2949	2791[e]
GDP (million current US$)	1360	3091
GDP (per capita current US$)	20624	46029
Gross fixed capital formation (% of GDP)	26.0	29.0
Telephone lines (per 100 inhabitants)	43.9	52.3[e]
Internet users, estimated (000s)	7.0	21.9

Total trade		Major trading partners			2005
	(million US$)[e]		(% of exports)[e]		(% of imports)[e]
Exports	122.8	Spain	52	Spain	51
Imports	1765.5	France	24	France	22
		Germany	15	Germany	5

Social indicators	2000-2006
Population growth rate 2000-2005 (% per annum)	0.4
Population aged 0-14 years (%)	16.0
Infant mortality rate 2000-2005 (per 1,000 births)	4[f]
Urban population (%)	91
Urban population growth rate 2000-2005 (% per annum)	—<
Rural population growth rate 2000-2005 (% per annum)	4.4
Foreign born (%)	23.7[g]
Primary-secondary gross enrolment ratio (w and m per 100)	92/92
Third-level students (women and men, % of total)	49/51
Television receivers (per 1,000 inhabitants)	462
Parliamentary seats (women and men, % of total)	29/71

Environment	2000-2006
Threatened species	8

a Prior to January 2002, Spanish peseta/French franc. b Data refer to non-commercial rates derived from the Operational Rates of Exchange for United Nations Programmes. c Spanish peseta/French franc. d October 2006. e 2004. f Rates based on 10 or fewer infant deaths. g 1989.

Angola

Region	Middle Africa
Largest urban agglom. (pop., 000s)	Luanda (2766)
Currency	kwanza
Population in 2006 (proj., 000s)	16400
Surface area (square kms)	1246700
Population density (per square km)	13
United Nations membership date	1 December 1976

Economic indicators	2000	2005
Exchange rate (national currency per US$)	16.82	80.37[a]
Consumer price index (2000=100)[b]	100	1846
Balance of payments, current account (million US$)	−150[c]	686[d]
Tourist arrivals (000s)	51	194[d]
GDP (million current US$)	9130	28853
GDP (per capita current US$)	660	1810
Gross fixed capital formation (% of GDP)	15.0	11.0
Labour force participation, adult female pop. (%)	73.4[e]	...
Agricultural production index (1999-2001=100)	100	113[d]
Food production index (1999-2001=100)	100	113[d]
Primary energy production (000s Mt oil equiv.)	37310	50274[d]
Motor vehicles (per 1,000 inhabitants)[f]	17.0	16.1[c]
Telephone lines (per 100 inhabitants)	0.5	0.6
Internet users, estimated (000s)	15.0	176.0

Social indicators	2000-2006
Population growth rate 2000-2005 (% per annum)	2.8
Population aged 0-14 years (%)	46.0
Population aged 60+ years (women and men, % of total)	4.0/4.0
Sex ratio (women per 100 men)	103
Life expectancy at birth 2000-2005 (women and men, years)	42/39
Infant mortality rate 2000-2005 (per 1,000 births)	139
Total fertility rate 2000-2005 (births per woman)	6.8
Contraceptive use (% of currently married women)	6
Urban population (%)	53
Urban population growth rate 2000-2005 (% per annum)	4.1
Rural population growth rate 2000-2005 (% per annum)	1.5
Foreign born (%)	0.4[g]
Refugees and others of concern to UNHCR[h]	68640
Government education expenditure (% of GNP)	3.2
Third-level students (women and men, % of total)	40/60
Television receivers (per 1,000 inhabitants)	21

Environment	2000-2006
Threatened species	88
Forested area (% of land area)	56
CO2 emissions (000s Mt of carbon dioxide/per capita)	8634/0.6
Energy consumption per capita (kilograms oil equiv.)	238
Precipitation (mm)	332[b]
Average minimum and maximum temperatures (centigrade)[b]	22.3/25.2

a September 2006. b Luanda. c 2002. d 2004. e 1995. f Source: World Automotive Market Report, Auto and Truck International (Illinois). g Estimated data. h Provisional.

Antigua and Barbuda

Region	Caribbean
Largest urban agglom. (pop., 000s)	St John's (32)
Currency	EC dollar
Population in 2006 (proj., 000s)	81[a]
Surface area (square kms)	442
Population density (per square km)	175
United Nations membership date	11 November 1981

Economic indicators	2000	2005
Exchange rate (national currency per US$) [b]	2.70	2.70[c]
Balance of payments, current account (million US$)	−79	−103[d]
Tourist arrivals (000s) [e]	207	245[f]
GDP (million current US$)	665	856
GDP (per capita current US$)	8698	10507
Gross fixed capital formation (% of GDP)	48.0	50.0
Employment in industrial sector (%)	18.6[g]	...
Employment in agricultural sector (%)	4.0[g]	...
Agricultural production index (1999-2001=100)	100	108[f]
Food production index (1999-2001=100)	100	108[f]
Motor vehicles (per 1,000 inhabitants)	324.8[hi]	...
Telephone lines (per 100 inhabitants)	50.1	47.2
Internet users, estimated (000s)	5.0	43.7

Total trade		Major trading partners			2005
	(million US$)	(% of exports)		(% of imports)	
Exports	120.6	Neth.Antilles	23	USA	49
Imports	525.2	UK	17	Trinidad Tbg	11
		St.Kitts & Nevis	10	Neth.Antilles	10

Social indicators	2000-2006
Population growth rate 2000-2005 (% per annum)	1.3
Population aged 0-14 years (%)	28.0
Sex ratio (women per 100 men)	109[jk]
Contraceptive use (% of currently married women)	53[lmn]
Urban population (%)	39
Urban population growth rate 2000-2005 (% per annum)	2.2
Rural population growth rate 2000-2005 (% per annum)	0.7
Foreign born (%)	22.1[o]
Government education expenditure (% of GNP)	4.0
Television receivers (per 1,000 inhabitants)	465
Parliamentary seats (women and men, % of total)	14/86

Environment	2000-2006
Threatened species	23
Forested area (% of land area)	21
CO2 emissions (000s Mt of carbon dioxide/per capita)	399/5.0
Energy consumption per capita (kilograms oil equiv.)	2024[o]
Precipitation (mm)	1052[p]
Average minimum and maximum temperatures (centigrade) [p]	23.9/29.6

a 2005. b Official rate. c October 2006. d 2002. e Including nationals residing abroad. f 2004. g 1991. h Including commercial vehicles. i 1998. j Refers to de facto population count. k 2001. l Including visiting unions. m For women aged 15-44 in union or married. n 1988. o Estimated data. p St. John's Antigua.

Argentina

Region	South America
Largest urban agglom. (pop., 000s)	Buenos Aires (12550)
Currency	peso
Population in 2006 (proj., 000s)	39134
Surface area (square kms)	2780400
Population density (per square km)	14
United Nations membership date	24 October 1945

Economic indicators	2000	2005
Exchange rate (national currency per US$)[a]	1.00	3.07[b]
Consumer price index (2000=100)[c]	100	182[d]
Industrial production index (1995=100)[e]	102	128[f]
Unemployment (percentage of labour force)[g]	15.0[h]	10.6[i]
Balance of payments, current account (million US$)	-8981	5705
Tourist arrivals (000s)[j]	2909	3353[k]
GDP (million current US$)	284346	183310
GDP (per capita current US$)	7707	4731
Gross fixed capital formation (% of GDP)	16.0	21.0
Labour force participation, adult female pop. (%)	44.1	45.8[l]
Labour force participation, adult male pop. (%)	73.2	71.5[l]
Employment in industrial sector (%)	22.7	23.5
Employment in agricultural sector (%)	0.7	1.1
Agricultural production index (1999-2001=100)	100	102[k]
Food production index (1999-2001=100)	100	102[k]
Primary energy production (000s Mt oil equiv.)	82024	85978[k]
Motor vehicles (per 1,000 inhabitants)	173.2	...
Telephone lines (per 100 inhabitants)[m]	21.5	22.8
Internet users, estimated (000s)[n]	2600.0	6863.5

Total trade		Major trading partners			2005
	(million US$)	(% of exports)		(% of imports)	
Exports	40106.4	Brazil	16	Brazil	37
Imports	28688.6	USA	11	USA	16
		Chile	11	China	5

Social indicators	2000-2006
Population growth rate 2000-2005 (% per annum)	1.0
Population aged 0-14 years (%)	26.0
Population aged 60+ years (women and men, % of total)	16.0/12.0
Sex ratio (women per 100 men)	104
Life expectancy at birth 2000-2005 (women and men, years)	78/71
Infant mortality rate 2000-2005 (per 1,000 births)	15
Total fertility rate 2000-2005 (births per woman)	2.4
Urban population (%)	90
Urban population growth rate 2000-2005 (% per annum)	1.2
Rural population growth rate 2000-2005 (% per annum)	-0.8
Foreign born (%)	4.2
Refugees and others of concern to UNHCR[o]	3899
Government education expenditure (% of GNP)	3.6
Primary-secondary gross enrolment ratio (w and m per 100)	101/98
Third-level students (women and men, % of total)	60/40
Newspaper circulation (per 1,000 inhabitants)	41
Television receivers (per 1,000 inhabitants)[n]	326
Intentional homicides (per 100,000 inhabitants)	7
Parliamentary seats (women and men, % of total)	37/63

Environment	2000-2006
Threatened species	203
Forested area (% of land area)	13
CO_2 emissions (000s Mt of carbon dioxide/per capita)	127728/3.4
Energy consumption per capita (kilograms oil equiv.)	1568
Precipitation (mm)	1215[c]
Average minimum and maximum temperatures (centigrade)[c]	13.5/22.5

a Official rate. b October 2006. c Buenos Aires. d September 2006. e Manufacturing.
f 2nd quarter 2006. g Persons aged 10 years and over, 28 urban agglomerations.
h May and October. i Second semester. j Excluding nationals residing abroad. k 2004.
l 2003. m From 1991, year ending 30 September. n Year ending 30 September.
o Provisional.

Armenia

Region	Western Asia
Largest urban agglom. (pop., 000s)	Yerevan (1103)
Currency	dram
Population in 2006 (proj., 000s)	3007
Surface area (square kms)	29800
Population density (per square km)	101
United Nations membership date	2 March 1992

Economic indicators	2000	2005
Exchange rate (national currency per US$)[a]	552.18	378.93[b]
Consumer price index (2000=100)	100	124[c]
Unemployment (percentage of labour force)[de]	11.7	8.1
Balance of payments, current account (million US$)	−278	−193
Tourist arrivals (000s)	45	263[f]
GDP (million current US$)	1912	4868
GDP (per capita current US$)	620	1614
Gross fixed capital formation (% of GDP)	18.0	28.0
Labour force participation, adult female pop. (%)	54.3[g]	44.8[h]
Labour force participation, adult male pop. (%)	68.7[g]	43.3[h]
Employment in industrial sector (%)	17.0[i]	16.5[j]
Employment in agricultural sector (%)	45.3[i]	46.0[j]
Agricultural production index (1999-2001=100)	96	140[f]
Food production index (1999-2001=100)	95	141[f]
Primary energy production (000s Mt oil equiv.)	281	361[f]
Telephone lines (per 100 inhabitants)	17.3	19.3
Internet users, estimated (000s)	40.0	161.0

Total trade		Major trading partners			2005
	(million US$)	(% of exports)			(% of imports)
Exports	937.0	Germany	16	Russian Fed.	15
Imports	1691.5	Netherlands	14	Belgium	9
		Belgium	13	USA	6

Social indicators	2000-2006
Population growth rate 2000-2005 (% per annum)	−0.4
Population aged 0-14 years (%)	21.0
Population aged 60+ years (women and men, % of total)	16.0/13.0
Sex ratio (women per 100 men)	115
Life expectancy at birth 2000-2005 (women and men, years)	75/68
Infant mortality rate 2000-2005 (per 1,000 births)	30
Total fertility rate 2000-2005 (births per woman)	1.3
Contraceptive use (% of currently married women)	61
Urban population (%)	64
Urban population growth rate 2000-2005 (% per annum)	−0.8
Rural population growth rate 2000-2005 (% per annum)	0.2
Foreign born (%)	8.9
Refugees and others of concern to UNHCR[k]	219620
Government education expenditure (% of GNP)	3.1
Primary-secondary gross enrolment ratio (w and m per 100)	95/93
Third-level students (women and men, % of total)	55/45
Television receivers (per 1,000 inhabitants)	293
Intentional homicides (per 100,000 inhabitants)	2
Parliamentary seats (women and men, % of total)	5/95

Environment	2000-2006
Threatened species	36
Forested area (% of land area)	12
CO2 emissions (000s Mt of carbon dioxide/per capita)	3432/1.1
Energy consumption per capita (kilograms oil equiv.)	598
Precipitation (mm)	227[l]
Average minimum and maximum temperatures (centigrade)[l]	5.5/18.2

a Official rate. b October 2006. c May 2006. d Persons aged 16 to 63 years. e Dec. of each year. f 2004. g 1997. h 2001. i 2002. j 2003. k Provisional. l Yerevan.

Aruba

Region	Caribbean
Largest urban agglom. (pop., 000s)	Oranjestad (30)
Currency	florin
Population in 2006 (proj., 000s)	99[a]
Surface area (square kms)	180
Population density (per square km)	503

Economic indicators	2000	2005
Exchange rate (national currency per US$)[b]	1.79	1.79[c]
Consumer price index (2000=100)	100	122[d]
Balance of payments, current account (million US$)	211	−259
Tourist arrivals (000s)	721	728[e]
GDP (million current US$)	1859	2258
GDP (per capita current US$)	20182	22696
Gross fixed capital formation (% of GDP)	24.0	32.0
Labour force participation, adult female pop. (%)	54.3[f]	...
Labour force participation, adult male pop. (%)	71.5[f]	...
Employment in industrial sector (%)	16.3[f]	...
Employment in agricultural sector (%)	0.5[f]	...
Primary energy production (000s Mt oil equiv.)	120	120[e]
Telephone lines (per 100 inhabitants)	41.4	39.7[gh]
Internet users, estimated (000s)	14.0	24.0[i]

Total trade		Major trading partners			2005
	(million US$)	(% of exports)		(% of imports)	
Exports	106.2	Netherlands	34	USA	56
Imports	1030.0	Panama	17	Netherlands	13
		Colombia	12	UK	4

Social indicators	2000-2006
Population growth rate 2000-2005 (% per annum)	1.5
Population aged 0-14 years (%)	22.0
Population aged 60+ years (women and men, % of total)	13.0/11.0
Sex ratio (women per 100 men)	111[jk]
Life expectancy at birth 2000-2005 (women and men, years)	76/70
Urban population (%)	47
Urban population growth rate 2000-2005 (% per annum)	1.5
Rural population growth rate 2000-2005 (% per annum)	1.6
Foreign born (%)	23.3
Government education expenditure (% of GNP)	4.0[l]
Primary-secondary gross enrolment ratio (w and m per 100)	106/109
Third-level students (women and men, % of total)	60/40
Television receivers (per 1,000 inhabitants)	269

Environment	2000-2006
Threatened species	21
CO_2 emissions (000s Mt of carbon dioxide/per capita)	2157/22.3
Energy consumption per capita (kilograms oil equiv.)	3519[m]
Precipitation (mm)	1007[n]

a 2005. b Official rate. c September 2006. d June 2006. e 2004. f 1997. g Estimate. h 2003. i 2001. j Refers to de facto population count. k 2000. l 1994. m Estimated data. n Juliana/St. Maarten.

Australia

Region	Oceania
Largest urban agglom. (pop., 000s)	Sydney (4331)
Currency	dollar
Population in 2006 (proj., 000s)	20366
Surface area (square kms)	7741220
Population density (per square km)	3
United Nations membership date	1 November 1945

Economic indicators	2000	2005
Exchange rate (national currency per US$)	1.81	1.30[a]
Consumer price index (2000=100)	100	120[b]
Industrial production index (1995=100)	112	119[c]
Unemployment (percentage of labour force)[d]	6.4	5.1
Balance of payments, current account (million US$)	−15306	−42084
Tourist arrivals (000s)[e]	4530	4774[f]
GDP (million current US$)	399658	709446
GDP (per capita current US$)	20956	35199
Gross fixed capital formation (% of GDP)	22.0	26.0
Labour force participation, adult female pop. (%)	54.6	55.3[f]
Labour force participation, adult male pop. (%)	71.7	70.4[f]
Employment in industrial sector (%)	21.7	21.1
Employment in agricultural sector (%)	5.0	3.6
Agricultural production index (1999-2001=100)	98	90[f]
Food production index (1999-2001=100)	97	92[f]
Primary energy production (000s Mt oil equiv.)	224313	247454[f]
Motor vehicles (per 1,000 inhabitants)[gh]	631.5[i]	667.7
Telephone lines (per 100 inhabitants)[j]	54.0	56.9
Internet users, estimated (000s)[j]	6600.0	14190.0

Total trade		Major trading partners			2005
	(million US$)	(% of exports)		(% of imports)	
Exports	105751.5	Japan	20	USA	14
Imports	118921.9	China	12	China	14
		Korea Rep.	8	Japan	11

Social indicators	2000-2006
Population growth rate 2000-2005 (% per annum)	1.1
Population aged 0-14 years (%)	20.0
Population aged 60+ years (women and men, % of total)	19.0/16.0
Sex ratio (women per 100 men)	102[h]
Life expectancy at birth 2000-2005 (women and men, years)	83/78[h]
Infant mortality rate 2000-2005 (per 1,000 births)	5
Total fertility rate 2000-2005 (births per woman)	1.8[h]
Contraceptive use (% of currently married women)	76[kl]
Urban population (%)[h]	88
Urban population growth rate 2000-2005 (% per annum)[h]	1.3
Rural population growth rate 2000-2005 (% per annum)[h]	−0.6
Foreign born (%)	24.6[m]
Refugees and others of concern to UNHCR[n]	66786
Government education expenditure (% of GNP)	4.9
Primary-secondary gross enrolment ratio (w and m per 100)	123/126
Third-level students (women and men, % of total)	54/46
Newspaper circulation (per 1,000 inhabitants)	162
Television receivers (per 1,000 inhabitants)[j]	665
Intentional homicides (per 100,000 inhabitants)	2
Parliamentary seats (women and men, % of total)	28/72

Environment	2000-2006
Threatened species	639
Forested area (% of land area)	20
CO2 emissions (000s Mt of carbon dioxide/per capita)[o]	371700/18.8
Energy consumption per capita (kilograms oil equiv.)	5169
Precipitation (mm)	1222[p]
Average minimum and maximum temperatures (centigrade)[p]	13.7/21.6

a October 2006. b June 2006. c 2nd quarter 2006. d Persons aged 15 years and over. e Arrivals by air. f 2004. g Data refer to fiscal years beginning 1 July. h Including Christmas Island, Cocos (Keeling) Islands and Norfolk Island. i 1999. j Year ending 30 June. k For women aged 20-49 in union or married. l 1986. m Estimated data. n Provisional. o Source: UNFCCC. p Sydney.

Austria

Region	Western Europe
Largest urban agglom. (pop., 000s)	Vienna (2260)
Currency	euro[a]
Population in 2006 (proj., 000s)	8205
Surface area (square kms)	83858
Population density (per square km)	98
United Nations membership date	14 December 1955

Economic indicators	2000	2005
Exchange rate (national currency per US$)	1.07	0.79[b]
Consumer price index (2000=100)	100	113[c]
Industrial production index (1995=100)[d]	134	156[c]
Unemployment (percentage of labour force)[e]	3.6[f]	4.9[g]
Balance of payments, current account (million US$)	−4864	4252
Tourist arrivals (000s)	17982	19373[g]
GDP (million current US$)	193838	306065
GDP (per capita current US$)	23942	37373
Gross fixed capital formation (% of GDP)	23.0	21.0
Labour force participation, adult female pop. (%)	48.4	50.8[g]
Labour force participation, adult male pop. (%)	68.4	67.0[g]
Employment in industrial sector (%)	30.3	27.5
Employment in agricultural sector (%)	5.8	5.5
Agricultural production index (1999-2001=100)	98	102[g]
Food production index (1999-2001=100)	98	102[g]
Primary energy production (000s Mt oil equiv.)	6879	6442[g]
Motor vehicles (per 1,000 inhabitants)	602.4	552.4[h]
Telephone lines (per 100 inhabitants)	49.9	45.3
Internet users, estimated (000s)	2700.0	4000.0

Total trade		Major trading partners			2005
	(million US$)	(% of exports)		(% of imports)	
Exports	114390.0	Germany	32	Germany	43
Imports	116626.5	Italy	8	Italy	7
		USA	6	France	4

Social indicators	2000-2006
Population growth rate 2000-2005 (% per annum)	0.2
Population aged 0-14 years (%)	16.0
Population aged 60+ years (women and men, % of total)	26.0/20.0
Sex ratio (women per 100 men)	104
Life expectancy at birth 2000-2005 (women and men, years)	82/76
Infant mortality rate 2000-2005 (per 1,000 births)	5
Total fertility rate 2000-2005 (births per woman)	1.4
Urban population (%)	66
Urban population growth rate 2000-2005 (% per annum)	0.3
Rural population growth rate 2000-2005 (% per annum)	0.1
Foreign born (%)	12.5
Refugees and others of concern to UNHCR[i]	62440
Government education expenditure (% of GNP)	5.6
Primary-secondary gross enrolment ratio (w and m per 100)	101/104
Third-level students (women and men, % of total)	53/47
Newspaper circulation (per 1,000 inhabitants)	309
Television receivers (per 1,000 inhabitants)	657
Intentional homicides (per 100,000 inhabitants)	1
Parliamentary seats (women and men, % of total)	31/69

Environment	2000-2006
Threatened species	73
Forested area (% of land area)	47
CO2 emissions (000s Mt of carbon dioxide/per capita)[j]	76210/9.4
Energy consumption per capita (kilograms oil equiv.)	3487
Precipitation (mm)	620[k]
Average minimum and maximum temperatures (centigrade)[k]	6.7/14.5

a Prior to 1 January 1999, schilling. b September 2006. c August 2006. d Monthly indices are adjusted for differences in the number of working days. e Persons aged 15 years and over. f May and November of each year. g 2004. h Including vehicles operated by police or other governmental security organizations. i Provisional. j Source: UNFCCC. k Vienna.

Azerbaijan

Region	Western Asia
Largest urban agglom. (pop., 000s)	Baku (1856)
Currency	manat
Population in 2006 (proj., 000s)	8471
Surface area (square kms)	86600
Population density (per square km)	98
United Nations membership date	2 March 1992

Economic indicators	2000	2005
Exchange rate (national currency per US$) [ab]	4565.00	0.88[c]
Consumer price index (2000=100)	100	134[d]
Unemployment (percentage of labour force) [ef]	1.2	1.4
Balance of payments, current account (million US$)	−168	167
Tourist arrivals (000s)	681	1349[g]
GDP (million current US$)	5273	12561
GDP (per capita current US$)	647	1493
Gross fixed capital formation (% of GDP)	23.0	46.0
Employment in industrial sector (%)	10.9	12.1
Employment in agricultural sector (%)	41.0	39.3
Agricultural production index (1999-2001=100)	101	120[g]
Food production index (1999-2001=100)	100	121[g]
Primary energy production (000s Mt oil equiv.)	19419	20457[g]
Motor vehicles (per 1,000 inhabitants)	57.2	72.0
Telephone lines (per 100 inhabitants)	9.8	13.0
Internet users, estimated (000s)	12.0	678.8

Total trade		Major trading partners			2005
	(million US$)	(% of exports)		(% of imports)	
Exports	4347.2	Italy	30	Russian Fed.	17
Imports	4211.2	France	9	UK	9
		Russian Fed.	7	Singapore	9

Social indicators	2000-2006
Population growth rate 2000-2005 (% per annum)	0.6
Population aged 0-14 years (%)	26.0
Population aged 60+ years (women and men, % of total)	10.0/8.0
Sex ratio (women per 100 men)	106
Life expectancy at birth 2000-2005 (women and men, years)	70/63
Infant mortality rate 2000-2005 (per 1,000 births)	76
Total fertility rate 2000-2005 (births per woman)	1.9
Contraceptive use (% of currently married women)	55[h]
Urban population (%)	52
Urban population growth rate 2000-2005 (% per annum)	0.9
Rural population growth rate 2000-2005 (% per annum)	0.4
Refugees and others of concern to UNHCR[i]	584292
Government education expenditure (% of GNP)	3.7
Primary-secondary gross enrolment ratio (w and m per 100)	86/89
Third-level students (women and men, % of total)	46/54
Television receivers (per 1,000 inhabitants)	331
Intentional homicides (per 100,000 inhabitants)	3
Parliamentary seats (women and men, % of total)	11/89

Environment	2000-2006
Threatened species	39
Forested area (% of land area)	13
CO2 emissions (000s Mt of carbon dioxide/per capita)	29223/3.5
Energy consumption per capita (kilograms oil equiv.)	1593
Precipitation (mm)	210[j]
Average minimum and maximum temperatures (centigrade)[j]	12.0/18.9

a Introduced a new currency in 2006, the new Manat. 1 new Manat (AZN) is equal to 5000 old Manats (AZM). b Official rate. c September 2006. d April 2006. e Employment office records. f 31st December of each year. g 2004. h For women aged 15-44 in union or married. i Provisional. j Baku.

Bahamas

Region	Caribbean
Largest urban agglom. (pop., 000s)	Nassau (233)
Currency	dollar
Population in 2006 (proj., 000s)	327
Surface area (square kms)	13878
Population density (per square km)	24
United Nations membership date	18 September 1973

Economic indicators	2000	2005
Exchange rate (national currency per US$)	1.00	1.00
Consumer price index (2000=100)	100	113[a]
Unemployment (percentage of labour force)[b]	7.8[c]	10.2
Balance of payments, current account (million US$)	−633	−203[d]
Tourist arrivals (000s)	1544	1561[d]
GDP (million current US$)	5004	5870
GDP (per capita current US$)	16599	18168
Gross fixed capital formation (% of GDP)	35.0	30.0
Labour force participation, adult female pop. (%)	70.9[c]	71.9[e]
Labour force participation, adult male pop. (%)	83.1[c]	81.7[e]
Employment in industrial sector (%)	16.6[c]	17.8
Employment in agricultural sector (%)	4.0[c]	3.5
Agricultural production index (1999-2001=100)	89	105[d]
Food production index (1999-2001=100)	89	105[d]
Motor vehicles (per 1,000 inhabitants)	346.0[f]	365.0[d]
Telephone lines (per 100 inhabitants)	37.9	43.9
Internet users, estimated (000s)	13.1	103.0

Total trade	Major trading partners		2005	
	(million US$)[g]	(% of exports)[g]		(% of imports)[g]
Exports	375.9	USA	78	USA	83
Imports	1927.3	France	6	Venezuela	5
		Germany	4	Neth.Antilles	3

Social indicators	2000-2006
Population growth rate 2000-2005 (% per annum)	1.4
Population aged 0-14 years (%)	28.0
Population aged 60+ years (women and men, % of total)	10.0/9.0
Sex ratio (women per 100 men)	106
Life expectancy at birth 2000-2005 (women and men, years)	73/66
Infant mortality rate 2000-2005 (per 1,000 births)	14
Total fertility rate 2000-2005 (births per woman)	2.3
Contraceptive use (% of currently married women)	62[hij]
Urban population (%)	90
Urban population growth rate 2000-2005 (% per annum)	1.7
Rural population growth rate 2000-2005 (% per annum)	−1.6
Foreign born (%)	9.9[k]
Government education expenditure (% of GNP)	3.8
Primary-secondary gross enrolment ratio (w and m per 100)	93/93
Television receivers (per 1,000 inhabitants)	248
Intentional homicides (per 100,000 inhabitants)	21
Parliamentary seats (women and men, % of total)	27/73

Environment	2000-2006
Threatened species	45
Forested area (% of land area)	84
CO2 emissions (000s Mt of carbon dioxide/per capita)	1873/6.0
Energy consumption per capita (kilograms oil equiv.)	2217[k]
Precipitation (mm)	1389[l]
Average minimum and maximum temperatures (centigrade)[l]	20.8/28.8

a May 2006. b Persons aged 15 years and over. c 1999. d 2004. e 2002. f Source: World Automotive Market Report, Auto and Truck International (Illinois). g 2001. h Including visiting unions. i For women aged 15-44 in union or married. j 1988. k Estimated data. l Nassau.

Bahrain

Region	Western Asia
Largest urban agglom. (pop., 000s)	Manama (162)
Currency	dinar
Population in 2006 (proj., 000s)	739
Surface area (square kms)	694
Population density (per square km)	1064
United Nations membership date	21 September 1971

Economic indicators	2000	2005
Exchange rate (national currency per US$)[a]	0.38	0.38[b]
Balance of payments, current account (million US$)	830	415[c]
Tourist arrivals (000s)	2420	3514[c]
GDP (million current US$)	7971	13348
GDP (per capita current US$)	11861	18370
Gross fixed capital formation (% of GDP)	14.0	21.0
Labour force participation, adult female pop. (%)	35.1	37.8[c]
Labour force participation, adult male pop. (%)	86.6	85.4[c]
Agricultural production index (1999-2001=100)	104	99[c]
Food production index (1999-2001=100)	104	99[c]
Primary energy production (000s Mt oil equiv.)	17562	16333[c]
Motor vehicles (per 1,000 inhabitants)	316.2	329.1[d]
Telephone lines (per 100 inhabitants)	25.4	27.1
Internet users, estimated (000s)	40.0	155.0

Total trade		Major trading partners	2005
	(million US$)	(% of exports)	(% of imports)
Exports	9732.9	...	...
Imports	7540.3	...	...
		...	

Social indicators	2000-2006
Population growth rate 2000-2005 (% per annum)	1.6
Population aged 0-14 years (%)	27.0
Population aged 60+ years (women and men, % of total)	5.0/4.0
Sex ratio (women per 100 men)	76
Life expectancy at birth 2000-2005 (women and men, years)	76/73
Infant mortality rate 2000-2005 (per 1,000 births)	14
Total fertility rate 2000-2005 (births per woman)	2.5
Urban population (%)	97
Urban population growth rate 2000-2005 (% per annum)	2.0
Rural population growth rate 2000-2005 (% per annum)	−7.1
Refugees and others of concern to UNHCR[e]	15
Primary-secondary gross enrolment ratio (w and m per 100)	103/100
Third-level students (women and men, % of total)	63/37
Television receivers (per 1,000 inhabitants)	414
Intentional homicides (per 100,000 inhabitants)	1
Parliamentary seats (women and men, % of total)	8/92

Environment	2000-2006
Threatened species	19
Forested area (% of land area)	−
CO2 emissions (000s Mt of carbon dioxide/per capita)	21912/31.0
Energy consumption per capita (kilograms oil equiv.)	10736
Precipitation (mm)	71[f]
Average minimum and maximum temperatures (centigrade)[f]	23.0/30.1

a Official rate. b October 2006. c 2004. d 2001. e Provisional. f Bahrain/Manama.

Bangladesh

Region	South-central Asia
Largest urban agglom. (pop., 000s)	Dhaka (12430)
Currency	taka
Population in 2006 (proj., 000s)	144437
Surface area (square kms)	143998
Population density (per square km)	1003
United Nations membership date	17 September 1974

Economic indicators	2000	2005
Exchange rate (national currency per US$)[a]	54.00	67.72[b]
Consumer price index (2000=100)[c]	100	131[d]
Industrial production index (1995=100)	132	216[e]
Unemployment (percentage of labour force)[f]	3.3	4.3[g]
Balance of payments, current account (million US$)	−306	−132
Tourist arrivals (000s)	199	271[h]
GDP (million current US$)	48626	64058
GDP (per capita current US$)	377	452
Gross fixed capital formation (% of GDP)	23.0	24.0
Labour force participation, adult female pop. (%)	55.9	26.1[g]
Labour force participation, adult male pop. (%)	87.2	87.4[g]
Employment in industrial sector (%)	10.3	13.7[g]
Employment in agricultural sector (%)	62.1	51.7[g]
Agricultural production index (1999-2001=100)	103	105[h]
Food production index (1999-2001=100)	103	105[h]
Primary energy production (000s Mt oil equiv.)	8380	11374[h]
Motor vehicles (per 1,000 inhabitants)	1.7[i]	...
Telephone lines (per 100 inhabitants)	0.4	0.8
Internet users, estimated (000s)	100.0	370.0

Total trade		Major trading partners			2005
	(million US$)[h]	(% of exports)[h]			(% of imports)[h]
Exports	8267.5	USA	26	China	14
Imports	11372.7	Germany	17	India	11
		UK	11	Japan	10

Social indicators	2000-2006
Population growth rate 2000-2005 (% per annum)	1.9
Population aged 0-14 years (%)	35.0
Population aged 60+ years (women and men, % of total)	6.0/5.0
Sex ratio (women per 100 men)	96
Life expectancy at birth 2000-2005 (women and men, years)	63/62
Infant mortality rate 2000-2005 (per 1,000 births)	59
Total fertility rate 2000-2005 (births per woman)	3.3
Contraceptive use (% of currently married women)	58
Urban population (%)	25
Urban population growth rate 2000-2005 (% per annum)	3.5
Rural population growth rate 2000-2005 (% per annum)	1.4
Foreign born (%)	0.7[i]
Refugees and others of concern to UNHCR[k]	271156
Government education expenditure (% of GNP)	2.1
Primary-secondary gross enrolment ratio (w and m per 100)	77/73
Third-level students (women and men, % of total)	32/68
Television receivers (per 1,000 inhabitants)	83
Parliamentary seats (women and men, % of total)	15/85

Environment	2000-2006
Threatened species	111
Forested area (% of land area)	10
CO2 emissions (000s Mt of carbon dioxide/per capita)	34690/0.3
Energy consumption per capita (kilograms oil equiv.)	108
Precipitation (mm)	2039
Average minimum and maximum temperatures (centigrade)	18.4/28.9

a Principal rate. b October 2006. c Government officials. d March 2006. e May 2006. f Persons aged 15 years and over. g 2003. h 2004. i 1998. j Estimated data. k Provisional.

Barbados

Region	Caribbean
Largest urban agglom. (pop., 000s)	Bridgetown (142)
Currency	dollar
Population in 2006 (proj., 000s)	270
Surface area (square kms)	430
Population density (per square km)	628
United Nations membership date	9 December 1966

Economic indicators	2000	2005
Exchange rate (national currency per US$)[a]	2.00	2.00[b]
Consumer price index (2000=100)	100	119[c]
Industrial production index (1995=100)	111	95[d]
Unemployment (percentage of labour force)[e]	9.4	9.8[f]
Balance of payments, current account (million US$)	−146	−337[f]
Tourist arrivals (000s)	545	552[f]
GDP (million current US$)	2559	2996
GDP (per capita current US$)	9616	11116
Gross fixed capital formation (% of GDP)	18.0	20.0
Labour force participation, adult female pop. (%)	62.8	64.4[f]
Labour force participation, adult male pop. (%)	74.8	75.4[f]
Employment in industrial sector (%)	20.4	17.3[f]
Employment in agricultural sector (%)	3.7	3.3[f]
Agricultural production index (1999-2001=100)	105	101[f]
Food production index (1999-2001=100)	105	101[f]
Primary energy production (000s Mt oil equiv.)	112	107[f]
Motor vehicles (per 1,000 inhabitants)[gh]	269.5[i]	...
Telephone lines (per 100 inhabitants)[j]	46.3	50.1
Internet users, estimated (000s)	10.0	160.0

Total trade	Major trading partners			2005
(million US$)	(% of exports)		(% of imports)	
Exports	361.2	...	USA	36
Imports	1671.9	...	Trinidad Tbg	21
		...	Japan	8

Social indicators	2000-2006
Population growth rate 2000-2005 (% per annum)	0.3
Population aged 0-14 years (%)	19.0
Population aged 60+ years (women and men, % of total)	16.0/10.0
Sex ratio (women per 100 men)	107
Life expectancy at birth 2000-2005 (women and men, years)	78/71
Infant mortality rate 2000-2005 (per 1,000 births)	11
Total fertility rate 2000-2005 (births per woman)	1.5
Contraceptive use (% of currently married women)	55[klm]
Urban population (%)	53
Urban population growth rate 2000-2005 (% per annum)	1.4
Rural population growth rate 2000-2005 (% per annum)	−0.9
Foreign born (%)	9.2[n]
Government education expenditure (% of GNP)	7.6
Primary-secondary gross enrolment ratio (w and m per 100)	108/109
Third-level students (women and men, % of total)	71/29
Television receivers (per 1,000 inhabitants)	289
Intentional homicides (per 100,000 inhabitants)	10
Parliamentary seats (women and men, % of total)	18/82

Environment	2000-2006
Threatened species	22
Forested area (% of land area)	5
CO2 emissions (000s Mt of carbon dioxide/per capita)	1192/4.4
Energy consumption per capita (kilograms oil equiv.)	1383
Precipitation (mm)	1273[o]
Average minimum and maximum temperatures (centigrade)[o]	25.1/27.1

a Official rate. b October 2006. c April 2006. d May 2006. e Persons aged 15 years and over. f 2004. g Including pick-ups. h Including buses and coaches. i 1999. j Until 2001: Data refer to BET and Bartel. From 2002: Cable and Wireless (Barbados) Limited. k Including visiting unions. l For women aged 15-44 in union or married. m 1988. n Estimated data. o Seawell airport.

Belarus

Region	Eastern Europe
Largest urban agglom. (pop., 000s)	Minsk (1778)
Currency	rubel
Population in 2006 (proj., 000s)	9700
Surface area (square kms)	207600
Population density (per square km)	47
United Nations membership date	24 October 1945

Economic indicators	2000	2005
Exchange rate (national currency per US$)[a]	1180.00	2141.00[b]
Consumer price index (2000=100)	100	408[c]
Unemployment (percentage of labour force)[d]	2.1	1.5
Balance of payments, current account (million US$)	−338	469
Tourist arrivals (000s)	60	67[e]
GDP (million current US$)	10418	29566
GDP (per capita current US$)	1039	3031
Gross fixed capital formation (% of GDP)	25.0	26.0
Labour force participation, adult female pop. (%)	52.7[f]	...
Labour force participation, adult male pop. (%)	65.7[f]	...
Employment in industrial sector (%)	34.9[g]	...
Employment in agricultural sector (%)	21.2[g]	...
Agricultural production index (1999-2001=100)	99	116[e]
Food production index (1999-2001=100)	98	116[e]
Primary energy production (000s Mt oil equiv.)[h]	2551	2498[e]
Motor vehicles (per 1,000 inhabitants)[h]	141.8	174.1[e]
Telephone lines (per 100 inhabitants)	27.4	33.7
Internet users, estimated (000s)	187.0	3394.4

Total trade		Major trading partners			2005
	(million US$)	(% of exports)		(% of imports)	
Exports	15979.3	Russian Fed.	36	Russian Fed.	61
Imports	16708.1	Netherlands	15	Germany	7
		UK	7	Ukraine	5

Social indicators	2000-2006
Population growth rate 2000-2005 (% per annum)	−0.6
Population aged 0-14 years (%)	15.0
Population aged 60+ years (women and men, % of total)	23.0/14.0
Sex ratio (women per 100 men)	114
Life expectancy at birth 2000-2005 (women and men, years)	74/62
Infant mortality rate 2000-2005 (per 1,000 births)	15
Total fertility rate 2000-2005 (births per woman)	1.2
Urban population (%)	72
Urban population growth rate 2000-2005 (% per annum)	0.1
Rural population growth rate 2000-2005 (% per annum)	−2.1
Refugees and others of concern to UNHCR[i]	13178
Government education expenditure (% of GNP)	5.8
Primary-secondary gross enrolment ratio (w and m per 100)	95/96
Third-level students (women and men, % of total)	57/43
Television receivers (per 1,000 inhabitants)	386
Intentional homicides (per 100,000 inhabitants)	11
Parliamentary seats (women and men, % of total)	30/70

Environment	2000-2006
Threatened species	19
Forested area (% of land area)	45
CO2 emissions (000s Mt of carbon dioxide/per capita)[j]	52590/5.3
Energy consumption per capita (kilograms oil equiv.)	2487
Precipitation (mm)	677[k]
Average minimum and maximum temperatures (centigrade)[k]	2.1/9.9

a Official rate. b September 2006. c April 2006. d Dec. of each year, Men aged 16 to 59 years, women aged 16 to 54 years. e 2004. f 1999. g 1994. h Passenger-cars only. i Provisional. j Source: UNFCCC. k Minsk.

Belgium

Region	Western Europe
Largest urban agglom. (pop., 000s)	Brussels (1012)
Currency	euro[a]
Population in 2006 (proj., 000s)	10437
Surface area (square kms)	30528
Population density (per square km)	342
United Nations membership date	27 December 1945

Economic indicators	2000	2005
Exchange rate (national currency per US$)	1.07	0.79[b]
Consumer price index (2000=100)	100	113[b]
Industrial production index (1995=100)	116	119[c]
Unemployment (percentage of labour force)[d]	7.0	8.4
Balance of payments, current account (million US$)	11360[e]	10198
Tourist arrivals (000s)	6457	6710[f]
GDP (million current US$)	231933	370815
GDP (per capita current US$)	22509	35590
Gross fixed capital formation (% of GDP)	21.0	20.0
Labour force participation, adult female pop. (%)	43.6	44.2[f]
Labour force participation, adult male pop. (%)	61.5	60.3[f]
Employment in industrial sector (%)	26.3	24.9[f]
Employment in agricultural sector (%)	1.8	2.0[f]
Agricultural production index (1999-2001=100)	...	102[f]
Primary energy production (000s Mt oil equiv.)	4516	4219[f]
Motor vehicles (per 1,000 inhabitants)	508.4	530.0[f]
Telephone lines (per 100 inhabitants)	49.1	46.0
Internet users, estimated (000s)	3000.0	4800.0

Total trade		Major trading partners			2005
	(million US$)	(% of exports)			(% of imports)
Exports	334106.3	Germany	19	Netherlands	18
Imports	320129.6	France	17	Germany	17
		Netherlands	12	France	11

Social indicators	2000-2006
Population growth rate 2000-2005 (% per annum)	0.2
Population aged 0-14 years (%)	17.0
Population aged 60+ years (women and men, % of total)	25.0/20.0
Sex ratio (women per 100 men)	104
Life expectancy at birth 2000-2005 (women and men, years)	82/76
Infant mortality rate 2000-2005 (per 1,000 births)	4
Total fertility rate 2000-2005 (births per woman)	1.7
Contraceptive use (% of currently married women)	78[gh]
Urban population (%)	97
Urban population growth rate 2000-2005 (% per annum)	0.2
Rural population growth rate 2000-2005 (% per annum)	−0.5
Foreign born (%)	8.6[i]
Refugees and others of concern to UNHCR[j]	34432
Government education expenditure (% of GNP)	6.1
Primary-secondary gross enrolment ratio (w and m per 100)	105/108
Third-level students (women and men, % of total)	54/46
Newspaper circulation (per 1,000 inhabitants)	152
Television receivers (per 1,000 inhabitants)	555
Parliamentary seats (women and men, % of total)	36/64

Environment	2000-2006
Threatened species	41
Forested area (% of land area)	22[e]
CO_2 emissions (000s Mt of carbon dioxide/per capita)[k]	126200/12.2
Energy consumption per capita (kilograms oil equiv.)	4340
Precipitation (mm)	820[l]
Average minimum and maximum temperatures (centigrade)[l]	6.7/13.9

a Prior to 1 January 1999, franc. b September 2006. c August 2006. d Persons aged 15 years and over. e Including Luxembourg. f 2004. g For women aged 21-39 in union or married. h 1991/92. i Estimated data. j Provisional. k Source: UNFCCC. l Brussels.

Belize

Region	Central America
Largest urban agglom. (pop., 000s)	Belize City (53)
Currency	dollar
Population in 2006 (proj., 000s)	275
Surface area (square kms)	22966
Population density (per square km)	12
United Nations membership date	25 September 1981

Economic indicators	2000	2005
Exchange rate (national currency per US$)[a]	2.00	2.00[b]
Consumer price index (2000=100)	100	113
Unemployment (percentage of labour force)[c]	12.8[d]	10.0
Balance of payments, current account (million US$)	−156	−182[e]
Tourist arrivals (000s)	196	231[e]
GDP (million current US$)	832	1105
GDP (per capita current US$)	3436	4097
Gross fixed capital formation (% of GDP)	29.0	23.0
Labour force participation, adult female pop. (%)	39.6[d]	...
Labour force participation, adult male pop. (%)	79.7[d]	...
Employment in industrial sector (%)	17.0[d]	...
Employment in agricultural sector (%)	27.5[d]	...
Agricultural production index (1999-2001=100)	103	117[e]
Food production index (1999-2001=100)	103	117[e]
Primary energy production (000s Mt oil equiv.)	8	9[e]
Motor vehicles (per 1,000 inhabitants)[fg]	134.2	159.4[h]
Telephone lines (per 100 inhabitants)[i]	14.9	12.3
Internet users, estimated (000s)[i]	15.0	38.0

Total trade		Major trading partners			2005
	(million US$)	(% of exports)			(% of imports)
Exports	207.5	USA	54	USA	40
Imports	438.6	UK	20	Cuba	15
		Trinidad Tbg	5	Mexico	12

Social indicators	2000-2006
Population growth rate 2000-2005 (% per annum)	2.1
Population aged 0-14 years (%)	37.0
Population aged 60+ years (women and men, % of total)	6.0/6.0
Sex ratio (women per 100 men)	98
Life expectancy at birth 2000-2005 (women and men, years)	75/69
Infant mortality rate 2000-2005 (per 1,000 births)	31
Total fertility rate 2000-2005 (births per woman)	3.2
Contraceptive use (% of currently married women)	47[jkl]
Urban population (%)	48
Urban population growth rate 2000-2005 (% per annum)	2.4
Rural population growth rate 2000-2005 (% per annum)	1.9
Refugees and others of concern to UNHCR[m]	638
Government education expenditure (% of GNP)	5.3
Primary-secondary gross enrolment ratio (w and m per 100)	106/105
Third-level students (women and men, % of total)	70/30
Television receivers (per 1,000 inhabitants)[i]	200
Intentional homicides (per 100,000 inhabitants)	23
Parliamentary seats (women and men, % of total)	12/88

Environment	2000-2006
Threatened species	69
Forested area (% of land area)	59
CO2 emissions (000s Mt of carbon dioxide/per capita)	780/3.0
Energy consumption per capita (kilograms oil equiv.)	1018[n]
Precipitation (mm)	2014[o]
Average minimum and maximum temperatures (centigrade)[o]	22.5/30.0

a Official rate. b October 2006. c Persons aged 14 years and over, April of each year. d 1999. e 2004. f Number of licensed vehicles. g Excluding government vehicles. h 2002. i Year beginning 1 April. j Including visiting unions. k For women aged 15-44 in union or married. l 1991. m Provisional. n Estimated data. o Belize City.

Benin

Region	Western Africa
Largest urban agglom. (pop., 000s)	Cotonou (719)
Currency	CFA franc
Population in 2006 (proj., 000s)	8703
Surface area (square kms)	112622
Population density (per square km)	77
United Nations membership date	20 September 1960

Economic indicators	2000	2005
Exchange rate (national currency per US$)[a]	704.95	516.66[b]
Consumer price index (2000=100)[c]	100	121[d]
Balance of payments, current account (million US$)	−111	−349[e]
Tourist arrivals (000s)	96	174[fg]
GDP (million current US$)	2359	4378
GDP (per capita current US$)	328	519
Gross fixed capital formation (% of GDP)	18.0	19.0
Labour force participation, adult female pop. (%)	...	68.7[h]
Labour force participation, adult male pop. (%)	...	66.6[h]
Agricultural production index (1999-2001=100)	103	132[g]
Food production index (1999-2001=100)	102	137[g]
Primary energy production (000s Mt oil equiv.)	35	20[g]
Motor vehicles (per 1,000 inhabitants)[i]	2.0[j]	19.6[e]
Telephone lines (per 100 inhabitants)	0.8	1.0
Internet users, estimated (000s)	15.0	425.0

Total trade		Major trading partners			2005
	(million US$)	(% of exports)			(% of imports)
Exports	288.2	China	36	France	18
Imports	898.7	India	7	China	9
		Nigeria	6	Ghana	7

Social indicators	2000-2006
Population growth rate 2000-2005 (% per annum)	3.2
Population aged 0-14 years (%)	44.0
Population aged 60+ years (women and men, % of total)	5.0/4.0
Sex ratio (women per 100 men)	98
Life expectancy at birth 2000-2005 (women and men, years)	55/53
Infant mortality rate 2000-2005 (per 1,000 births)	105
Total fertility rate 2000-2005 (births per woman)	5.9
Contraceptive use (% of currently married women)	19
Urban population (%)	40
Urban population growth rate 2000-2005 (% per annum)	4.1
Rural population growth rate 2000-2005 (% per annum)	2.6
Foreign born (%)	2.1
Refugees and others of concern to UNHCR[k]	31989
Government education expenditure (% of GNP)	3.3
Primary-secondary gross enrolment ratio (w and m per 100)	52/74
Third-level students (women and men, % of total)	20/80
Television receivers (per 1,000 inhabitants)	59
Parliamentary seats (women and men, % of total)	7/93

Environment	2000-2006
Threatened species	45
Forested area (% of land area)	24
CO2 emissions (000s Mt of carbon dioxide/per capita)	2045/0.3
Energy consumption per capita (kilograms oil equiv.)	100
Precipitation (mm)	1308[c]
Average minimum and maximum temperatures (centigrade)[c]	24.3/30.1

a Official rate. b October 2006. c Cotonou. d May 2006. e 2003. f Country estimates. g 2004. h 2001. i Source: World Automotive Market Report, Auto and Truck International (Illinois). j 1998. k Provisional.

Bermuda

Region	Northern America
Largest urban agglom. (pop., 000s)	Hamilton (1)
Currency	dollar
Population in 2006 (proj., 000s)	64[a]
Surface area (square kms)	53
Population density (per square km)	1171

Economic indicators	2000	2005
Exchange rate (national currency per US$)	1.00	1.00
Consumer price index (2000=100)	100	118[b]
Tourist arrivals (000s)[c]	332	272[d]
GDP (million current US$)	3522	4090
GDP (per capita current US$)	56026	63731
Gross fixed capital formation (% of GDP)	20.0	19.0
Labour force participation, adult female pop. (%)	67.9	...
Labour force participation, adult male pop. (%)	79.5[e]	...
Employment in industrial sector (%)	9.5[f]	...
Employment in agricultural sector (%)	1.4[f]	...
Agricultural production index (1999-2001=100)	100	98[g]
Food production index (1999-2001=100)	100	98[g]
Motor vehicles (per 1,000 inhabitants)	420.0	422.2[d]
Telephone lines (per 100 inhabitants)[h]	89.2	88.3[di]
Internet users, estimated (000s)[h]	27.0	42.0

Social indicators	2000-2006
Population growth rate 2000-2005 (% per annum)	0.4
Population aged 0-14 years (%)	19.0
Population aged 60+ years (women and men, % of total)	17.0/13.0
Sex ratio (women per 100 men)	108[ejk]
Life expectancy at birth 2000-2005 (women and men, years)	78/71[i]
Infant mortality rate 2000-2005 (per 1,000 births)	4[m]
Total fertility rate 2000-2005 (births per woman)	1.6
Urban population (%)	100
Urban population growth rate 2000-2005 (% per annum)	0.4
Rural population growth rate 2000-2005 (% per annum)	—
Foreign born (%)	21.7[n]
Primary-secondary gross enrolment ratio (w and m per 100)	99/92
Third-level students (women and men, % of total)	55/45
Newspaper circulation (per 1,000 inhabitants)	239
Television receivers (per 1,000 inhabitants)[h]	1099
Intentional homicides (per 100,000 inhabitants)	—

Environment	2000-2006
Threatened species	49
Forested area (% of land area)	—
CO2 emissions (000s Mt of carbon dioxide/per capita)	498/7.8
Energy consumption per capita (kilograms oil equiv.)	2811[n]

a 2005. b February 2006. c Arrivals by air. d 2004. e 2000. f 1996. g 2003. h Year beginning 1 April. i Estimate. j Including armed forces stationed in the area. k Refers to de jure population count. l 1994. m Rates based on 10 or fewer infant deaths. n Estimated data.

Bhutan

Region	South-central Asia
Largest urban agglom. (pop., 000s)	Thimphu (85)
Currency	ngultrum
Population in 2006 (proj., 000s)	2211
Surface area (square kms)	47000
Population density (per square km)	47
United Nations membership date	21 September 1971

Economic indicators	2000	2005
Exchange rate (national currency per US$)[a]	46.75	45.03[b]
Consumer price index (2000=100)	100	117
Tourist arrivals (000s)	8	9[c]
GDP (million current US$)	447	917
GDP (per capita current US$)	231	424
Gross fixed capital formation (% of GDP)	60.0	70.0
Agricultural production index (1999-2001=100)	90	95[c]
Food production index (1999-2001=100)	90	95[c]
Primary energy production (000s Mt oil equiv.)	191	204[c]
Telephone lines (per 100 inhabitants)	2.2	3.9[c]
Internet users, estimated (000s)	2.3	25.0

Social indicators	2000-2006
Population growth rate 2000-2005 (% per annum)	2.2
Population aged 0-14 years (%)	38.0
Population aged 60+ years (women and men, % of total)	7.0/7.0
Sex ratio (women per 100 men)	97
Life expectancy at birth 2000-2005 (women and men, years)	64/61
Infant mortality rate 2000-2005 (per 1,000 births)	56
Total fertility rate 2000-2005 (births per woman)	4.4
Urban population (%)	11
Urban population growth rate 2000-2005 (% per annum)	5.1
Rural population growth rate 2000-2005 (% per annum)	1.9
Government education expenditure (% of GNP)	5.9
Third-level students (women and men, % of total)	34/66
Television receivers (per 1,000 inhabitants)	32
Parliamentary seats (women and men, % of total)	9/91

Environment	2000-2006
Threatened species	54
Forested area (% of land area)	64
CO_2 emissions (000s Mt of carbon dioxide/per capita)	387/0.2
Energy consumption per capita (kilograms oil equiv.)	65[d]
Precipitation (mm)	799[e]
Average minimum and maximum temperatures (centigrade)[e]	14.0/24.2

a Official rate. b October 2006. c 2004. d Estimated data. e Wangdi Phodrang .

Bolivia

Region	South America
Largest urban agglom. (pop., 000s)	La Paz (1527)
Currency	boliviano
Population in 2006 (proj., 000s)	9354
Surface area (square kms)	1098581
Population density (per square km)	9
United Nations membership date	14 November 1945

Economic indicators	2000	2005
Exchange rate (national currency per US$)	6.39	8.00[a]
Consumer price index (2000=100)[b]	100	122[c]
Industrial production index (1995=100)[d]	104	99[e]
Unemployment (percentage of labour force)[f]	4.8	5.5[g]
Balance of payments, current account (million US$)	−446	498
Tourist arrivals (000s)[h]	319	405[i]
GDP (million current US$)	8398	9728
GDP (per capita current US$)	1010	1059
Gross fixed capital formation (% of GDP)	18.0	12.0
Labour force participation, adult female pop. (%)	59.6	60.2[j]
Labour force participation, adult male pop. (%)	82.2	75.9[j]
Employment in industrial sector (%)	28.2	...
Employment in agricultural sector (%)	4.9	...
Agricultural production index (1999-2001=100)	104	111[i]
Food production index (1999-2001=100)	104	110[i]
Primary energy production (000s Mt oil equiv.)	4738	10617[i]
Motor vehicles (per 1,000 inhabitants)	55.8	51.9[i]
Telephone lines (per 100 inhabitants)	6.1	7.0
Internet users, estimated (000s)	120.0	480.0

Total trade	(million US$)	Major trading partners	(% of exports)		2005 (% of imports)
Exports	2797.4	Brazil	36	Brazil	22
Imports	2343.3	USA	15	Argentina	17
		Argentina	9	USA	14

Social indicators	2000-2006
Population growth rate 2000-2005 (% per annum)	2.0
Population aged 0-14 years (%)	38.0
Population aged 60+ years (women and men, % of total)	7.0/6.0
Sex ratio (women per 100 men)	101
Life expectancy at birth 2000-2005 (women and men, years)	66/62
Infant mortality rate 2000-2005 (per 1,000 births)	56
Total fertility rate 2000-2005 (births per woman)	4.0
Contraceptive use (% of currently married women)	58[k]
Urban population (%)	64
Urban population growth rate 2000-2005 (% per annum)	2.7
Rural population growth rate 2000-2005 (% per annum)	0.7
Refugees and others of concern to UNHCR[l]	538
Government education expenditure (% of GNP)	6.7
Primary-secondary gross enrolment ratio (w and m per 100)	101/103
Third-level students (women and men, % of total)	35/65[m]
Television receivers (per 1,000 inhabitants)	134
Parliamentary seats (women and men, % of total)	15/85

Environment	2000-2006
Threatened species	154
Forested area (% of land area)	49
CO2 emissions (000s Mt of carbon dioxide/per capita)	7908/0.9
Energy consumption per capita (kilograms oil equiv.)	350
Precipitation (mm)	484[n]
Average minimum and maximum temperatures (centigrade)[n]	12.3/18.6

a October 2006. b Urban areas. c August 2006. d Calculated by the Statistics Division of the United Nations from component national indices. e 1st quarter 2005. f Persons aged 10 years and over. g 2002. h Data based on surveys. As from 2000 a new survey was applied. i 2004. j 2001. k 2003/04. l Provisional. m 1995. n Cochabamba.

Bosnia and Herzegovina

Region	Southern Europe
Largest urban agglom. (pop., 000s)	Sarajevo (380)
Currency	convertible mark
Population in 2006 (proj., 000s)	3912
Surface area (square kms)	51197
Population density (per square km)	76
United Nations membership date	22 May 1992

Economic indicators	2000	2005
Exchange rate (national currency per US$)	2.10	1.54[a]
Balance of payments, current account (million US$)	−396	−2087
Tourist arrivals (000s)	171	190[b]
GDP (million current US$)	4527	9132
GDP (per capita current US$)	1177	2337
Gross fixed capital formation (% of GDP)	21.0	20.0
Agricultural production index (1999-2001=100)	88	98[b]
Food production index (1999-2001=100)	88	98[b]
Primary energy production (000s Mt oil equiv.)	5503	3181[b]
Telephone lines (per 100 inhabitants)	20.6	24.8
Internet users, estimated (000s)	40.0	806.4

Total trade		Major trading partners			2005
	(million US$)	(% of exports)		(% of imports)	
Exports	2388.5	Croatia	20	Croatia	17
Imports	7053.8	Serbia, Mtneg	17	Germany	14
		Italy	13	Serbia, Mtneg	12

Social indicators	2000-2006
Population growth rate 2000-2005 (% per annum)	0.3
Population aged 0-14 years (%)	17.0
Population aged 60+ years (women and men, % of total)	21.0/17.0
Sex ratio (women per 100 men)	106
Life expectancy at birth 2000-2005 (women and men, years)	77/71
Infant mortality rate 2000-2005 (per 1,000 births)	14
Total fertility rate 2000-2005 (births per woman)	1.3
Contraceptive use (% of currently married women)	48
Urban population (%)	46
Urban population growth rate 2000-2005 (% per annum)	1.4
Rural population growth rate 2000-2005 (% per annum)	−0.6
Refugees and others of concern to UNHCR[c]	199967
Television receivers (per 1,000 inhabitants)	249
Parliamentary seats (women and men, % of total)	12/88

Environment	2000-2006
Threatened species	56
Forested area (% of land area)	45
CO2 emissions (000s Mt of carbon dioxide/per capita)	19161/4.9
Energy consumption per capita (kilograms oil equiv.)	1170
Precipitation (mm)	931[d]
Average minimum and maximum temperatures (centigrade)[d]	4.8/15.0

a October 2006. b 2004. c Provisional. d Sarajevo.

Botswana

Region	Southern Africa
Largest urban agglom. (pop., 000s)	Gaborone (210)
Currency	pula
Population in 2006 (proj., 000s)	1760
Surface area (square kms)	581730
Population density (per square km)	3
United Nations membership date	17 October 1966

Economic indicators	2000	2005
Exchange rate (national currency per US$)[a]	5.36	6.35[b]
Consumer price index (2000=100)	100	166[c]
Unemployment (percentage of labour force)[d]	15.8	23.8[e]
Balance of payments, current account (million US$)	545	483[e]
Tourist arrivals (000s)	1104	975[e]
GDP (million current US$)	4889	8850
GDP (per capita current US$)	2787	5014
Gross fixed capital formation (% of GDP)	27.0	26.0
Labour force participation, adult female pop. (%)	48.5	45.8[f]
Labour force participation, adult male pop. (%)	67.6	65.0[f]
Employment in industrial sector (%)	20.9	22.0[g]
Employment in agricultural sector (%)	19.7	22.6[g]
Agricultural production index (1999-2001=100)	99	104[g]
Food production index (1999-2001=100)	99	104[g]
Motor vehicles (per 1,000 inhabitants)	75.8	109.9
Telephone lines (per 100 inhabitants)[h]	8.3	7.5
Internet users, estimated (000s)[h]	50.0	60.0[g]

Total trade		Major trading partners			2005
	(million US$)[e]	(% of exports)[e]		(% of imports)[e]	
Exports	3801.6	UK	79	South Africa	85
Imports	3964.0	Norway	8	Sweden	3
		South Africa	8	UK	3

Social indicators	2000-2006
Population growth rate 2000-2005 (% per annum)	0.1
Population aged 0-14 years (%)	38.0
Population aged 60+ years (women and men, % of total)	6.0/4.0
Sex ratio (women per 100 men)	103
Life expectancy at birth 2000-2005 (women and men, years)	37/36
Infant mortality rate 2000-2005 (per 1,000 births)	51
Total fertility rate 2000-2005 (births per woman)	3.2
Contraceptive use (% of currently married women)	40
Urban population (%)	57
Urban population growth rate 2000-2005 (% per annum)	1.6
Rural population growth rate 2000-2005 (% per annum)	−1.7
Refugees and others of concern to UNHCR[i]	3156
Government education expenditure (% of GNP)	2.3
Primary-secondary gross enrolment ratio (w and m per 100)	93/92
Third-level students (women and men, % of total)	46/54
Television receivers (per 1,000 inhabitants)[h]	44
Parliamentary seats (women and men, % of total)	11/89

Environment	2000-2006
Threatened species	17
Forested area (% of land area)	22
CO_2 emissions (000s Mt of carbon dioxide/per capita)	4123/2.3

a Official rate. b October 2006. c September 2006. d Persons aged 12 years and over. e 2003. f 2001. g 2004. h Year beginning 1 April. i Provisional.

Brazil

Region	South America
Largest urban agglom. (pop., 000s)	Sao Paulo (18333)
Currency	real
Population in 2006 (proj., 000s)	188883
Surface area (square km)	8514047
Population density (per square km)	22
United Nations membership date	24 October 1945

Economic indicators	2000	2005
Exchange rate (national currency per US$)	1.95	2.14[a]
Consumer price index (2000=100)	100	158[b]
Industrial production index (1995=100)	110	142[b]
Unemployment (percentage of labour force)[c]	9.6[d]	8.9[e]
Balance of payments, current account (million US$)	−24225	14199
Tourist arrivals (000s)	5313	4794[e]
GDP (million current US$)	601732	799413
GDP (per capita current US$)	3461	4289
Gross fixed capital formation (% of GDP)	19.0	19.0
Labour force participation, adult female pop. (%)	49.3	55.9[f]
Labour force participation, adult male pop. (%)	78.4	80.7[e]
Employment in industrial sector (%)	19.3[d]	21.0[e]
Employment in agricultural sector (%)	24.2[d]	21.0[e]
Agricultural production index (1999-2001=100)	99	126[e]
Food production index (1999-2001=100)	99	124[e]
Primary energy production (000s Mt oil equiv.)	109740	129451[e]
Motor vehicles (per 1,000 inhabitants)	104.2[g]	118.0[h]
Telephone lines (per 100 inhabitants)	17.8	23.0
Internet users, estimated (000s)	5000.0	36356.0

Total trade		Major trading partners			2005
	(million US$)	(% of exports)			(% of imports)
Exports	116128.8	USA	20	USA	17
Imports	76435.6	Argentina	9	Argentina	9
		China	6	Germany	8

Social indicators	2000-2006
Population growth rate 2000-2005 (% per annum)	1.4
Population aged 0-14 years (%)	28.0
Population aged 60+ years (women and men, % of total)	10.0/8.0
Sex ratio (women per 100 men)	103
Life expectancy at birth 2000-2005 (women and men, years)	74/66
Infant mortality rate 2000-2005 (per 1,000 births)	27
Total fertility rate 2000-2005 (births per woman)	2.4
Urban population (%)	84
Urban population growth rate 2000-2005 (% per annum)	2.1
Rural population growth rate 2000-2005 (% per annum)	−2.1
Foreign born (%)	0.3[i]
Refugees and others of concern to UNHCR[j]	7744
Government education expenditure (% of GNP)	4.3
Primary-secondary gross enrolment ratio (w and m per 100)	118/114
Third-level students (women and men, % of total)	56/44
Newspaper circulation (per 1,000 inhabitants)	45
Television receivers (per 1,000 inhabitants)	358
Intentional homicides (per 100,000 inhabitants)	27
Parliamentary seats (women and men, % of total)	9/91

Environment	2000-2006
Threatened species	721
Forested area (% of land area)	64
CO2 emissions (000s Mt of carbon dioxide/per capita)	298902/1.6
Energy consumption per capita (kilograms oil equiv.)	817
Precipitation (mm)	1455[k]
Average minimum and maximum temperatures (centigrade)[k]	15.5/24.9

a October 2006. b August 2006. c Persons aged 10 years and over, excl. rural population of Rondônia, Acre, Amazonas, Roraima, Pará and Amapá. d 1999. e 2004. f 2003. g Source: World Automotive Market Report, Auto and Truck International (Illinois). h 2002. i Estimated data. j Provisional. k Sao Paulo.

British Virgin Islands

Region	Caribbean
Largest urban agglom. (pop., 000s)	Road Town (13)
Currency	US dollar
Population in 2006 (proj., 000s)	22[a]
Surface area (square kms)	151
Population density (per square km)	137

Economic indicators	2000	2005
Exchange rate (national currency per US$)	1.00	1.00
Consumer price index (2000=100)	100	112[b]
Tourist arrivals (000s)	272	305[c]
GDP (million current US$)	784	972
GDP (per capita current US$)	38203	44150
Gross fixed capital formation (% of GDP)	24.0	24.0
Employment in industrial sector (%)	21.3[d]	...
Employment in agricultural sector (%)	2.0[d]	...
Agricultural production index (1999-2001=100)	100	100[e]
Food production index (1999-2001=100)	100	100[e]
Motor vehicles (per 1,000 inhabitants)	369.9	519.6[c]
Telephone lines (per 100 inhabitants)	...	53.2[f]
Internet users, estimated (000s)	...	4.0[f]

Social indicators	2000-2006
Population growth rate 2000-2005 (% per annum)	1.4
Population aged 0-14 years (%)	26.3
Population aged 60+ years (women and men, % of total)	7.3/7.0
Sex ratio (women per 100 men)	94[gh]
Urban population (%)	61
Urban population growth rate 2000-2005 (% per annum)	2.5
Rural population growth rate 2000-2005 (% per annum)	−0.2
Primary-secondary gross enrolment ratio (w and m per 100)	102/103
Television receivers (per 1,000 inhabitants)	326

Environment	2000-2006
Threatened species	31
CO2 emissions (000s Mt of carbon dioxide/per capita)	77/3.6
Energy consumption per capita (kilograms oil equiv.)	1311[i]

a 2005. b February 2006. c 2004. d 1991. e 2003. f 2002. g Refers to de facto population count. h 2001. i Estimated data.

Brunei Darussalam

Region	South-eastern Asia
Largest urban agglom. (pop., 000s)	Bandar Seri Begawan (64)
Currency	dollar
Population in 2006 (proj., 000s)	382
Surface area (square kms)	5765
Population density (per square km)	66
United Nations membership date	21 September 1984

Economic indicators	2000	2005
Exchange rate (national currency per US$)	1.73	1.56[a]
Consumer price index (2000=100)	100	101
Tourist arrivals (000s)	984	840[b]
GDP (million current US$)	4316	6280
GDP (per capita current US$)	12944	16800
Gross fixed capital formation (% of GDP)	9.0	9.0
Employment in industrial sector (%)	...	21.4[b]
Employment in agricultural sector (%)	...	1.4[b]
Agricultural production index (1999-2001=100)	99	121[c]
Food production index (1999-2001=100)	99	121[c]
Primary energy production (000s Mt oil equiv.)	20978	22366[c]
Motor vehicles (per 1,000 inhabitants)	605.8	701.3
Telephone lines (per 100 inhabitants)	24.3	25.6[cd]
Internet users, estimated (000s)	30.0	135.0

Total trade		Major trading partners			2005
	(million US$)[e]	(% of exports)[e]		(% of imports)[e]	
Exports	4144.3	Japan	41	Malaysia	20
Imports	1243.6	Korea Rep.	11	Singapore	20
		Thailand	9	USA	11

Social indicators	2000-2006
Population growth rate 2000-2005 (% per annum)	2.3
Population aged 0-14 years (%)	30.0
Population aged 60+ years (women and men, % of total)	4.0/5.0
Sex ratio (women per 100 men)	93
Life expectancy at birth 2000-2005 (women and men, years)	79/74
Infant mortality rate 2000-2005 (per 1,000 births)	6
Total fertility rate 2000-2005 (births per woman)	2.5
Urban population (%)	74
Urban population growth rate 2000-2005 (% per annum)	2.9
Rural population growth rate 2000-2005 (% per annum)	0.6
Primary-secondary gross enrolment ratio (w and m per 100)	102/100
Third-level students (women and men, % of total)	67/33
Television receivers (per 1,000 inhabitants)	644

Environment	2000-2006
Threatened species	168
Forested area (% of land area)	84
CO2 emissions (000s Mt of carbon dioxide/per capita)	4558/12.7
Energy consumption per capita (kilograms oil equiv.)	8962
Precipitation (mm)	2913[f]
Average minimum and maximum temperatures (centigrade)[f]	23.3/31.8

a October 2006. b 2001. c 2004. d Estimate. e 2003. f Bandar Seri Begawan.

Bulgaria

Region	Eastern Europe
Largest urban agglom. (pop., 000s)	Sofia (1093)
Currency	lev
Population in 2006 (proj., 000s)	7671
Surface area (square kms)	110912
Population density (per square km)	69
United Nations membership date	14 December 1955

Economic indicators	2000	2005
Exchange rate (national currency per US$)	2.10	1.54[a]
Consumer price index (2000=100)	100	140[b]
Industrial production index (1995=100)	78	128[c]
Unemployment (percentage of labour force)[d]	17.9[e]	10.7[f]
Balance of payments, current account (million US$)	−703	−3133
Tourist arrivals (000s)[g]	2785	4630[h]
GDP (million current US$)	12600	26419
GDP (per capita current US$)	1576	3420
Gross fixed capital formation (% of GDP)	16.0	23.0
Labour force participation, adult female pop. (%)	42.9	44.2[i]
Labour force participation, adult male pop. (%)	52.4	55.3[h]
Employment in industrial sector (%)	28.3	34.2
Employment in agricultural sector (%)	26.2	8.9
Agricultural production index (1999-2001=100)	98	109[h]
Food production index (1999-2001=100)	98	108[h]
Primary energy production (000s Mt oil equiv.)	6269	6421[h]
Motor vehicles (per 1,000 inhabitants)	286.9	358.9[h]
Telephone lines (per 100 inhabitants)	35.4	32.1
Internet users, estimated (000s)	430.0	1591.7

Total trade		Major trading partners			2005
	(million US$)	(% of exports)		(% of imports)	
Exports	11725.1	Italy	12	Germany	14
Imports	18180.3	Turkey	10	Italy	9
		Germany	10	Turkey	6

Social indicators	2000-2006
Population growth rate 2000-2005 (% per annum)	−0.7
Population aged 0-14 years (%)	14.0
Population aged 60+ years (women and men, % of total)	25.0/20.0
Sex ratio (women per 100 men)	107
Life expectancy at birth 2000-2005 (women and men, years)	76/69
Infant mortality rate 2000-2005 (per 1,000 births)	13
Total fertility rate 2000-2005 (births per woman)	1.2
Urban population (%)	70
Urban population growth rate 2000-2005 (% per annum)	−0.4
Rural population growth rate 2000-2005 (% per annum)	−1.4
Refugees and others of concern to UNHCR[j]	5218
Government education expenditure (% of GNP)	4.4
Primary-secondary gross enrolment ratio (w and m per 100)	101/105
Third-level students (women and men, % of total)	52/48
Newspaper circulation (per 1,000 inhabitants)	175
Television receivers (per 1,000 inhabitants)	453
Intentional homicides (per 100,000 inhabitants)	3
Parliamentary seats (women and men, % of total)	22/78

Environment	2000-2006
Threatened species	47
Forested area (% of land area)	33
CO2 emissions (000s Mt of carbon dioxide/per capita)[k]	53320/6.8
Energy consumption per capita (kilograms oil equiv.)	1872
Precipitation (mm)	571[l]
Average minimum and maximum temperatures (centigrade)[l]	5.0/15.1

a October 2006. b May 2006. c August 2006. d Dec. of each year. e Men aged 16 to 60 years, women aged 16 to 55 years. f Men aged 16 to 62.5 years, women aged 16 to 57 years. g Excluding children without own passports. h 2004. i 2003. j Provisional. k Source: UNFCCC. l Sofia.

Burkina Faso

Region	Western Africa
Largest urban agglom. (pop., 000s)	Ouagadougou (926)
Currency	CFA franc
Population in 2006 (proj., 000s)	13634
Surface area (square kms)	274000
Population density (per square km)	50
United Nations membership date	20 September 1960

Economic indicators	2000	2005
Exchange rate (national currency per US$)[a]	704.95	516.66[b]
Consumer price index (2000=100)[c]	100	122[d]
Balance of payments, current account (million US$)	−392	−381[e]
Tourist arrivals (000s)	126	222[f]
GDP (million current US$)	2415	5397
GDP (per capita current US$)	214	408
Gross fixed capital formation (% of GDP)	27.0	24.0
Agricultural production index (1999-2001=100)	86	125[f]
Food production index (1999-2001=100)	85	115[f]
Primary energy production (000s Mt oil equiv.)	9	9[f]
Motor vehicles (per 1,000 inhabitants)[g]	4.3	4.0[h]
Telephone lines (per 100 inhabitants)	0.5	0.7
Internet users, estimated (000s)	9.0	64.6

Total trade		Major trading partners			2005
	(million US$)[f]		(% of exports)[f]		(% of imports)[f]
Exports	396.5	Ghana	61	Côte d'Ivoire	18
Imports	1267.2	France	11	France	14
		Côte d'Ivoire	4	Japan	13

Social indicators	2000-2006
Population growth rate 2000-2005 (% per annum)	3.2
Population aged 0-14 years (%)	47.0
Population aged 60+ years (women and men, % of total)	5.0/4.0
Sex ratio (women per 100 men)	99
Life expectancy at birth 2000-2005 (women and men, years)	48/47
Infant mortality rate 2000-2005 (per 1,000 births)	121
Total fertility rate 2000-2005 (births per woman)	6.7
Contraceptive use (% of currently married women)	14
Urban population (%)	18
Urban population growth rate 2000-2005 (% per annum)	5.2
Rural population growth rate 2000-2005 (% per annum)	2.7
Foreign born (%)	9.4[i]
Refugees and others of concern to UNHCR[j]	1295
Primary-secondary gross enrolment ratio (w and m per 100)	29/38
Third-level students (women and men, % of total)	22/78
Television receivers (per 1,000 inhabitants)	12
Parliamentary seats (women and men, % of total)	12/88

Environment	2000-2006
Threatened species	15
Forested area (% of land area)	26
CO2 emissions (000s Mt of carbon dioxide/per capita)	1041/0.1
Energy consumption per capita (kilograms oil equiv.)	28[i]

a Official rate. b October 2006. c Ouagadougou. d July 2006. e 2001. f 2004. g Source: World Automotive Market Report, Auto and Truck International (Illinois). h 2003. i Estimated data. j Provisional.

Burundi

Region	Eastern Africa
Largest urban agglom. (pop., 000s)	Bujumbura (447)
Currency	franc
Population in 2006 (proj., 000s)	7834
Surface area (square kms)	27834
Population density (per square km)	281
United Nations membership date	18 September 1962

Economic indicators	2000	2005
Exchange rate (national currency per US$)[a]	778.20	1059.10[b]
Unemployment (percentage of labour force)	14.0[c]	...
Balance of payments, current account (million US$)	−54	−263
Tourist arrivals (000s)[d]	29	36[e]
GDP (million current US$)	711	845
GDP (per capita current US$)	110	112
Gross fixed capital formation (% of GDP)	10.0	13.0
Agricultural production index (1999-2001=100)	95	104[f]
Food production index (1999-2001=100)	96	104[f]
Primary energy production (000s Mt oil equiv.)	10	13[f]
Motor vehicles (per 1,000 inhabitants)[g]	2.5	2.3[h]
Telephone lines (per 100 inhabitants)	0.3	0.4
Internet users, estimated (000s)	5.0	40.0

Total trade		Major trading partners			2005
	(million US$)[f]	(% of exports)[f]		(% of imports)[f]	
Exports	82.7	Switzerland	56	Kenya	15
Imports	172.7	Belgium	10	Japan	14
		Untd Arab Em	7	Belgium	12

Social indicators	2000-2006
Population growth rate 2000-2005 (% per annum)	3.0
Population aged 0-14 years (%)	45.0
Population aged 60+ years (women and men, % of total)	5.0/3.0
Sex ratio (women per 100 men)	105
Life expectancy at birth 2000-2005 (women and men, years)	44/42
Infant mortality rate 2000-2005 (per 1,000 births)	106
Total fertility rate 2000-2005 (births per woman)	6.8
Contraceptive use (% of currently married women)	16
Urban population (%)	10
Urban population growth rate 2000-2005 (% per annum)	6.1
Rural population growth rate 2000-2005 (% per annum)	2.7
Foreign born (%)	1.2[i]
Refugees and others of concern to UNHCR[j]	120329
Government education expenditure (% of GNP)	5.3
Primary-secondary gross enrolment ratio (w and m per 100)	41/50
Third-level students (women and men, % of total)	28/72
Television receivers (per 1,000 inhabitants)	40
Parliamentary seats (women and men, % of total)[k]	31/69

Environment	2000-2006
Threatened species	54
Forested area (% of land area)	4
CO2 emissions (000s Mt of carbon dioxide/per capita)	236/0.0
Energy consumption per capita (kilograms oil equiv.)	12[i]
Precipitation (mm)	838
Average minimum and maximum temperatures (centigrade)	22.1/24.6

a Official rate. b October 2006. c 1999, Bujumbura. d Including nationals residing abroad. e 2001. f 2004. g Source: World Automotive Market Report, Auto and Truck International (Illinois). h 2003. i Estimated data. j Provisional. k 1993.

Cambodia

Region	South-eastern Asia
Largest urban agglom. (pop., 000s)	Phnom-Penh (1364)
Currency	riel
Population in 2006 (proj., 000s)	14351
Surface area (square kms)	181035
Population density (per square km)	79
United Nations membership date	14 December 1955

Economic indicators	2000	2005
Exchange rate (national currency per US$)	3905.00	4110.00[a]
Consumer price index (2000=100)[b]	100	121[c]
Unemployment (percentage of labour force)[d]	2.5	1.8[e]
Balance of payments, current account (million US$)	−138	−356
Tourist arrivals (000s)	466	1055[f]
GDP (million current US$)	3668	5397
GDP (per capita current US$)	288	384
Gross fixed capital formation (% of GDP)	18.0	22.0
Labour force participation, adult female pop. (%)	76.3	81.6[e]
Labour force participation, adult male pop. (%)	80.8	85.1[e]
Employment in industrial sector (%)	8.4	10.5[e]
Employment in agricultural sector (%)	73.7	70.2[e]
Agricultural production index (1999-2001=100)	99	106[f]
Food production index (1999-2001=100)	99	105[f]
Primary energy production (000s Mt oil equiv.)	1	3[f]
Motor vehicles (per 1,000 inhabitants)	0.9	...
Telephone lines (per 100 inhabitants)	0.2	0.3
Internet users, estimated (000s)	6.0	41.0[f]

Total trade		Major trading partners			2005
	(million US$)[f]	(% of exports)[f]		(% of imports)[f]	
Exports	2797.5	USA	47	China, HK SAR	20
Imports	2062.7	China, HK SAR	22	China	17
		Germany	8	Thailand	11

Social indicators	2000-2005
Population growth rate 2000-2005 (% per annum)	2.0
Population aged 0-14 years (%)	37.0
Population aged 60+ years (women and men, % of total)	7.0/4.0
Sex ratio (women per 100 men)	107
Life expectancy at birth 2000-2005 (women and men, years)	60/52
Infant mortality rate 2000-2005 (per 1,000 births)	95
Total fertility rate 2000-2005 (births per woman)	4.1
Contraceptive use (% of currently married women)	24
Urban population (%)	20
Urban population growth rate 2000-2005 (% per annum)	5.1
Rural population growth rate 2000-2005 (% per annum)	1.3
Foreign born (%)	1.6[g]
Refugees and others of concern to UNHCR[h]	226
Government education expenditure (% of GNP)	2.2
Primary-secondary gross enrolment ratio (w and m per 100)	76/87
Third-level students (women and men, % of total)	31/69
Television receivers (per 1,000 inhabitants)	8
Parliamentary seats (women and men, % of total)	10/90

Environment	2000-2006
Threatened species	122
Forested area (% of land area)	53
CO2 emissions (000s Mt of carbon dioxide/per capita)	535/0.0
Energy consumption per capita (kilograms oil equiv.)	13[g]
Precipitation (mm)	1636[i]
Average minimum and maximum temperatures (centigrade)[b]	23.8/32.5

a June 2006. b Phnom Penh. c July 2006. d Persons aged 10 years and over. e 2001. f 2004. g Estimated data. h Provisional. i Phnom Penh-Pochentong.

Cameroon

Region	Middle Africa
Largest urban agglom. (pop., 000s)	Douala (1761)
Currency	CFA franc
Population in 2006 (proj., 000s)	16601
Surface area (square kms)	475442
Population density (per square km)	35
United Nations membership date	20 September 1960

Economic indicators	2000	2005
Exchange rate (national currency per US$)[a]	704.95	516.66[b]
Consumer price index (2000=100)	100	114[c]
Unemployment (percentage of labour force)[d]	8.1[e]	7.5[f]
Tourist arrivals (000s)	277	190[g]
GDP (million current US$)	9287	16823
GDP (per capita current US$)	625	1031
Gross fixed capital formation (% of GDP)	16.0	16.0
Labour force participation, adult female pop. (%)	61.5[e]	66.1[f]
Employment in industrial sector (%)	9.1[h]	...
Employment in agricultural sector (%)	60.6[h]	...
Agricultural production index (1999-2001=100)	100	105[g]
Food production index (1999-2001=100)	100	105[g]
Primary energy production (000s Mt oil equiv.)	7642	6974[g]
Motor vehicles (per 1,000 inhabitants)	11.0	14.6[i]
Telephone lines (per 100 inhabitants)	0.6	0.6
Internet users, estimated (000s)	40.0	250.0

Total trade		Major trading partners			2005
	(million US$)	(% of exports)		(% of imports)	
Exports	2447.0	Spain	19	Nigeria	22
Imports	2737.4	Italy	13	France	19
		France	12	China	5

Social indicators	2000-2006
Population growth rate 2000-2005 (% per annum)	1.9
Population aged 0-14 years (%)	41.0
Population aged 60+ years (women and men, % of total)	6.0/5.0
Sex ratio (women per 100 men)	101
Life expectancy at birth 2000-2005 (women and men, years)	47/45
Infant mortality rate 2000-2005 (per 1,000 births)	94
Total fertility rate 2000-2005 (births per woman)	4.7
Contraceptive use (% of currently married women)	26
Urban population (%)	55
Urban population growth rate 2000-2005 (% per annum)	3.6
Rural population growth rate 2000-2005 (% per annum)	−0.1
Foreign born (%)	1.0[j]
Refugees and others of concern to UNHCR[k]	58808
Government education expenditure (% of GNP)	4.0
Primary-secondary gross enrolment ratio (w and m per 100)	71/88
Third-level students (women and men, % of total)	39/61
Television receivers (per 1,000 inhabitants)	45
Parliamentary seats (women and men, % of total)	9/91

Environment	2000-2006
Threatened species	514
Forested area (% of land area)	51
CO2 emissions (000s Mt of carbon dioxide/per capita)	3543/0.2
Energy consumption per capita (kilograms oil equiv.)	80
Precipitation (mm)	1628[l]
Average minimum and maximum temperatures (centigrade)[l]	18.6/28.4

a Official rate. b October 2006. c March 2006. d Persons aged 15 years and over. e 1996. f 2001. g 2004. h 1990. i 2003. j Estimated data. k Provisional. l Yaounde.

Canada

Region	Northern America
Largest urban agglom. (pop., 000s)	Toronto (5312)
Currency	dollar
Population in 2006 (proj., 000s)	32566
Surface area (square kms)	9970610
Population density (per square km)	3
United Nations membership date	9 November 1945

Economic indicators	2000	2005
Exchange rate (national currency per US$)	1.50	1.12[a]
Consumer price index (2000=100)	100	115[b]
Industrial production index (1995=100)[c]	126	129[d]
Unemployment (percentage of labour force)[ef]	6.8	6.8
Balance of payments, current account (million US$)	19622	26555
Tourist arrivals (000s)	19627	19095[g]
GDP (million current US$)	714454	1131760
GDP (per capita current US$)	23280	35073
Gross fixed capital formation (% of GDP)	19.0	21.0
Labour force participation, adult female pop. (%)	59.4	62.1[g]
Labour force participation, adult male pop. (%)	72.3	73.3[g]
Employment in industrial sector (%)	22.6	22.0
Employment in agricultural sector (%)	3.3	2.7
Agricultural production index (1999-2001=100)	102	101[g]
Food production index (1999-2001=100)	103	102[g]
Primary energy production (000s Mt oil equiv.)[h]	371513	393509[g]
Motor vehicles (per 1,000 inhabitants)	571.2	583.6
Telephone lines (per 100 inhabitants)	67.7	56.6
Internet users, estimated (000s)	12971.0	16800.0

Total trade		Major trading partners			2005
	(million US$)		(% of exports)		(% of imports)
Exports	360135.7	USA	84	USA	57
Imports	314436.0	Japan	2	China	8
		UK	2	Japan	4

Social indicators	2000-2006
Population growth rate 2000-2005 (% per annum)	1.0
Population aged 0-14 years (%)	18.0
Population aged 60+ years (women and men, % of total)	20.0/16.0
Sex ratio (women per 100 men)	102
Life expectancy at birth 2000-2005 (women and men, years)	82/77
Infant mortality rate 2000-2005 (per 1,000 births)	5
Total fertility rate 2000-2005 (births per woman)	1.5
Urban population (%)	80
Urban population growth rate 2000-2005 (% per annum)	1.2
Rural population growth rate 2000-2005 (% per annum)	0.3
Foreign born (%)	19.1
Refugees and others of concern to UNHCR[i]	167723
Government education expenditure (% of GNP)	5.4
Primary-secondary gross enrolment ratio (w and m per 100)	104/105
Third-level students (women and men, % of total)	56/44
Newspaper circulation (per 1,000 inhabitants)	168
Television receivers (per 1,000 inhabitants)	706
Intentional homicides (per 100,000 inhabitants)	2
Parliamentary seats (women and men, % of total)	24/76

Environment	2000-2006
Threatened species	82
Forested area (% of land area)	27
CO2 emissions (000s Mt of carbon dioxide/per capita)[j]	586070/18.5
Energy consumption per capita (kilograms oil equiv.)	8535
Precipitation (mm)	834[k]
Average minimum and maximum temperatures (centigrade)[k]	5.6/12.7

a October 2006. b July 2006. c Monthly indices are adjusted for differences in the number of working days. d August 2006. e Persons aged 15 years and over, excl. residents of the territories and indigenous persons living on reserves. f Excl. full-time members of the armed forces. g 2004. h Including vehicles operated by police or other governmental security organizations. i Provisional. j Source: UNFCCC. k Toronto.

Cape Verde

Region	Western Africa
Largest urban agglom. (pop., 000s)	Praia (117)
Currency	escudo
Population in 2006 (proj., 000s)	519
Surface area (square kms)	4033
Population density (per square km)	129
United Nations membership date	16 September 1975

Economic indicators	2000	2005
Exchange rate (national currency per US$)[a]	118.51	86.02[b]
Consumer price index (2000=100)	100	111[c]
Balance of payments, current account (million US$)	−58	−77[d]
Tourist arrivals (000s)	115	157[e]
GDP (million current US$)	539	1038
GDP (per capita current US$)	1197	2048
Gross fixed capital formation (% of GDP)	30.0	22.0
Agricultural production index (1999-2001=100)	97	92[e]
Food production index (1999-2001=100)	97	92[e]
Primary energy production (000s Mt oil equiv.)	1	1[e]
Motor vehicles (per 1,000 inhabitants)	37.7[f]	
Telephone lines (per 100 inhabitants)	12.6	14.1
Internet users, estimated (000s)	8.0	29.0

Total trade		Major trading partners			2005
	(million US$)[e]	(% of exports)[e]		(% of imports)[e]	
Exports	15.2	Portugal	80	Portugal	38
Imports	429.2	USA	18	USA	14
		...		Netherlands	8

Social indicators	2000-2006
Population growth rate 2000-2005 (% per annum)	2.4
Population aged 0-14 years (%)	40.0
Population aged 60+ years (women and men, % of total)	7.0/4.0
Sex ratio (women per 100 men)	108
Life expectancy at birth 2000-2005 (women and men, years)	73/67
Infant mortality rate 2000-2005 (per 1,000 births)	30
Total fertility rate 2000-2005 (births per woman)	3.8
Urban population (%)	57
Urban population growth rate 2000-2005 (% per annum)	3.8
Rural population growth rate 2000-2005 (% per annum)	0.6
Foreign born (%)	2.4[g]
Government education expenditure (% of GNP)	7.4
Primary-secondary gross enrolment ratio (w and m per 100)	88/88
Third-level students (women and men, % of total)	53/47
Television receivers (per 1,000 inhabitants)	105
Parliamentary seats (women and men, % of total)	15/85

Environment	2000-2006
Threatened species	28
Forested area (% of land area)	21
CO2 emissions (000s Mt of carbon dioxide/per capita)	144/0.3
Energy consumption per capita (kilograms oil equiv.)	219[g]
Precipitation (mm)	70[h]
Average minimum and maximum temperatures (centigrade)[h]	23.5/5.8

a Official rate. b August 2006. c July 2006. d 2003. e 2004. f 1999. g Estimated data.
h Sal.

Cayman Islands

Region	Caribbean
Largest urban agglom. (pop., 000s)	George Town (26)
Currency	dollar
Population in 2006 (proj., 000s)	45[a]
Surface area (square kms)	264
Population density (per square km)	1479

Economic indicators	2000	2005
Exchange rate (national currency per US$)[b]	...	0.81[c]
Consumer price index (2000=100)	100	115[d]
Tourist arrivals (000s)	354	260[e]
GDP (million current US$)	1323	1593
GDP (per capita current US$)	33319	35381
Gross fixed capital formation (% of GDP)	22.0	22.0
Labour force participation, adult female pop. (%)	82.4[f]	...
Labour force participation, adult male pop. (%)	89.0[f]	...
Agricultural production index (1999-2001=100)	100	100[g]
Food production index (1999-2001=100)	100	100[g]
Motor vehicles (per 1,000 inhabitants)	609.5	651.5[h]

Social indicators	2000-2006
Population growth rate 2000-2005 (% per annum)	2.5
Urban population (%)	100
Urban population growth rate 2000-2005 (% per annum)	2.5
Rural population growth rate 2000-2005 (% per annum)	−
Foreign born (%)	47.9[i]
Primary-secondary gross enrolment ratio (w and m per 100)	96/94
Third-level students (women and men, % of total)	75/25
Television receivers (per 1,000 inhabitants)	278
Intentional homicides (per 100,000 inhabitants)	10

Environment	2000-2006
Threatened species	23
CO2 emissions (000s Mt of carbon dioxide/per capita)	304/7.1
Energy consumption per capita (kilograms oil equiv.)	2407[j]
Precipitation (mm)	1435
Average minimum and maximum temperatures (centigrade)	21.2/32.0

a 2005. b Data refer to non-commercial rates derived from the Operational Rates of Exchange for United Nations Programmes. c November 2006. d March 2006. e 2004. f 1999. g 2003. h 2002. i 1989. j Estimated data.

Central African Republic

Region	Middle Africa
Largest urban agglom. (pop., 000s)	Bangui (541)
Currency	CFA franc
Population in 2006 (proj., 000s)	4093
Surface area (square kms)	622984
Population density (per square km)	7
United Nations membership date	20 September 1960

Economic indicators	2000	2005
Exchange rate (national currency per US$)[a]	704.95	516.66[b]
Consumer price index (2000=100)[c]	100	119[d]
Tourist arrivals (000s)	11	6[e]
GDP (million current US$)	906	1325
GDP (per capita current US$)	240	328
Gross fixed capital formation (% of GDP)	9.0	5.0
Agricultural production index (1999-2001=100)	102	105[e]
Food production index (1999-2001=100)	102	108[e]
Primary energy production (000s Mt oil equiv.)	7	7[e]
Motor vehicles (per 1,000 inhabitants)[f]	3.1	3.0[g]
Telephone lines (per 100 inhabitants)	0.3	0.3
Internet users, estimated (000s)	2.0	11.0

Total trade		Major trading partners		2005
	(million US$)	(% of exports)		(% of imports)
Exports	469.5	Belgium	29	...
Imports	186.5	France	19	...
		Switzerland	14	...

Social indicators	2000-2006
Population growth rate 2000-2005 (% per annum)	1.3
Population aged 0-14 years (%)	43.0
Population aged 60+ years (women and men, % of total)	7.0/5.0
Sex ratio (women per 100 men)	105
Life expectancy at birth 2000-2005 (women and men, years)	40/39
Infant mortality rate 2000-2005 (per 1,000 births)	98
Total fertility rate 2000-2005 (births per woman)	5.0
Contraceptive use (% of currently married women)	28
Urban population (%)	38
Urban population growth rate 2000-2005 (% per annum)	1.6
Rural population growth rate 2000-2005 (% per annum)	1.2
Foreign born (%)	1.6[h]
Refugees and others of concern to UNHCR[i]	26603
Third-level students (women and men, % of total)	16/84
Television receivers (per 1,000 inhabitants)	6
Parliamentary seats (women and men, % of total)	11/89

Environment	2000-2006
Threatened species	31
Forested area (% of land area)	37
CO2 emissions (000s Mt of carbon dioxide/per capita)	252/0.1
Energy consumption per capita (kilograms oil equiv.)	22[h]
Precipitation (mm)	1560
Average minimum and maximum temperatures (centigrade)	25.1/27.4

a Official rate. b October 2006. c Excluding rent, Bangui. d May 2006. e 2004. f Source: World Automotive Market Report, Auto and Truck International (Illinois). g 2002. h Estimated data. i Provisional.

Chad

Region	Middle Africa
Largest urban agglom. (pop., 000s)	N'Djamena (888)
Currency	CFA franc
Population in 2006 (proj., 000s)	10032
Surface area (square kms)	1284000
Population density (per square km)	8
United Nations membership date	20 September 1960

Economic indicators	2000	2005
Exchange rate (national currency per US$)[a]	704.95	516.66[b]
Consumer price index (2000=100)[c]	100	138[d]
Tourist arrivals (000s)	43	21[e]
GDP (million current US$)	1386	4942
GDP (per capita current US$)	169	507
Gross fixed capital formation (% of GDP)	21.0	25.0
Labour force participation, adult female pop. (%)	64.7[f]	...
Labour force participation, adult male pop. (%)	81.0[f]	...
Agricultural production index (1999-2001=100)	93	113[g]
Food production index (1999-2001=100)	93	112[g]
Telephone lines (per 100 inhabitants)	0.1	0.2
Internet users, estimated (000s)	3.0	40.0

Social indicators	2000-2006
Population growth rate 2000-2005 (% per annum)	3.4
Population aged 0-14 years (%)	47.0
Population aged 60+ years (women and men, % of total)	5.0/4.0
Sex ratio (women per 100 men)	102
Life expectancy at birth 2000-2005 (women and men, years)	45/43
Infant mortality rate 2000-2005 (per 1,000 births)	116
Total fertility rate 2000-2005 (births per woman)	6.7
Contraceptive use (% of currently married women)	3
Urban population (%)	25
Urban population growth rate 2000-2005 (% per annum)	5.0
Rural population growth rate 2000-2005 (% per annum)	2.9
Refugees and others of concern to UNHCR[h]	276927
Government education expenditure (% of GNP)	1.7[i]
Primary-secondary gross enrolment ratio (w and m per 100)	37/62
Third-level students (women and men, % of total)	13/87
Television receivers (per 1,000 inhabitants)	6
Parliamentary seats (women and men, % of total)	7/93

Environment	2000-2006
Threatened species	23
Forested area (% of land area)	10
CO2 emissions (000s Mt of carbon dioxide/per capita)	117/0.0
Energy consumption per capita (kilograms oil equiv.)	4[j]
Precipitation (mm)	510[c]
Average minimum and maximum temperatures (centigrade)[c]	20.8/30.5

a Official rate. b October 2006. c N'Djamena. d May 2006. e 2003. f 1993. g 2004.
h Provisional. i 1999. j Estimated data.

Chile

Region	South America
Largest urban agglom. (pop., 000s)	Santiago (5683)
Currency	peso
Population in 2006 (proj., 000s)	16465
Surface area (square kms)	756626
Population density (per square km)	22
United Nations membership date	24 October 1945

Economic indicators	2000	2005
Exchange rate (national currency per US$)[a]	572.68	525.99[b]
Consumer price index (2000=100)[c]	100	119[d]
Industrial production index (1995=100)[e]	130	157[d]
Unemployment (percentage of labour force)[f]	8.3	6.9
Balance of payments, current account (million US$)	−898	703
Tourist arrivals (000s)	1742	1785[g]
GDP (million current US$)	75196	111339
GDP (per capita current US$)	4879	6833
Gross fixed capital formation (% of GDP)	21.0	22.0
Labour force participation, adult female pop. (%)	35.0	37.8[g]
Labour force participation, adult male pop. (%)	72.9	71.2[g]
Employment in industrial sector (%)	23.4	23.0
Employment in agricultural sector (%)	14.4	13.2
Agricultural production index (1999-2001=100)	99	113[g]
Food production index (1999-2001=100)	99	113[g]
Primary energy production (000s Mt oil equiv.)	4324	4143[g]
Motor vehicles (per 1,000 inhabitants)	131.2	140.2[g]
Telephone lines (per 100 inhabitants)	21.7	22.0
Internet users, estimated (000s)	2537.3	4300.0[g]

Total trade		Major trading partners			2005
	(million US$)	(% of exports)			(% of imports)
Exports	38595.6	USA	16	Argentina	16
Imports	29857.2	Japan	12	USA	16
		China	11	Brazil	13

Social indicators	2000-2006
Population growth rate 2000-2005 (% per annum)	1.1
Population aged 0-14 years (%)	25.0
Population aged 60+ years (women and men, % of total)	13.0/10.0
Sex ratio (women per 100 men)	102
Life expectancy at birth 2000-2005 (women and men, years)	81/75
Infant mortality rate 2000-2005 (per 1,000 births)	8
Total fertility rate 2000-2005 (births per woman)	2.0
Urban population (%)	88
Urban population growth rate 2000-2005 (% per annum)	1.5
Rural population growth rate 2000-2005 (% per annum)	−1.4
Refugees and others of concern to UNHCR[h]	913
Government education expenditure (% of GNP)	4.1
Primary-secondary gross enrolment ratio (w and m per 100)	95/97
Third-level students (women and men, % of total)	48/52
Television receivers (per 1,000 inhabitants)	321
Intentional homicides (per 100,000 inhabitants)	5
Parliamentary seats (women and men, % of total)	13/87

Environment	2000-2006
Threatened species	132
Forested area (% of land area)	21
CO2 emissions (000s Mt of carbon dioxide/per capita)	58591/3.7
Energy consumption per capita (kilograms oil equiv.)	1511
Precipitation (mm)	313[c]
Average minimum and maximum temperatures (centigrade)[c]	8.3/22.5

a Principal rate. b October 2006. c Santiago. d August 2006. e Calculated by the Statistics Division of the United Nations from component national indices. f Persons aged 15 years and over. g 2004. h Provisional.

China[a]

Region	Eastern Asia
Largest urban agglom. (pop., 000s)	Shanghai (14503)
Currency	yuan
Population in 2006 (proj., 000s)	1323636
Surface area (square kms)	9596961
Population density (per square km)	138
United Nations membership date	24 October 1945

Economic indicators	2000	2005
Exchange rate (national currency per US$)[b]	8.28	7.88[c]
Consumer price index (2000=100)	100	109[d]
Unemployment (percentage of labour force)[ef]	3.1	4.2
Balance of payments, current account (million US$)	20518	160818
Tourist arrivals (000s)	31229	41761[g]
GDP (million current US$)	1079192	1981648
GDP (per capita current US$)	862	1533
Gross fixed capital formation (% of GDP)	37.0	42.0
Labour force participation, adult female pop. (%)	73.7[h]	...
Labour force participation, adult male pop. (%)	85.6[h]	...
Employment in industrial sector (%)	17.3	17.7[i]
Employment in agricultural sector (%)	46.3	44.1[i]
Agricultural production index (1999-2001=100)[j]	100	118[g]
Food production index (1999-2001=100)[j]	100	118[g]
Primary energy production (000s Mt oil equiv.)	863675	1242421[g]
Motor vehicles (per 1,000 inhabitants)	10.2[k]	20.1[g]
Telephone lines (per 100 inhabitants)	11.2	26.6
Internet users, estimated (000s)	22500.0	113162.6

Total trade		Major trading partners			2005
	(million US$)	(% of exports)		(% of imports)	
Exports	761953.4	USA	21	Japan	15
Imports	659952.8	China, HK SAR	16	Korea Rep.	12
		Japan	11	China	8

Social indicators	2000-2006
Population growth rate 2000-2005 (% per annum)	0.6
Population aged 0-14 years (%)	21.0
Population aged 60+ years (women and men, % of total)	12.0/10.0
Sex ratio (women per 100 men)	95
Life expectancy at birth 2000-2005 (women and men, years)	73/70
Infant mortality rate 2000-2005 (per 1,000 births)	35
Total fertility rate 2000-2005 (births per woman)	1.7
Urban population (%)	40
Urban population growth rate 2000-2005 (% per annum)	3.1
Rural population growth rate 2000-2005 (% per annum)	-0.9
Foreign born (%)	<[l]
Refugees and others of concern to UNHCR[m]	301125
Government education expenditure (% of GNP)	2.1[n]
Primary-secondary gross enrolment ratio (w and m per 100)	92/92
Third-level students (women and men, % of total)	44/56
Newspaper circulation (per 1,000 inhabitants)	59
Television receivers (per 1,000 inhabitants)	382
Parliamentary seats (women and men, % of total)	20/80

Environment	2000-2006
Threatened species	804
Forested area (% of land area)	18
CO2 emissions (000s Mt of carbon dioxide/per capita)	4151412/3.2
Energy consumption per capita (kilograms oil equiv.)	970
Precipitation (mm)	578[o]
Average minimum and maximum temperatures (centigrade)[o]	6.5/17.7

a For statistical purposes the data for China do not include those for Hong Kong Special Administrative Region (Hong Kong SAR), Macao Special Administrative Region (Macao SAR) and Taiwan Province of China, except in statistics relating to population of Taiwan Province. b Principal rate. c October 2006. d July 2006. e Unemployed in urban areas, persons aged 15 years and over. f Dec. of each year. g 2004. h 1995. i 2002. j Data generally include those for Taiwan Province of China. k 1998. l Estimated data. m Provisional. n 1999. o Beijing.

China, Hong Kong SAR[a]

Region	Eastern Asia
Largest urban agglom. (pop., 000s)	Hong Kong (7041)
Currency	dollar
Population in 2006 (proj., 000s)	7118
Surface area (square kms)	1099
Population density (per square km)	6477

Economic indicators	2000	2005
Exchange rate (national currency per US$)	7.80	7.78[b]
Consumer price index (2000=100)	100	96[c]
Industrial production index (1995=100)[d]	85	77[e]
Unemployment (percentage of labour force)[f]	4.9	5.6[g]
Balance of payments, current account (million US$)	6993	20284
Tourist arrivals (000s)	8814	13655[h]
GDP (million current US$)	168754	172649
GDP (per capita current US$)	25426	24521
Gross fixed capital formation (% of GDP)	26.0	22.0
Labour force participation, adult female pop. (%)	49.1	51.9[h]
Labour force participation, adult male pop. (%)	...	72.0[ij]
Employment in industrial sector (%)	20.3	15.2
Employment in agricultural sector (%)	0.3	0.3
Motor vehicles (per 1,000 inhabitants)	72.9	70.9[h]
Telephone lines (per 100 inhabitants)[k]	58.9	53.9
Internet users, estimated (000s)[k]	1855.2	3526.2

Total trade		Major trading partners			2005
	(million US$)	(% of exports)		(% of imports)	
Exports	292118.7	China	45	China	45
Imports	300160.4	USA	16	Japan	11
		Japan	5	Singapore	6

Social indicators	2000-2006
Population growth rate 2000-2005 (% per annum)	1.2
Population aged 0-14 years (%)	14.0
Population aged 60+ years (women and men, % of total)	15.0/16.0
Sex ratio (women per 100 men)	113
Life expectancy at birth 2000-2005 (women and men, years)	85/79
Infant mortality rate 2000-2005 (per 1,000 births)	4
Total fertility rate 2000-2005 (births per woman)	0.9
Contraceptive use (% of currently married women)	86
Urban population (%)	100
Urban population growth rate 2000-2005 (% per annum)	1.2
Rural population growth rate 2000-2005 (% per annum)	—
Foreign born (%)	40.3
Refugees and others of concern to UNHCR[g]	3031
Government education expenditure (% of GNP)	4.6
Primary-secondary gross enrolment ratio (w and m per 100)	93/97
Third-level students (women and men, % of total)	51/49
Newspaper circulation (per 1,000 inhabitants)	223
Television receivers (per 1,000 inhabitants)[k]	549
Intentional homicides (per 100,000 inhabitants)	1[j]

Environment	2000-2006
Threatened species	43
CO2 emissions (000s Mt of carbon dioxide/per capita)	37865/5.5
Energy consumption per capita (kilograms oil equiv.)	1772
Precipitation (mm)	2214[l]
Average minimum and maximum temperatures (centigrade)	20.8/25.7

a Pursuant to a Joint Declaration signed on 19 December 1984, the United Kingdom restored Hong Kong to the People's Republic of China with effect from 1 July 1997; the People's Republic of China resumed the exercise of sovereignty over the territory with effect from that date. b October 2006. c August 2006. d Calculated by the Statistics Division of the United Nations from component national indices. e 2nd quarter 2006. f Persons aged 15 years and over, excluding marine, military and institutional populations. g Provisional. h 2004. i 2003. j Hong Kong Census and Statistics Department. k Year beginning 1 April. l Hong Kong.

China, Macao SAR[a]

Region	Eastern Asia
Largest urban agglom. (pop., 000s)	Macao (460)
Currency	pataca
Population in 2006 (proj., 000s)	463
Surface area (square kms)	26
Population density (per square km)	17817

Economic indicators	2000	2005
Exchange rate (national currency per US$)[b]	8.03	7.99[c]
Consumer price index (2000=100)	100	104[d]
Unemployment (percentage of labour force)[e]	6.8	4.1
Balance of payments, current account (million US$)	2739[f]	4163[g]
Tourist arrivals (000s)	5197	8324[g]
GDP (million current US$)	6102	10992
GDP (per capita current US$)	13757	23887
Gross fixed capital formation (% of GDP)	11.0	23.0
Labour force participation, adult female pop. (%)	54.9	54.8[g]
Labour force participation, adult male pop. (%)	73.3	70.1[g]
Employment in industrial sector (%)	28.2	25.2
Employment in agricultural sector (%)	0.2	0.1
Motor vehicles (per 1,000 inhabitants)	126.3	150.9[gh]
Telephone lines (per 100 inhabitants)	40.9	37.9
Internet users, estimated (000s)	60.0	170.0

Total trade		Major trading partners			2005
	(million US$)[g]	(% of exports)[g]		(% of imports)[g]	
Exports	2812.1	USA	49	China	44
Imports	3477.8	China	14	China, HK SAR	11
		Germany	8	Japan	10

Social indicators	2000-2006
Population growth rate 2000-2005 (% per annum)	0.7
Population aged 0-14 years (%)	16.0
Population aged 60+ years (women and men, % of total)	11.0/10.0
Sex ratio (women per 100 men)	108
Life expectancy at birth 2000-2005 (women and men, years)	82/78
Infant mortality rate 2000-2005 (per 1,000 births)	8
Total fertility rate 2000-2005 (births per woman)	0.8
Urban population (%)	100
Urban population growth rate 2000-2005 (% per annum)	0.7
Rural population growth rate 2000-2005 (% per annum)	−0.3
Foreign born (%)	56.1
Government education expenditure (% of GNP)	3.0
Primary-secondary gross enrolment ratio (w and m per 100)	99/101
Third-level students (women and men, % of total)	41/59
Newspaper circulation (per 1,000 inhabitants)	377
Television receivers (per 1,000 inhabitants)	292
Intentional homicides (per 100,000 inhabitants)	4[ij]

Environment	2000-2006
Threatened species	5[g]
CO_2 emissions (000s Mt of carbon dioxide/per capita)	1868/4.1
Energy consumption per capita (kilograms oil equiv.)	1594
Precipitation (mm)	2123[k]
Average minimum and maximum temperatures (centigrade)	20.2/25.2

a Pursuant to a Joint Declaration signed on 13 April 1987, Portugal restored Macao to the People's Republic of China with effect from 20 December 1999; the People's Republic of China resumed the exercise of sovereignty over the territory with effect from that date. b Data refer to non-commercial rates derived from the Operational Rates of Exchange for United Nations Programmes. c October 2006. d August 2006. e Persons aged 14 years and over. f 2002. g 2004. h Including special-purpose vehicles. i Based on 30 or fewer deaths. j 1994. k Macao.

Colombia

Region	South America
Largest urban agglom. (pop., 000s)	Bogotá (7747)
Currency	peso
Population in 2006 (proj., 000s)	46279
Surface area (square kms)	1138914
Population density (per square km)	41
United Nations membership date	5 November 1945

Economic indicators	2000	2005
Exchange rate (national currency per US$)[a]	2187.02	2315.38[b]
Consumer price index (2000=100)[c]	100	146[d]
Industrial production index (1995=100)[e]	93	106[f]
Unemployment (percentage of labour force)	20.5[g]	11.8[h]
Balance of payments, current account (million US$)	765	−1978
Tourist arrivals (000s)	557	791[i]
GDP (million current US$)	83766	121878
GDP (per capita current US$)	1989	2673
Gross fixed capital formation (% of GDP)	13.0	19.0
Labour force participation, adult female pop. (%)	57.1	49.3[i]
Labour force participation, adult male pop. (%)	74.0	74.3[i]
Employment in industrial sector (%)	25.5	18.8
Employment in agricultural sector (%)	1.1	22.4
Agricultural production index (1999-2001=100)	101	110[i]
Food production index (1999-2001=100)	101	110[i]
Primary energy production (000s Mt oil equiv.)	69310	71559[i]
Motor vehicles (per 1,000 inhabitants)	28.8[j]	54.8[i]
Telephone lines (per 100 inhabitants)	17.0	16.8
Internet users, estimated (000s)	878.0	4738.5

Total trade		Major trading partners			2005
	(million US$)	(% of exports)		(% of imports)	
Exports	21190.4	USA	42	USA	28
Imports	21204.2	Venezuela	10	Mexico	8
		Ecuador	6	China	8

Social indicators	2000-2006
Population growth rate 2000-2005 (% per annum)	1.6
Population aged 0-14 years (%)	31.0
Population aged 60+ years (women and men, % of total)	8.0/7.0
Sex ratio (women per 100 men)	102
Life expectancy at birth 2000-2005 (women and men, years)	75/69
Infant mortality rate 2000-2005 (per 1,000 births)	26
Total fertility rate 2000-2005 (births per woman)	2.6
Contraceptive use (% of currently married women)	78[k]
Urban population (%)	73
Urban population growth rate 2000-2005 (% per annum)	2.0
Rural population growth rate 2000-2005 (% per annum)	0.5
Foreign born (%)	0.3[l]
Refugees and others of concern to UNHCR[m]	2000210
Government education expenditure (% of GNP)	5.1
Primary-secondary gross enrolment ratio (w and m per 100)	93/90
Third-level students (women and men, % of total)	51/49
Television receivers (per 1,000 inhabitants)	251
Parliamentary seats (women and men, % of total)	17/83

Environment	2000-2006
Threatened species	614
Forested area (% of land area)	48
CO2 emissions (000s Mt of carbon dioxide/per capita)	55631/1.3
Energy consumption per capita (kilograms oil equiv.)	467
Precipitation (mm)	799[n]
Average minimum and maximum temperatures (centigrade)[n]	7.4/16.0

a Principal rate. b October 2006. c Low income group. d July 2006. e Manufacturing. f February 2006. g Persons aged 10 years and over. h Persons aged 12 years and over. i 2004. j Source: World Automotive Market Report, Auto and Truck International (Illinois). k 2005. l Estimated data. m Provisional. n Bogota.

Comoros

Region	Eastern Africa
Largest urban agglom. (pop., 000s)	Moroni (44)
Currency	franc
Population in 2006 (proj., 000s)	798[a]
Surface area (square kms)	2235
Population density (per square km)	366
United Nations membership date	12 November 1975

Economic indicators	2000	2005
Exchange rate (national currency per US$)[b]	528.71	387.50[c]
Tourist arrivals (000s)	24	18[d]
GDP (million current US$)	204	380
GDP (per capita current US$)	292	477
Gross fixed capital formation (% of GDP)	10.0	9.0
Agricultural production index (1999-2001=100)	100	105[d]
Food production index (1999-2001=100)	100	105[d]
Motor vehicles (per 1,000 inhabitants)	10.3[e]	...
Telephone lines (per 100 inhabitants)	1.0	2.1
Internet users, estimated (000s)	1.5	20.0

Total trade		Major trading partners			2005
	(million US$)[f]	(% of exports)[f]		(% of imports)[f]	
Exports	6.9	France	43	South Africa	54
Imports	71.9	USA	16	France	19
		Singapore	16	Pakistan	7

Social indicators	2000-2006
Population growth rate 2000-2005 (% per annum)	2.6[g]
Population aged 0-14 years (%)	42.0
Population aged 60+ years (women and men, % of total)	5.0/4.0
Sex ratio (women per 100 men)	99[g]
Life expectancy at birth 2000-2005 (women and men, years)	65/61[g]
Infant mortality rate 2000-2005 (per 1,000 births)	58
Total fertility rate 2000-2005 (births per woman)	4.9[g]
Contraceptive use (% of currently married women)	26
Urban population (%)[h]	37
Urban population growth rate 2000-2005 (% per annum)	4.4
Rural population growth rate 2000-2005 (% per annum)[h]	1.7
Foreign born (%)	2.6[i]
Refugees and others of concern to UNHCR[j]	1
Government education expenditure (% of GNP)	3.9
Primary-secondary gross enrolment ratio (w and m per 100)	55/65
Third-level students (women and men, % of total)	43/57
Television receivers (per 1,000 inhabitants)	27
Parliamentary seats (women and men, % of total)	3/97

Environment	2000-2006
Threatened species	29
Forested area (% of land area)	4
CO2 emissions (000s Mt of carbon dioxide/per capita)	89/0.1
Energy consumption per capita (kilograms oil equiv.)	46[i]
Precipitation (mm)	2700[k]
Average minimum and maximum temperatures (centigrade)[k]	21.2/29.5

a 2005. b Official rate. c October 2006. d 2004. e 1987. f 2000. g Including Mayotte. h Including island of Mayotte. i Estimated data. j Provisional. k Moroni.

Congo

Region	Middle Africa
Largest urban agglom. (pop., 000s)	Brazzaville (1173)
Currency	CFA franc
Population in 2006 (proj., 000s)	4117
Surface area (square kms)	342000
Population density (per square km)	12
United Nations membership date	20 September 1960

Economic indicators	2000	2005
Exchange rate (national currency per US$)[a]	704.95	516.66[b]
Consumer price index (2000=100)[c]	100	115[d]
Balance of payments, current account (million US$)	648	903
Tourist arrivals (000s)	19	22[e]
GDP (million current US$)	3220	5528
GDP (per capita current US$)	937	1382
Gross fixed capital formation (% of GDP)	18.0	23.0
Agricultural production index (1999-2001=100)	100	109[f]
Food production index (1999-2001=100)	100	109[f]
Primary energy production (000s Mt oil equiv.)	13839	11360[f]
Motor vehicles (per 1,000 inhabitants)	15.4	14.9[g]
Telephone lines (per 100 inhabitants)	0.8	0.4
Internet users, estimated (000s)	0.8	50.0

Total trade		Major trading partners		2005
	(million US$)[h]	(% of exports)		(% of imports)[h]
Exports	1722.0	...	Netherlands	38
Imports	681.5	...	France	28
		...	USA	10

Social indicators	2000-2006
Population growth rate 2000-2005 (% per annum)	3.0
Population aged 0-14 years (%)	47.0
Population aged 60+ years (women and men, % of total)	5.0/4.0
Sex ratio (women per 100 men)	102
Life expectancy at birth 2000-2005 (women and men, years)	53/51
Infant mortality rate 2000-2005 (per 1,000 births)	72
Total fertility rate 2000-2005 (births per woman)	6.3
Urban population (%)	60
Urban population growth rate 2000-2005 (% per annum)	3.6
Rural population growth rate 2000-2005 (% per annum)	2.1
Foreign born (%)	5.7[i]
Refugees and others of concern to UNHCR[j]	69907
Government education expenditure (% of GNP)	4.4
Primary-secondary gross enrolment ratio (w and m per 100)	61/68
Third-level students (women and men, % of total)	16/84
Television receivers (per 1,000 inhabitants)	13
Parliamentary seats (women and men, % of total)	11/89

Environment	2000-2006
Threatened species	73
Forested area (% of land area)	65
CO2 emissions (000s Mt of carbon dioxide/per capita)	1380/0.4
Energy consumption per capita (kilograms oil equiv.)	137
Precipitation (mm)	1371

a Official rate. b October 2006. c Brazzaville. d May 2006. e 2002. f 2004. g 2001. h 2003. i Estimated data. j Provisional.

Congo, Democratic Republic of the

Region	Middle Africa
Largest urban agglom. (pop., 000s)	Kinshasa (6049)
Currency	franc congolais
Population in 2006 (proj., 000s)	59320
Surface area (square kms)	2344858
Population density (per square km)	25
United Nations membership date	20 September 1960

Economic indicators	2000	2005
Exchange rate (national currency per US$)	50.00	449.66[a]
Tourist arrivals (000s)	103	35[b]
GDP (million current US$)	5256	7212
GDP (per capita current US$)	105	125
Gross fixed capital formation (% of GDP)	10.0	12.0
Agricultural production index (1999-2001=100)	100	97[c]
Food production index (1999-2001=100)	100	98[c]
Primary energy production (000s Mt oil equiv.)	1739	1696[c]
Motor vehicles (per 1,000 inhabitants)	4.2[d]	...
Telephone lines (per 100 inhabitants)	0.1	<[e]
Internet users, estimated (000s)	3.0	140.6

Social indicators	2000-2006
Population growth rate 2000-2005 (% per annum)	2.8
Population aged 0-14 years (%)	47.0
Population aged 60+ years (women and men, % of total)	5.0/4.0
Sex ratio (women per 100 men)	102
Life expectancy at birth 2000-2005 (women and men, years)	44/42
Infant mortality rate 2000-2005 (per 1,000 births)	119
Total fertility rate 2000-2005 (births per woman)	6.7
Contraceptive use (% of currently married women)	31
Urban population (%)	32
Urban population growth rate 2000-2005 (% per annum)	4.3
Rural population growth rate 2000-2005 (% per annum)	2.1
Foreign born (%)	1.5[f]
Refugees and others of concern to UNHCR[g]	243529
Government education expenditure (% of GNP)	1.0[h]
Primary-secondary gross enrolment ratio (w and m per 100)	37/51
Television receivers (per 1,000 inhabitants)	4
Parliamentary seats (women and men, % of total)	7/93

Environment	2000-2006
Threatened species	193
Forested area (% of land area)	60
CO2 emissions (000s Mt of carbon dioxide/per capita)	1789/0.0
Energy consumption per capita (kilograms oil equiv.)	19
Precipitation (mm)	1371

a June 2006. b 2003. c 2004. d 1999. e 2002. f Estimated data. g Provisional. h 1985.

Cook Islands

Region	Oceania-Polynesia
Largest urban agglom. (pop., 000s)	Avarua (13)
Currency	New Zealand dollar
Population in 2006 (proj., 000s)	18[a]
Surface area (square kms)	236
Population density (per square km)	76

Economic indicators	2000	2005
Exchange rate (national currency per US$)	2.27	1.50[b]
Consumer price index (2000=100)[c]	100	122[d]
Tourist arrivals (000s)	73	83[e]
GDP (million current US$)	81	183
GDP (per capita current US$)	4291	10201
Gross fixed capital formation (% of GDP)	11.0	11.0
Labour force participation, adult female pop. (%)	...	65.2[f]
Labour force participation, adult male pop. (%)	64.9[g]	81.8[f]
Agricultural production index (1999-2001=100)	102	60[h]
Food production index (1999-2001=100)	102	60[h]
Telephone lines (per 100 inhabitants)	...	34.5[i]
Internet users, estimated (000s)	2.8	5.0

Total trade		Major trading partners			2005
	(million US$)[e]	(% of exports)[e]		(% of imports)[e]	
Exports	7.1	Japan	39	New Zealand	74
Imports	67.4	New Zealand	18	Australia	12
		China	15	Japan	3

Social indicators	2000-2006
Population growth rate 2000-2005 (% per annum)	−1.0
Population aged 0-14 years (%)	30.0
Sex ratio (women per 100 men)	94[fj]
Life expectancy at birth 2000-2005 (women and men, years)	74/21[ik]
Total fertility rate 2000-2005 (births per woman)	2.9[kl]
Urban population (%)	70
Urban population growth rate 2000-2005 (% per annum)	0.6
Rural population growth rate 2000-2005 (% per annum)	−4.2
Foreign born (%)	12.8[m]
Government education expenditure (% of GNP)	0.4
Primary-secondary gross enrolment ratio (w and m per 100)	73/73
Television receivers (per 1,000 inhabitants)	147

Environment	2000-2006
Threatened species	25
CO2 emissions (000s Mt of carbon dioxide/per capita)	31/1.7
Energy consumption per capita (kilograms oil equiv.)	484[m]

a 2005. b October 2006. c Rarotonga. d 2nd quarter 2006. e 2004. f 2001. g 1996.
h 2003. i 2002. j Refers to de facto population count. k Published by the Secretariat of
the Pacific Community. l 1996-2001. m Estimated data.

Costa Rica

Region	Central America
Largest urban agglom. (pop., 000s)	San José (1217)
Currency	colón
Population in 2006 (proj., 000s)	4399
Surface area (square kms)	51100
Population density (per square km)	86
United Nations membership date	2 November 1945

Economic indicators	2000	2005
Exchange rate (national currency per US$)	318.02	517.37[a]
Consumer price index (2000=100)[b]	100	189[c]
Industrial production index (1995=100)[d]	144	190[e]
Unemployment (percentage of labour force)[f]	5.2	6.6
Balance of payments, current account (million US$)	−707	−959
Tourist arrivals (000s)	1088	1453[g]
GDP (million current US$)	15947	19818
GDP (per capita current US$)	4059	4580
Gross fixed capital formation (% of GDP)	18.0	19.0
Labour force participation, adult female pop. (%)	36.5	39.9[g]
Labour force participation, adult male pop. (%)	80.0	78.8[g]
Employment in industrial sector (%)	22.3	21.6
Employment in agricultural sector (%)	20.4	15.2
Agricultural production index (1999-2001=100)	100	98[g]
Food production index (1999-2001=100)	100	99[g]
Primary energy production (000s Mt oil equiv.)	588	668[g]
Motor vehicles (per 1,000 inhabitants)	132.3	192.8[g]
Telephone lines (per 100 inhabitants)	23.5	32.1
Internet users, estimated (000s)	228.0	922.5

Total trade		Major trading partners			2005
	(million US$)	(% of exports)		(% of imports)	
Exports	7150.7	USA	43	USA	41
Imports	9173.3	China, HK SAR	7	Japan	6
		Netherlands	6	Venezuela	5

Social indicators	2000-2006
Population growth rate 2000-2005 (% per annum)	1.9
Population aged 0-14 years (%)	28.0
Population aged 60+ years (women and men, % of total)	9.0/8.0
Sex ratio (women per 100 men)	97
Life expectancy at birth 2000-2005 (women and men, years)	81/76
Infant mortality rate 2000-2005 (per 1,000 births)	10
Total fertility rate 2000-2005 (births per woman)	2.3
Urban population (%)	62
Urban population growth rate 2000-2005 (% per annum)	2.8
Rural population growth rate 2000-2005 (% per annum)	0.6
Foreign born (%)	7.8
Refugees and others of concern to UNHCR[h]	11476[g]
Government education expenditure (% of GNP)	5.1
Primary-secondary gross enrolment ratio (w and m per 100)	97/95
Third-level students (women and men, % of total)	54/46
Newspaper circulation (per 1,000 inhabitants)	70
Television receivers (per 1,000 inhabitants)	251
Intentional homicides (per 100,000 inhabitants)	6
Parliamentary seats (women and men, % of total)	39/61

Environment	2000-2006
Threatened species	240
Forested area (% of land area)	39
CO2 emissions (000s Mt of carbon dioxide/per capita)	6340/1.5
Energy consumption per capita (kilograms oil equiv.)	535
Precipitation (mm)	1866[i]
Average minimum and maximum temperatures (centigrade)[i]	16.2/24.9

a October 2006. b Central area. c June 2006. d Calculated by the Statistics Division of the United Nations from component national indices. e 2nd quarter 2006. f Persons aged 12 years and over. g 2004. h Provisional. i San Jose.

Côte d'Ivoire

Region	Western Africa
Largest urban agglom. (pop., 000s)	Abidjan (3577)
Currency	CFA franc
Population in 2006 (proj., 000s)	18454
Surface area (square kms)	322463
Population density (per square km)	57
United Nations membership date	20 September 1960

Economic indicators	2000	2005
Exchange rate (national currency per US$)[a]	704.95	516.66[b]
Consumer price index (2000=100)[c]	100	121[d]
Balance of payments, current account (million US$)	−241	303[e]
Tourist arrivals (000s)	301[f]	...
GDP (million current US$)	10682	16785
GDP (per capita current US$)	638	925
Gross fixed capital formation (% of GDP)	12.0	10.0
Agricultural production index (1999-2001=100)	104	98[e]
Food production index (1999-2001=100)	102	101[e]
Primary energy production (000s Mt oil equiv.)	1689	2292[e]
Motor vehicles (per 1,000 inhabitants)[g]	10.1	9.7[h]
Telephone lines (per 100 inhabitants)	1.8	1.5
Internet users, estimated (000s)	40.0	200.0

Total trade		Major trading partners			2005
	(million US$)[i]	(% of exports)[i]		(% of imports)[i]	
Exports	5493.4	France	19	France	33
Imports	3536.3	Netherlands	18	Nigeria	14
		USA	7	UK	7

Social indicators	2000-2006
Population growth rate 2000-2005 (% per annum)	1.6
Population aged 0-14 years (%)	42.0
Population aged 60+ years (women and men, % of total)	5.0/5.0
Sex ratio (women per 100 men)	97
Life expectancy at birth 2000-2005 (women and men, years)	47/45
Infant mortality rate 2000-2005 (per 1,000 births)	118
Total fertility rate 2000-2005 (births per woman)	5.1
Contraceptive use (% of currently married women)	15[j]
Urban population (%)	45
Urban population growth rate 2000-2005 (% per annum)	2.5
Rural population growth rate 2000-2005 (% per annum)	0.9
Foreign born (%)	14.8[k]
Refugees and others of concern to UNHCR[l]	82111
Government education expenditure (% of GNP)	4.8
Primary-secondary gross enrolment ratio (w and m per 100)	40/59
Television receivers (per 1,000 inhabitants)	52
Parliamentary seats (women and men, % of total)	9/91

Environment	2000-2006
Threatened species	181
Forested area (% of land area)	22
CO2 emissions (000s Mt of carbon dioxide/per capita)	5723/0.3
Energy consumption per capita (kilograms oil equiv.)	101

a Official rate. b October 2006. c Abidjan. d July 2006. e 2004. f 1998. g Source: World Automotive Market Report, Auto and Truck International (Illinois). h 2002. i 2003. j 1998/99. k Estimated data. l Provisional.

Croatia

Region	Southern Europe
Largest urban agglom. (pop., 000s)	Zagreb (689)
Currency	kuna
Population in 2006 (proj., 000s)	4556
Surface area (square kms)	56538
Population density (per square km)	81
United Nations membership date	22 May 1992

Economic indicators	2000	2005
Exchange rate (national currency per US$)	8.16	5.79[a]
Consumer price index (2000=100)	100	118[b]
Industrial production index (1995=100)	115	175[c]
Unemployment (percentage of labour force)[d]	16.1	12.7
Balance of payments, current account (million US$)	−471	−2541
Tourist arrivals (000s)	5831	7912[e]
GDP (million current US$)	18428	36947
GDP (per capita current US$)	4090	8118
Gross fixed capital formation (% of GDP)	22.0	29.0
Labour force participation, adult female pop. (%)	44.0	42.3[e]
Labour force participation, adult male pop. (%)	58.4	55.7[e]
Employment in industrial sector (%)	28.8	28.6
Employment in agricultural sector (%)	14.5	17.3
Agricultural production index (1999-2001=100)	95	97[e]
Food production index (1999-2001=100)	95	97[e]
Primary energy production (000s Mt oil equiv.)	3557	3860[e]
Motor vehicles (per 1,000 inhabitants)	277.9	341.2[f]
Telephone lines (per 100 inhabitants)	38.5	41.5
Internet users, estimated (000s)	299.4	1472.4

Total trade		Major trading partners			2005
	(million US$)	(% of exports)		(% of imports)	
Exports	8772.6	Italy	21	Italy	16
Imports	18560.4	Bosnia/Herzeg	14	Germany	15
		Germany	11	Russian Fed.	9

Social indicators	2000-2006
Population growth rate 2000-2005 (% per annum)	0.2
Population aged 0-14 years (%)	16.0
Population aged 60+ years (women and men, % of total)	25.0/18.0
Sex ratio (women per 100 men)	108
Life expectancy at birth 2000-2005 (women and men, years)	78/71
Infant mortality rate 2000-2005 (per 1,000 births)	7
Total fertility rate 2000-2005 (births per woman)	1.4
Urban population (%)	56
Urban population growth rate 2000-2005 (% per annum)	0.5
Rural population growth rate 2000-2005 (% per annum)	−0.2
Refugees and others of concern to UNHCR[g]	15756
Government education expenditure (% of GNP)	4.9
Primary-secondary gross enrolment ratio (w and m per 100)	90/90
Third-level students (women and men, % of total)	53/47
Newspaper circulation (per 1,000 inhabitants)	132
Television receivers (per 1,000 inhabitants)	320
Intentional homicides (per 100,000 inhabitants)	2
Parliamentary seats (women and men, % of total)	22/78

Environment	2000-2006
Threatened species	75
Forested area (% of land area)	32
CO2 emissions (000s Mt of carbon dioxide/per capita)[h]	23000/5.1
Energy consumption per capita (kilograms oil equiv.)	2013
Precipitation (mm)	856[i]
Average minimum and maximum temperatures (centigrade)[i]	5.4/15.7

a October 2006. b April 2006. c June 2006. d Persons aged 15 years and over. e 2004.
f Work vehicles are included. g Provisional. h Source: UNFCCC. i Zagreb.

Cuba

Region	Caribbean
Largest urban agglom. (pop., 000s)	Havana (2189)
Currency	peso
Population in 2006 (proj., 000s)	11294
Surface area (square kms)	110861
Population density (per square km)	102
United Nations membership date	24 October 1945

Economic indicators	2000	2005
Exchange rate (national currency per US$)	1.00	1.00
Unemployment (percentage of labour force)[a]	5.4	1.9
Tourist arrivals (000s)	1741	2017[b]
GDP (million current US$)	32685	46932
GDP (per capita current US$)	2938	4165
Gross fixed capital formation (% of GDP)	11.0	9.0
Labour force participation, adult female pop. (%)	...	37.9[b]
Employment in industrial sector (%)	19.1	19.4[b]
Employment in agricultural sector (%)	27.1	21.2[b]
Agricultural production index (1999-2001=100)	105	109[b]
Food production index (1999-2001=100)	105	110[b]
Primary energy production (000s Mt oil equiv.)	3270	3948[b]
Motor vehicles (per 1,000 inhabitants)	0.9	2.1[bc]
Telephone lines (per 100 inhabitants)	4.4	7.5
Internet users, estimated (000s)	60.0	190.0

Total trade		Major trading partners			2005
	(million US$)[b]	(% of exports)[b]		(% of imports)[b]	
Exports	2332.1	Netherlands	28	Venezuela	20
Imports	5609.6	Canada	21	Spain	11
		Venezuela	16	China	11

Social indicators	2000-2006
Population growth rate 2000-2005 (% per annum)	0.3
Population aged 0-14 years (%)	19.0
Population aged 60+ years (women and men, % of total)	16.0/14.0
Sex ratio (women per 100 men)	100
Life expectancy at birth 2000-2005 (women and men, years)	79/75
Infant mortality rate 2000-2005 (per 1,000 births)	6
Total fertility rate 2000-2005 (births per woman)	1.6
Contraceptive use (% of currently married women)	73
Urban population (%)	76
Urban population growth rate 2000-2005 (% per annum)	0.2
Rural population growth rate 2000-2005 (% per annum)	0.3
Refugees and others of concern to UNHCR[d]	739
Government education expenditure (% of GNP)	8.7
Primary-secondary gross enrolment ratio (w and m per 100)	95/97
Third-level students (women and men, % of total)	49/51
Newspaper circulation (per 1,000 inhabitants)	54
Television receivers (per 1,000 inhabitants)	267
Intentional homicides (per 100,000 inhabitants)	5
Parliamentary seats (women and men, % of total)	36/64

Environment	2000-2006
Threatened species	277
Forested area (% of land area)	21
CO2 emissions (000s Mt of carbon dioxide/per capita)	25295/2.3
Energy consumption per capita (kilograms oil equiv.)	743
Precipitation (mm)	1189[e]
Average minimum and maximum temperatures (centigrade)[e]	21.6/28.8

a Men aged 17 to 60 years, women aged 17 to 55 years. b 2004. c Including specialized enterprises only. d Provisional. e Havana.

Cyprus

Region	Western Asia
Largest urban agglom. (pop., 000s)	Nicosia (211)
Currency	pound
Population in 2006 (proj., 000s)	845
Surface area (square kms)	9251
Population density (per square km)	91
United Nations membership date	20 September 1960

Economic indicators	2000	2005
Exchange rate (national currency per US$)[a]	0.62	0.45[b]
Consumer price index (2000=100)	100	118[c]
Industrial production index (1995=100)[d]	105	123[e]
Unemployment (percentage of labour force)[fg]	4.9	5.3
Balance of payments, current account (million US$)	–488	–962
Tourist arrivals (000s)	2686	2349[h]
GDP (million current US$)	9124	16723
GDP (per capita current US$)	13155	22432
Gross fixed capital formation (% of GDP)	17.0	19.0
Labour force participation, adult female pop. (%)	48.2	53.3[h]
Labour force participation, adult male pop. (%)	73.9	73.4[h]
Employment in industrial sector (%)	23.3	24.0[h]
Employment in agricultural sector (%)	5.3	4.8[h]
Agricultural production index (1999-2001=100)	102	105[h]
Food production index (1999-2001=100)	102	105[h]
Motor vehicles (per 1,000 inhabitants)	492.4	575.8
Telephone lines (per 100 inhabitants)[i]	64.8	50.3
Internet users, estimated (000s)	120.0	326.0

Total trade		Major trading partners			2005
	(million US$)	(% of exports)		(% of imports)	
Exports	1546.4	France	17	Greece	17
Imports	6382.1	UK	17	Italy	10
		Greece	11	UK	9

Social indicators	2000-2006
Population growth rate 2000-2005 (% per annum)	1.2
Population aged 0-14 years (%)	20.0
Population aged 60+ years (women and men, % of total)	18.0/16.0
Sex ratio (women per 100 men)	106
Life expectancy at birth 2000-2005 (women and men, years)	81/76
Infant mortality rate 2000-2005 (per 1,000 births)	6
Total fertility rate 2000-2005 (births per woman)	1.6
Urban population (%)	69
Urban population growth rate 2000-2005 (% per annum)	1.4
Rural population growth rate 2000-2005 (% per annum)	0.8
Foreign born (%)	12.7
Refugees and others of concern to UNHCR[j]	13769
Government education expenditure (% of GNP)	7.6
Primary-secondary gross enrolment ratio (w and m per 100)	98/97
Third-level students (women and men, % of total)	48/52
Newspaper circulation (per 1,000 inhabitants)	111
Television receivers (per 1,000 inhabitants)	384
Parliamentary seats (women and men, % of total)	14/86

Environment	2000-2006
Threatened species	36
Forested area (% of land area)	19
CO_2 emissions (000s Mt of carbon dioxide/per capita)	7291/8.9
Energy consumption per capita (kilograms oil equiv.)	2822
Precipitation (mm)	320[k]
Average minimum and maximum temperatures (centigrade)[l]	13.7/24.3

a Official rate. b October 2006. c September 2006. d For government controlled areas. e June 2006. f Persons aged 15 years and over, the data relate to the government-controlled areas. g Second quarter of each year. h 2004. i Excluding 8,960 main lines in the occupied areas. j Provisional. k Larnaca. l Lanarca airport.

Czech Republic

Region	Eastern Europe
Largest urban agglom. (pop., 000s)	Prague (1171)
Currency	koruna
Population in 2006 (proj., 000s)	10209
Surface area (square kms)	78866
Population density (per square km)	129
United Nations membership date	19 January 1993

Economic indicators	2000	2005
Exchange rate (national currency per US$)[a]	37.81	22.23[b]
Consumer price index (2000=100)	100	115[c]
Industrial production index (1995=100)	106	144[d]
Unemployment (percentage of labour force)[e]	8.8	7.9
Balance of payments, current account (million US$)	−2690	−2495
Tourist arrivals (000s)	4666	6061[f]
GDP (million current US$)	55703	122345
GDP (per capita current US$)	5425	11972
Gross fixed capital formation (% of GDP)	28.0	26.0
Labour force participation, adult female pop. (%)	51.6	50.5[f]
Labour force participation, adult male pop. (%)	69.8	68.4[f]
Employment in industrial sector (%)	39.5	39.5
Employment in agricultural sector (%)	5.1	4.0
Agricultural production index (1999-2001=100)	97	105[f]
Food production index (1999-2001=100)	97	105[f]
Primary energy production (000s Mt oil equiv.)	25472	26595[f]
Motor vehicles (per 1,000 inhabitants)	368.0[g]	416.6[fh]
Telephone lines (per 100 inhabitants)	37.7	31.5
Internet users, estimated (000s)	1000.0	2758.0

Total trade	(million US$)	Major trading partners	(% of exports)		2005 (% of imports)
Exports	78208.5	Germany	34	Germany	30
Imports	76527.3	Slovakia	9	Russian Fed.	6
		Austria	6	Slovakia	5

Social indicators	2000-2006
Population growth rate 2000-2005 (% per annum)	−0.1
Population aged 0-14 years (%)	15.0
Population aged 60+ years (women and men, % of total)	23.0/17.0
Sex ratio (women per 100 men)	105
Life expectancy at birth 2000-2005 (women and men, years)	79/72
Infant mortality rate 2000-2005 (per 1,000 births)	6
Total fertility rate 2000-2005 (births per woman)	1.2
Urban population (%)	74
Urban population growth rate 2000-2005 (% per annum)	−0.2
Rural population growth rate 2000-2005 (% per annum)	0.3
Foreign born (%)	4.5
Refugees and others of concern to UNHCR[i]	2726
Government education expenditure (% of GNP)	4.8
Primary-secondary gross enrolment ratio (w and m per 100)	98/98
Third-level students (women and men, % of total)	51/49
Television receivers (per 1,000 inhabitants)	603
Intentional homicides (per 100,000 inhabitants)	1
Parliamentary seats (women and men, % of total)	15/85

Environment	2000-2006
Threatened species	50
Forested area (% of land area)	34
CO2 emissions (000s Mt of carbon dioxide/per capita)[j]	127120/12.4
Energy consumption per capita (kilograms oil equiv.)	3575
Precipitation (mm)	526[k]
Average minimum and maximum temperatures (centigrade)[k]	3.6/12.5

a Principal rate. b October 2006. c September 2006. d August 2006. e Persons aged 15 years and over, excl. persons on child-care leave. f 2004. g Including vans. h Including special-purpose commercial vehicles and farm tractors. i Provisional. j Source: UNFCCC. k Prague.

Denmark

Region	Northern Europe
Largest urban agglom. (pop., 000s)	Copenhagen (1088)
Currency	krone
Population in 2006 (proj., 000s)	5446
Surface area (square kms)	43094
Population density (per square km)	126
United Nations membership date	24 October 1945

Economic indicators	2000	2005
Exchange rate (national currency per US$)	8.02	5.87[a]
Consumer price index (2000=100)	100	113[b]
Industrial production index (1995=100)	116	126[c]
Unemployment (percentage of labour force)[d]	4.6	5.0
Balance of payments, current account (million US$)	2262	5941[e]
Tourist arrivals (000s)	3535	3358[e]
GDP (million current US$)	160082	258718
GDP (per capita current US$)	29980	47641
Gross fixed capital formation (% of GDP)	20.0	21.0
Labour force participation, adult female pop. (%)	60.2	60.4[e]
Labour force participation, adult male pop. (%)	71.1	71.7[e]
Employment in industrial sector (%)	26.1	23.8
Employment in agricultural sector (%)	3.3	2.9
Agricultural production index (1999-2001=100)	100	101[e]
Food production index (1999-2001=100)	100	101[e]
Primary energy production (000s Mt oil equiv.)	26382	29269[e]
Motor vehicles (per 1,000 inhabitants)[fg]	421.9	437.0[e]
Telephone lines (per 100 inhabitants)	72.0	61.7
Internet users, estimated (000s)	2090.0	2854.0

Total trade		Major trading partners			2005
	(million US$)		(% of exports)		(% of imports)
Exports	82415.4	Germany	16	Germany	21
Imports	74265.1	Sweden	13	Sweden	14
		UK	8	Netherlands	7

Social indicators	2000-2006
Population growth rate 2000-2005 (% per annum)	0.3
Population aged 0-14 years (%)	19.0
Population aged 60+ years (women and men, % of total)	23.0/19.0
Sex ratio (women per 100 men)	102
Life expectancy at birth 2000-2005 (women and men, years)	79/75
Infant mortality rate 2000-2005 (per 1,000 births)	5
Total fertility rate 2000-2005 (births per woman)	1.8
Contraceptive use (% of currently married women)	78[hi]
Urban population (%)	86
Urban population growth rate 2000-2005 (% per annum)	0.5
Rural population growth rate 2000-2005 (% per annum)	−0.3
Refugees and others of concern to UNHCR[j]	45329
Government education expenditure (% of GNP)	8.5
Primary-secondary gross enrolment ratio (w and m per 100)	113/110
Third-level students (women and men, % of total)	58/42
Newspaper circulation (per 1,000 inhabitants)	282
Television receivers (per 1,000 inhabitants)	975
Intentional homicides (per 100,000 inhabitants)	1
Parliamentary seats (women and men, % of total)	37/63

Environment	2000-2006
Threatened species	40
Forested area (% of land area)	11
CO2 emissions (000s Mt of carbon dioxide/per capita)[k]	60750/11.3
Energy consumption per capita (kilograms oil equiv.)	3303
Precipitation (mm)	525[l]
Average minimum and maximum temperatures (centigrade)[l]	5.0/11.0

a October 2006. b September 2006. c August 2006. d Persons aged 15 to 66 years.
e 2004. f Excluding Faeroe Islands. g Including vehicles operated by police or other
governmental security organizations. h For women aged 15-44 in union or married.
i 1988. j Provisional. k Source: UNFCCC. l Copenhagen.

Djibouti

Region	Eastern Africa
Largest urban agglom. (pop., 000s)	Djibouti (555)
Currency	franc
Population in 2006 (proj., 000s)	807
Surface area (square kms)	23200
Population density (per square km)	35
United Nations membership date	20 September 1977

Economic indicators	2000	2005
Exchange rate (national currency per US$)[a]	177.72	177.72[b]
Tourist arrivals (000s)	20	26[c]
GDP (million current US$)	553	705
GDP (per capita current US$)	774	889
Gross fixed capital formation (% of GDP)	12.0	14.0
Employment in industrial sector (%)	7.9[d]	...
Agricultural production index (1999-2001=100)	100	110[c]
Food production index (1999-2001=100)	100	110[c]
Motor vehicles (per 1,000 inhabitants)	30.8[e]	...
Telephone lines (per 100 inhabitants)	1.5	1.6
Internet users, estimated (000s)	1.4	10.0

Social indicators	2000-2006
Population growth rate 2000-2005 (% per annum)	2.1
Population aged 0-14 years (%)	41.0
Population aged 60+ years (women and men, % of total)	5.0/4.0
Sex ratio (women per 100 men)	100
Life expectancy at birth 2000-2005 (women and men, years)	54/51
Infant mortality rate 2000-2005 (per 1,000 births)	93
Total fertility rate 2000-2005 (births per woman)	5.1
Urban population (%)	86
Urban population growth rate 2000-2005 (% per annum)	2.7
Rural population growth rate 2000-2005 (% per annum)	−1.5
Foreign born (%)	4.2[f]
Refugees and others of concern to UNHCR[g]	10475
Government education expenditure (% of GNP)	5.6
Primary-secondary gross enrolment ratio (w and m per 100)	26/35
Third-level students (women and men, % of total)	45/55
Television receivers (per 1,000 inhabitants)	78
Parliamentary seats (women and men, % of total)	11/89

Environment	2000-2006
Threatened species	29
Forested area (% of land area)	<
CO2 emissions (000s Mt of carbon dioxide/per capita)	366/0.5
Energy consumption per capita (kilograms oil equiv.)	158[f]
Precipitation (mm)	164[h]
Average minimum and maximum temperatures (centigrade)[h]	25.9/33.9

a Official rate. b October 2006. c 2004. d 1991. e 1994. f Estimated data. g Provisional. h Djibouti.

Dominica

Region	Caribbean
Largest urban agglom. (pop., 000s)	Roseau (14)
Currency	EC dollar
Population in 2006 (proj., 000s)	79[a]
Surface area (square kms)	751
Population density (per square km)	93
United Nations membership date	18 December 1978

Economic indicators	2000	2005
Exchange rate (national currency per US$)[b]	2.70	2.70[c]
Unemployment (percentage of labour force)[d]	15.7[e]	11.0[f]
Balance of payments, current account (million US$)	−53	−38[g]
Tourist arrivals (000s)	70	79[h]
GDP (million current US$)	269	283
GDP (per capita current US$)	3457	3580
Gross fixed capital formation (% of GDP)	28.0	25.0
Employment in industrial sector (%)	18.2[i]	...
Employment in agricultural sector (%)	23.7[i]	...
Agricultural production index (1999-2001=100)	102	98[h]
Food production index (1999-2001=100)	102	98[h]
Primary energy production (000s Mt oil equiv.)	3	3[h]
Motor vehicles (per 1,000 inhabitants)[j]	157.2[k]	...
Telephone lines (per 100 inhabitants)	31.7	29.4
Internet users, estimated (000s)	6.0	26.0

Total trade		Major trading partners			2005
	(million US$)	(% of exports)		(% of imports)	
Exports	41.8	UK	17	USA	37
Imports	165.3	Jamaica	13	Trinidad Tbg	21
		Antigua,Barb	11	UK	7

Social indicators	2000-2006
Population growth rate 2000-2005 (% per annum)	0.3
Population aged 60+ years (women and men, % of total)	15.0/12.0
Sex ratio (women per 100 men)	99[lm]
Infant mortality rate 2000-2005 (per 1,000 births)	20[n]
Contraceptive use (% of currently married women)	50[op]
Urban population (%)	73
Urban population growth rate 2000-2005 (% per annum)	0.8
Rural population growth rate 2000-2005 (% per annum)	−1.0
Foreign born (%)	4.8[q]
Government education expenditure (% of GNP)	5.5[e]
Primary-secondary gross enrolment ratio (w and m per 100)	100/100
Television receivers (per 1,000 inhabitants)	252
Parliamentary seats (women and men, % of total)	13/87

Environment	2000-2006
Threatened species	37
Forested area (% of land area)	61
CO2 emissions (000s Mt of carbon dioxide/per capita)	138/1.8
Energy consumption per capita (kilograms oil equiv.)	558
Precipitation (mm)	2575[r]
Average minimum and maximum temperatures (centigrade)[r]	23.1/29.2

a 2005. b Official rate. c October 2006. d Persons aged 15 years and over. e 1999.
f Estimates based on 2001 Population Census reults. g 2002. h 2004. i 1997. j Including
large public service excavators and trench diggers. k 1998. l Refers to de facto population count. m 2001. n Data tabulated by date of registration rather than occurrence.
o For women aged 15-44 in union or married. p 1987. q Estimated data. r Melville
Hall airport.

Dominican Republic

Region	Caribbean	
Largest urban agglom. (pop., 000s)	Santo Domingo (2022)	
Currency	peso	
Population in 2006 (proj., 000s)	9021	
Surface area (square kms)	48511	
Population density (per square km)	186	
United Nations membership date	24 October 1945	

Economic indicators	2000	2005
Exchange rate (national currency per US$)[a]	16.67	33.79[b]
Consumer price index (2000=100)	100	251[c]
Industrial production index (1995=100)	133	150[d]
Unemployment (percentage of labour force)[e]	13.9	17.9
Balance of payments, current account (million US$)	−1027	−143
Tourist arrivals (000s)[fg]	2978	3450[h]
GDP (million current US$)	19772	29101
GDP (per capita current US$)	2392	3272
Gross fixed capital formation (% of GDP)	23.0	20.0
Labour force participation, adult female pop. (%)	46.2	52.8[i]
Labour force participation, adult male pop. (%)	80.1	74.8[i]
Employment in industrial sector (%)	23.8	21.1[i]
Employment in agricultural sector (%)	15.9	15.9[i]
Agricultural production index (1999-2001=100)	99	103[h]
Food production index (1999-2001=100)	98	103[h]
Primary energy production (000s Mt oil equiv.)	66	170[h]
Motor vehicles (per 1,000 inhabitants)	89.4	110.7[h]
Telephone lines (per 100 inhabitants)	11.2	10.0
Internet users, estimated (000s)	327.1	1500.0

Social indicators	2000-2006
Population growth rate 2000-2005 (% per annum)	1.5
Population aged 0-14 years (%)	33.0
Population aged 60+ years (women and men, % of total)	6.0/6.0
Sex ratio (women per 100 men)	98
Life expectancy at birth 2000-2005 (women and men, years)	71/64
Infant mortality rate 2000-2005 (per 1,000 births)	35
Total fertility rate 2000-2005 (births per woman)	2.7
Contraceptive use (% of currently married women)	70
Urban population (%)	67
Urban population growth rate 2000-2005 (% per annum)	2.8
Rural population growth rate 2000-2005 (% per annum)	−1.0
Foreign born (%)	1.2
Government education expenditure (% of GNP)	1.2
Primary-secondary gross enrolment ratio (w and m per 100)	92/88
Third-level students (women and men, % of total)	61/39
Newspaper circulation (per 1,000 inhabitants)	28
Television receivers (per 1,000 inhabitants)	225
Parliamentary seats (women and men, % of total)	17/83

Environment	2000-2006
Threatened species	108
Forested area (% of land area)	28
CO2 emissions (000s Mt of carbon dioxide/per capita)	21347/2.5
Energy consumption per capita (kilograms oil equiv.)	669
Precipitation (mm)	1447[j]
Average minimum and maximum temperatures (centigrade)[j]	21.5/30.4

a Principal rate. b October 2006. c July 2006. d 2nd quarter 2006. e Persons aged 10 years and over. f Arrivals by air. g Including nationals residing abroad. h 2004. i 2002. j Santo Domingo.

Ecuador

Region	South America
Largest urban agglom. (pop., 000s)	Guayaquil (2387)
Currency	dollar[ab]
Population in 2006 (proj., 000s)	13419
Surface area (square kms)	283561
Population density (per square km)	47
United Nations membership date	21 December 1945

Economic indicators	2000	2005
Exchange rate (national currency per US$)[c]	1.00	1.00
Consumer price index (2000=100)	100	182[d]
Unemployment (percentage of labour force)[e]	9.0	7.9
Balance of payments, current account (million US$)	921	−157[f]
Tourist arrivals (000s)[g]	627	819[h]
GDP (million current US$)	15934	33062
GDP (per capita current US$)	1295	2499
Gross fixed capital formation (% of GDP)	20.0	22.0
Labour force participation, adult female pop. (%)	2000.0	53.9[h]
Labour force participation, adult male pop. (%)	82.4	80.7[h]
Employment in industrial sector (%)	23.9	21.2
Employment in agricultural sector (%)	8.5	8.3
Agricultural production index (1999-2001=100)	99	106[f]
Food production index (1999-2001=100)	99	107[f]
Primary energy production (000s Mt oil equiv.)	22020	28590[f]
Motor vehicles (per 1,000 inhabitants)	50.1	55.0[f]
Telephone lines (per 100 inhabitants)	9.7	12.9
Internet users, estimated (000s)	180.0	968.0

Total trade		Major trading partners			2005
	(million US$)		(% of exports)		(% of imports)
Exports	9869.4	USA	50	USA	19
Imports	9608.7	Peru	9	Colombia	14
		Panama	7	Brazil	7

Social indicators	2000-2006
Population growth rate 2000-2005 (% per annum)	1.4
Population aged 0-14 years (%)	32.0
Population aged 60+ years (women and men, % of total)	9.0/8.0
Sex ratio (women per 100 men)	99
Life expectancy at birth 2000-2005 (women and men, years)	77/71
Infant mortality rate 2000-2005 (per 1,000 births)	25
Total fertility rate 2000-2005 (births per woman)	2.8
Urban population (%)	63
Urban population growth rate 2000-2005 (% per annum)	2.3
Rural population growth rate 2000-2005 (% per annum)	0.2
Foreign born (%)	0.7[i]
Refugees and others of concern to UNHCR[j]	262552
Government education expenditure (% of GNP)	1.1
Primary-secondary gross enrolment ratio (w and m per 100)	89/90
Third-level students (women and men, % of total)	44/56[k]
Newspaper circulation (per 1,000 inhabitants)	99
Television receivers (per 1,000 inhabitants)	250
Intentional homicides (per 100,000 inhabitants)	17
Parliamentary seats (women and men, % of total)	16/84

Environment	2000-2006
Threatened species	2180
Forested area (% of land area)	38
CO_2 emissions (000s Mt of carbon dioxide/per capita)	23245/1.8
Energy consumption per capita (kilograms oil equiv.)	558
Precipitation (mm)	1014[l]

a 2005. b Adopted US dollar as the national currency effective 15 September 2000.
c On 20 March 2000, the Central Bank of Ecuador started to exchange the existing
local currency (sucres) for U.S. dollars at the fixed exchange rate of 25,000 sucres per
U.S. dollar. d February 2006. e Persons aged 10 years and over, Nov. of each year.
f 2004. g Excluding nationals residing abroad. h 2003. i Estimated data. j Provisional.
k 1996. l Quito.

Egypt

Region	Northern Africa
Largest urban agglom. (pop., 000s)	Cairo (11128)
Currency	pound
Population in 2006 (proj., 000s)	75437
Surface area (square kms)	1001449
Population density (per square km)	75
United Nations membership date	24 October 1945

Economic indicators	2000	2005
Exchange rate (national currency per US$)[a]	3.69	5.73[b]
Consumer price index (2000=100)	100	137[c]
Unemployment (percentage of labour force)[d]	9.0	11.0
Balance of payments, current account (million US$)	−971	3922[e]
Tourist arrivals (000s)	5116	7795[e]
GDP (million current US$)	99601	101406
GDP (per capita current US$)	1480	1370
Gross fixed capital formation (% of GDP)	17.0	18.0
Labour force participation, adult female pop. (%)	20.4[f]	18.4[g]
Labour force participation, adult male pop. (%)	73.7[f]	68.7[g]
Employment in industrial sector (%)	21.3	19.8[h]
Employment in agricultural sector (%)	29.6	29.9[h]
Agricultural production index (1999-2001=100)	102	111[e]
Food production index (1999-2001=100)	103	111[e]
Primary energy production (000s Mt oil equiv.)	61817	71456[e]
Motor vehicles (per 1,000 inhabitants)	34.2	38.1
Telephone lines (per 100 inhabitants)	8.6	14.0
Internet users, estimated (000s)	450.0	5000.0

Total trade		Major trading partners			2005
	(million US$)[e]	(% of exports)[e]		(% of imports)[e]	
Exports	7912.8	Italy	12	USA	10
Imports	13331.8	USA	7	Germany	7
		Spain	6	China	5

Social indicators	2000-2006
Population growth rate 2000-2005 (% per annum)	1.9
Population aged 0-14 years (%)	34.0
Population aged 60+ years (women and men, % of total)	8.0/7.0
Sex ratio (women per 100 men)	100
Life expectancy at birth 2000-2005 (women and men, years)	72/67
Infant mortality rate 2000-2005 (per 1,000 births)	37
Total fertility rate 2000-2005 (births per woman)	3.3
Contraceptive use (% of currently married women)	59[i]
Urban population (%)	43
Urban population growth rate 2000-2005 (% per annum)	2.0
Rural population growth rate 2000-2005 (% per annum)	1.8
Foreign born (%)	0.2[j]
Refugees and others of concern to UNHCR[k]	100047
Primary-secondary gross enrolment ratio (w and m per 100)	91/96
Third-level students (women and men, % of total)	58/43[l]
Television receivers (per 1,000 inhabitants)	250
Intentional homicides (per 100,000 inhabitants)	<
Parliamentary seats (women and men, % of total)	4/96

Environment	2000-2006
Threatened species	63
Forested area (% of land area)	<
CO2 emissions (000s Mt of carbon dioxide/per capita)	139893/2.0
Energy consumption per capita (kilograms oil equiv.)	842
Precipitation (mm)	25[m]
Average minimum and maximum temperatures (centigrade)[m]	15.8/27.7

a Principal rate. b October 2006. c February 2006. d Persons aged 15 to 64 years, January, April, July and October. e 2004. f 1999. g 2002. h 2003. i 2005. j 1986. k Provisional. l 1996. m Cairo.

El Salvador

Region	Central America
Largest urban agglom. (pop., 000s)	San Salvador (1517)
Currency	colón
Population in 2006 (proj., 000s)	6999
Surface area (square kms)	21041
Population density (per square km)	333
United Nations membership date	24 October 1945

Economic indicators	2000	2005
Exchange rate (national currency per US$)[a]	8.76	8.75[b]
Consumer price index (2000=100)[c]	100	125[d]
Industrial production index (1995=100)	126	145[e]
Unemployment (percentage of labour force)[f]	7.0	6.8[g]
Balance of payments, current account (million US$)	−431	−612[g]
Tourist arrivals (000s)[h]	795	966[g]
GDP (million current US$)	13134	16980
GDP (per capita current US$)	2091	2468
Gross fixed capital formation (% of GDP)	17.0	16.0
Labour force participation, adult female pop. (%)	44.5[i]	44.1[g]
Labour force participation, adult male pop. (%)	78.8[i]	77.5[g]
Employment in industrial sector (%)	24.4	23.7[g]
Employment in agricultural sector (%)	20.7	19.1[g]
Agricultural production index (1999-2001=100)	100	99[g]
Food production index (1999-2001=100)	102	105[g]
Primary energy production (000s Mt oil equiv.)	169	200[g]
Motor vehicles (per 1,000 inhabitants)	63.5	...
Telephone lines (per 100 inhabitants)	10.0	14.1
Internet users, estimated (000s)	70.0	637.1

Total trade	(million US$)	Major trading partners			2005
		(% of exports)		(% of imports)	
Exports	1657.6	Guatemala	25	USA	29
Imports	5380.3	USA	20	Guatemala	10
		Honduras	15	Mexico	10

Social indicators	2000-2006
Population growth rate 2000-2005 (% per annum)	1.8
Population aged 0-14 years (%)	34.0
Population aged 60+ years (women and men, % of total)	8.0/7.0
Sex ratio (women per 100 men)	103
Life expectancy at birth 2000-2005 (women and men, years)	74/68
Infant mortality rate 2000-2005 (per 1,000 births)	26
Total fertility rate 2000-2005 (births per woman)	2.9
Contraceptive use (% of currently married women)	67[jk]
Urban population (%)	60
Urban population growth rate 2000-2005 (% per annum)	2.3
Rural population growth rate 2000-2005 (% per annum)	1.2
Foreign born (%)	0.4[l]
Refugees and others of concern to UNHCR[m]	50
Government education expenditure (% of GNP)	2.9
Primary-secondary gross enrolment ratio (w and m per 100)	88/90
Third-level students (women and men, % of total)	54/46
Television receivers (per 1,000 inhabitants)	236
Parliamentary seats (women and men, % of total)	17/83

Environment	2000-2006
Threatened species	59
Forested area (% of land area)	6
CO2 emissions (000s Mt of carbon dioxide/per capita)	6553/1.0
Energy consumption per capita (kilograms oil equiv.)	315
Precipitation (mm)	1734[n]
Average minimum and maximum temperatures (centigrade)[n]	18.4/30.1

a Principal rate. b October 2006. c Urban areas. d July 2006. e 2nd quarter 2006.
f Persons aged 10 years and over. g 2004. h Excluding nationals residing abroad. i 1999.
j 2002/03. k For women aged 15-44 in union or married. l Estimated data. m Provisional. n San Salvador.

Equatorial Guinea

Region	Middle Africa
Largest urban agglom. (pop., 000s)	Malabo (96)
Currency	CFA franc
Population in 2006 (proj., 000s)	515
Surface area (square kms)	28051
Population density (per square km)	18
United Nations membership date	12 November 1968

Economic indicators	2000	2005
Exchange rate (national currency per US$)[a]	704.95	516.66[b]
Consumer price index (2000=100)	100	138[c]
GDP (million current US$)	1216	5651
GDP (per capita current US$)	2709	11222
Gross fixed capital formation (% of GDP)	61.0	26.0
Agricultural production index (1999-2001=100)	100	94[d]
Food production index (1999-2001=100)	100	93[d]
Primary energy production (000s Mt oil equiv.)	5905	7901[d]
Telephone lines (per 100 inhabitants)	1.4	2.0
Internet users, estimated (000s)	0.7	7.0

Social indicators	2000-2006
Population growth rate 2000-2005 (% per annum)	2.3
Population aged 0-14 years (%)	44.0
Population aged 60+ years (women and men, % of total)	6.0/5.0
Sex ratio (women per 100 men)	102
Life expectancy at birth 2000-2005 (women and men, years)	44/43
Infant mortality rate 2000-2005 (per 1,000 births)	102
Total fertility rate 2000-2005 (births per woman)	5.9
Urban population (%)	39
Urban population growth rate 2000-2005 (% per annum)	2.4
Rural population growth rate 2000-2005 (% per annum)	2.3
Foreign born (%)	0.3[e]
Refugees and others of concern to UNHCR[f]	—
Government education expenditure (% of GNP)	2.2
Primary-secondary gross enrolment ratio (w and m per 100)	68/82
Third-level students (women and men, % of total)	30/70
Parliamentary seats (women and men, % of total)	18/82

Environment	2000-2006
Threatened species	104
Forested area (% of land area)	63
CO2 emissions (000s Mt of carbon dioxide/per capita)	166/0.3
Energy consumption per capita (kilograms oil equiv.)	957[e]

a Official rate. b October 2006. c April 2005. d 2004. e Estimated data. f Provisional.

Eritrea

Region	Eastern Africa
Largest urban agglom. (pop., 000s)	Asmara (551)
Currency	nakfa
Population in 2006 (proj., 000s)	4560
Surface area (square kms)	117600
Population density (per square km)	39
United Nations membership date	28 May 1993

Economic indicators	2000	2005
Exchange rate (national currency per US$)[a]	10.20	15.38[b]
Balance of payments, current account (million US$)	−105	...
Tourist arrivals (000s)[c]	70	87[d]
GDP (million current US$)	634	1077
GDP (per capita current US$)	178	245
Gross fixed capital formation (% of GDP)	32.0	24.0
Employment in industrial sector (%)	18.7[e]	...
Agricultural production index (1999-2001=100)	92	86[d]
Food production index (1999-2001=100)	92	86[d]
Telephone lines (per 100 inhabitants)	0.8	0.9
Internet users, estimated (000s)	5.0	80.0

Total trade		Major trading partners			2005
	(million US$)[f]	(% of exports)[f]		(% of imports)[f]	
Exports	6.6	Sudan	20	USA	16
Imports	432.8	Italy	12	Untd Arab Em	12
		Singapore	12	Italy	12

Social indicators	2000-2006
Population growth rate 2000-2005 (% per annum)	4.3
Population aged 0-14 years (%)	45.0
Population aged 60+ years (women and men, % of total)	5.0/3.0
Sex ratio (women per 100 men)	104
Life expectancy at birth 2000-2005 (women and men, years)	55/52
Infant mortality rate 2000-2005 (per 1,000 births)	65
Total fertility rate 2000-2005 (births per woman)	5.5
Contraceptive use (% of currently married women)	8
Urban population (%)	19
Urban population growth rate 2000-2005 (% per annum)	6.0
Rural population growth rate 2000-2005 (% per annum)	3.9
Foreign born (%)	0.4[g]
Refugees and others of concern to UNHCR[h]	6041
Government education expenditure (% of GNP)	3.8
Primary-secondary gross enrolment ratio (w and m per 100)	39/54
Third-level students (women and men, % of total)	13/87
Television receivers (per 1,000 inhabitants)	59

Environment	2000-2006
Threatened species	42
Forested area (% of land area)	14
CO_2 emissions (000s Mt of carbon dioxide/per capita)	702/0.2
Energy consumption per capita (kilograms oil equiv.)	57
Precipitation (mm)	533[i]
Average minimum and maximum temperatures (centigrade)[i]	8.9/23.2

a Official rate. b September 2006. c Including nationals residing abroad. d 2004.
e 1984. f 2003. g Estimated data. h Provisional. i Asmara.

Estonia

Region	Northern Europe
Largest urban agglom. (pop., 000s)	Tallin (392)
Currency	kroon
Population in 2006 (proj., 000s)	1325
Surface area (square kms)	45100
Population density (per square km)	29
United Nations membership date	17 September 1991

Economic indicators	2000	2005
Exchange rate (national currency per US$)[a]	16.82	12.33[b]
Consumer price index (2000=100)	100	126[c]
Industrial production index (1995=100)	136	237[d]
Unemployment (percentage of labour force)[e]	13.6	7.9
Balance of payments, current account (million US$)	−299	−1445
Tourist arrivals (000s)[f]	1220	1750[g]
GDP (million current US$)	5477	12762
GDP (per capita current US$)	4007	9598
Gross fixed capital formation (% of GDP)	26.0	28.0
Labour force participation, adult female pop. (%)	57.6	58.1[g]
Labour force participation, adult male pop. (%)	71.0	68.4[g]
Employment in industrial sector (%)	33.3	34.0
Employment in agricultural sector (%)	7.2	5.3
Agricultural production index (1999-2001=100)	98	102[g]
Food production index (1999-2001=100)	98	102[g]
Primary energy production (000s Mt oil equiv.)	2724	3220[g]
Motor vehicles (per 1,000 inhabitants)	403.9	417.2[g]
Telephone lines (per 100 inhabitants)	36.3	33.3
Internet users, estimated (000s)	391.6	690.0

Total trade		Major trading partners			2005
	(million US$)	(% of exports)		(% of imports)	
Exports	7710.3	Finland	27	Germany	14
Imports	10164.5	Sweden	13	Finland	14
		Latvia	9	Russian Fed.	10

Social indicators	2000-2006
Population growth rate 2000-2005 (% per annum)	−0.6
Population aged 0-14 years (%)	15.0
Population aged 60+ years (women and men, % of total)	26.0/16.0
Sex ratio (women per 100 men)	118
Life expectancy at birth 2000-2005 (women and men, years)	77/65
Infant mortality rate 2000-2005 (per 1,000 births)	10
Total fertility rate 2000-2005 (births per woman)	1.4
Contraceptive use (% of currently married women)	70[hi]
Urban population (%)	69
Urban population growth rate 2000-2005 (% per annum)	−0.6
Rural population growth rate 2000-2005 (% per annum)	−0.4
Foreign born (%)	18.4
Refugees and others of concern to UNHCR[j]	136015
Government education expenditure (% of GNP)	6.1
Primary-secondary gross enrolment ratio (w and m per 100)	99/99
Third-level students (women and men, % of total)	62/38
Newspaper circulation (per 1,000 inhabitants)	192
Television receivers (per 1,000 inhabitants)	507
Intentional homicides (per 100,000 inhabitants)	12
Parliamentary seats (women and men, % of total)	19/81

Environment	2000-2006
Threatened species	15
Forested area (% of land area)	49
CO$_2$ emissions (000s Mt of carbon dioxide/per capita)[k]	19110/14.2
Energy consumption per capita (kilograms oil equiv.)	3843
Precipitation (mm)	675[l]
Average minimum and maximum temperatures (centigrade)[l]	1.8/8.9

a Official rate. b October 2006. c September 2006. d August 2006. e Persons aged 15 to 74 years. f Arrivals of non-resident tourists at national borders (excluding same-day visitors). g 2004. h For women aged 20-49 in union or married. i 1994. j Provisional. k Source: UNFCCC. l Tallinn.

Ethiopia

Region	Eastern Africa
Largest urban agglom. (pop., 000s)	Addis Ababa (2893)
Currency	birr
Population in 2006 (proj., 000s)	79289
Surface area (square kms)	1104300
Population density (per square km)	72
United Nations membership date	13 November 1945

Economic indicators	2000	2005
Exchange rate (national currency per US$)[a]	8.31	8.69[b]
Consumer price index (2000=100)[c]	107	112[de]
Unemployment (percentage of labour force)[f]	...	5.0
Balance of payments, current account (million US$)	13	−668[g]
Tourist arrivals (000s)	136	210[g]
GDP (million current US$)	6473	9297
GDP (per capita current US$)	94	120
Gross fixed capital formation (% of GDP)	16.0	27.0
Labour force participation, adult female pop. (%)	71.9[h]	57.3[g]
Labour force participation, adult male pop. (%)	89.7[h]	70.6[g]
Employment in industrial sector (%)	2.6[i]	...
Employment in agricultural sector (%)	93.0[i]	...
Agricultural production index (1999-2001=100)	98	112[g]
Food production index (1999-2001=100)	98	112[g]
Primary energy production (000s Mt oil equiv.)	144	217[g]
Motor vehicles (per 1,000 inhabitants)[j]	1.6[h]	1.7[k]
Telephone lines (per 100 inhabitants)[l]	0.4	0.8
Internet users, estimated (000s)[l]	10.0	164.0

Total trade		Major trading partners			2005
	(million US$)[m]	(% of exports)[m]			(% of imports)[m]
Exports	512.7	Djibouti	19	USA	14
Imports	2685.9	Germany	11	China	12
		Japan	9	Italy	9

Social indicators	2000-2006
Population growth rate 2000-2005 (% per annum)	2.4
Population aged 0-14 years (%)	45.0
Population aged 60+ years (women and men, % of total)	5.0/4.0
Sex ratio (women per 100 men)	101
Life expectancy at birth 2000-2005 (women and men, years)	49/47
Infant mortality rate 2000-2005 (per 1,000 births)	100
Total fertility rate 2000-2005 (births per woman)	5.9
Contraceptive use (% of currently married women)	8
Urban population (%)	16
Urban population growth rate 2000-2005 (% per annum)	3.8
Rural population growth rate 2000-2005 (% per annum)	2.2
Refugees and others of concern to UNHCR[n]	101173
Government education expenditure (% of GNP)	4.6
Primary-secondary gross enrolment ratio (w and m per 100)	47/61
Third-level students (women and men, % of total)	25/75
Television receivers (per 1,000 inhabitants)	8
Parliamentary seats (women and men, % of total)	19/81

Environment	2000-2006
Threatened species	111
Forested area (% of land area)	4
CO2 emissions (000s Mt of carbon dioxide/per capita)	7347/0.1
Energy consumption per capita (kilograms oil equiv.)	27
Precipitation (mm)	1055[e]
Average minimum and maximum temperatures (centigrade)[e]	15.9/23.2

a Official rate. b June 2006. c Base: 2001=100. d April 2005. e Addis Ababa. f Persons aged 10 years and over, March. g 2004. h 1999. i 1994. j Data refer to fiscal years ending 7 July. k 2002. l Year ending 30 June. m 2003. n Provisional.

Fiji

Region	Oceania-Melanesia
Largest urban agglom. (pop., 000s)	Greater Suva (219)
Currency	dollar
Population in 2006 (proj., 000s)	854
Surface area (square kms)	18274
Population density (per square km)	47
United Nations membership date	13 October 1970

Economic indicators	2000	2005
Exchange rate (national currency per US$)[a]	2.19	1.73[b]
Consumer price index (2000=100)	100	118[b]
Industrial production index (1995=100)	123	122[c]
Unemployment (percentage of labour force)[d]	5.4[e]	...
Tourist arrivals (000s)[f]	294	499[g]
GDP (million current US$)	1686	2998
GDP (per capita current US$)	2080	3536
Gross fixed capital formation (% of GDP)	15.0	18.0
Labour force participation, adult female pop. (%)	39.4[h]	...
Labour force participation, adult male pop. (%)	79.2[h]	...
Agricultural production index (1999-2001=100)	102	96[g]
Food production index (1999-2001=100)	102	96[g]
Primary energy production (000s Mt oil equiv.)	37	37[g]
Motor vehicles (per 1,000 inhabitants)	126.3[i]	164.1[g]
Telephone lines (per 100 inhabitants)	10.7	12.4
Internet users, estimated (000s)	12.0	70.0

Total trade		Major trading partners			2005
	(million US$)		(% of exports)		(% of imports)
Exports	701.7	Australia	20	Singapore	30
Imports	1607.3	Singapore	20	Australia	25
		USA	15	New Zealand	18

Social indicators	2000-2006
Population growth rate 2000-2005 (% per annum)	0.9
Population aged 0-14 years (%)	32.0
Population aged 60+ years (women and men, % of total)	7.0/6.0
Sex ratio (women per 100 men)	97
Life expectancy at birth 2000-2005 (women and men, years)	70/66
Infant mortality rate 2000-2005 (per 1,000 births)	22
Total fertility rate 2000-2005 (births per woman)	2.9
Urban population (%)	51
Urban population growth rate 2000-2005 (% per annum)	1.9
Rural population growth rate 2000-2005 (% per annum)	−0.1
Foreign born (%)	1.8[j]
Government education expenditure (% of GNP)	6.8
Primary-secondary gross enrolment ratio (w and m per 100)	97/96
Third-level students (women and men, % of total)	53/47
Television receivers (per 1,000 inhabitants)	117
Parliamentary seats (women and men, % of total)	13/87

Environment	2000-2006
Threatened species	102
Forested area (% of land area)	45
CO2 emissions (000s Mt of carbon dioxide/per capita)	1120/1.3
Energy consumption per capita (kilograms oil equiv.)	426[k]
Precipitation (mm)	3040[l]
Average minimum and maximum temperatures (centigrade)[l]	22.2/28.7

a Official rate. b September 2006. c 1st quarter 2006. d Persons aged 15 years and over. e 1995. f Excluding nationals residing abroad. g 2004. h 1996. i 1998. j 1986. k Estimated data. l Suva.

Finland

Region	Northern Europe
Largest urban agglom. (pop., 000s)	Helsinki (1091)
Currency	euro[a]
Population in 2006 (proj., 000s)	5262[b]
Surface area (square kms)	338145
Population density (per square km)	16
United Nations membership date	14 December 1955

Economic indicators	2000	2005
Exchange rate (national currency per US$)	1.07	0.79[c]
Consumer price index (2000=100)	100	108[d]
Industrial production index (1995=100)	145	165[d]
Unemployment (percentage of labour force)[e]	9.7	8.3
Balance of payments, current account (million US$)	8975	4983
Tourist arrivals (000s)	2714	2840[f]
GDP (million current US$)	120563	193155
GDP (per capita current US$)	23290	36798
Gross fixed capital formation (% of GDP)	20.0	19.0
Labour force participation, adult female pop. (%)	63.2	63.3[f]
Labour force participation, adult male pop. (%)	69.5	68.4[f]
Employment in industrial sector (%)	27.2	25.6
Employment in agricultural sector (%)	6.0	4.8
Agricultural production index (1999-2001=100)	103	104[f]
Food production index (1999-2001=100)	103	104[f]
Primary energy production (000s Mt oil equiv.)	4431	4070[f]
Motor vehicles (per 1,000 inhabitants)	473.1	534.4
Telephone lines (per 100 inhabitants)[g]	55.0	40.4
Internet users, estimated (000s)	1927.0	2800.0

Total trade		Major trading partners			2005
	(million US$)	(% of exports)		(% of imports)	
Exports	65238.3	Russian Fed.	11	Germany	15
Imports	58472.5	Sweden	11	Russian Fed.	14
		Germany	10	Sweden	10

Social indicators	2000-2006
Population growth rate 2000-2005 (% per annum)	0.3[b]
Population aged 0-14 years (%)	17.0
Population aged 60+ years (women and men, % of total)	24.0/18.0
Sex ratio (women per 100 men)	104[b]
Life expectancy at birth 2000-2005 (women and men, years)	82/75[b]
Infant mortality rate 2000-2005 (per 1,000 births)	4
Total fertility rate 2000-2005 (births per woman)	1.7[b]
Contraceptive use (% of currently married women)	77[hi]
Urban population (%)[b]	61
Urban population growth rate 2000-2005 (% per annum)	0.3
Rural population growth rate 2000-2005 (% per annum)[b]	0.3
Foreign born (%)	2.6
Refugees and others of concern to UNHCR[j]	12535
Government education expenditure (% of GNP)	6.6
Primary-secondary gross enrolment ratio (w and m per 100)	106/104
Third-level students (women and men, % of total)	53/47
Newspaper circulation (per 1,000 inhabitants)	445
Television receivers (per 1,000 inhabitants)	679
Intentional homicides (per 100,000 inhabitants)	3
Parliamentary seats (women and men, % of total)	38/62

Environment	2000-2006
Threatened species	29
Forested area (% of land area)	72
CO2 emissions (000s Mt of carbon dioxide/per capita)[k]	73190/14.0
Energy consumption per capita (kilograms oil equiv.)	5280
Precipitation (mm)	650[l]
Average minimum and maximum temperatures (centigrade)[l]	1.0/8.7

a Prior to 1 January 1999, markka. b Including Åland Islands. c September 2006.
d August 2006. e Persons aged 15 to 74 years. f 2004. g Telephone subscribers . h For
women aged 25-49 in union or married. i 1989. j Provisional. k Source: UNFCCC.
l Helsinki-Vantaa.

France

Region	Western Europe
Largest urban agglom. (pop., 000s)	Paris (9820)
Currency	euro[a]
Population in 2006 (proj., 000s)	60723
Surface area (square kms)	551500
Population density (per square km)	110
United Nations membership date	24 October 1945

Economic indicators	2000	2005
Exchange rate (national currency per US$)	1.07	0.79[b]
Consumer price index (2000=100)	100	112[c]
Industrial production index (1995=100)	115	88[d]
Unemployment (percentage of labour force)[e]	10.0	9.9
Balance of payments, current account (million US$)	19	−33
Tourist arrivals (000s)[f]	77190	75121[g]
GDP (million current US$)[h]	1327962	2126578
GDP (per capita current US$)[h]	21776	34128
Gross fixed capital formation (% of GDP)	19.0	20.0
Labour force participation, adult female pop. (%)	48.2	49.2[g]
Labour force participation, adult male pop. (%)	61.9	62.0[g]
Employment in industrial sector (%)	30.1[i]	24.6[g]
Employment in agricultural sector (%)	6.4[i]	4.0[g]
Agricultural production index (1999-2001=100)	101	102[g]
Food production index (1999-2001=100)	101	102[g]
Primary energy production (000s Mt oil equiv.)[j]	47939	47450[g]
Motor vehicles (per 1,000 inhabitants)	573.5	492.9[gk]
Telephone lines (per 100 inhabitants)	57.7	59.0
Internet users, estimated (000s)	8460.0	26154.0

Total trade		Major trading partners			2005
	(million US$)[j]	(% of exports)[j]		(% of imports)[j]	
Exports	434424.9	Germany	15	Germany	17
Imports	475999.3	Spain	10	Italy	9
		Italy	9	Belgium	8

Social indicators	2000-2006
Population growth rate 2000-2005 (% per annum)	0.4
Population aged 0-14 years (%)	18.0
Population aged 60+ years (women and men, % of total)	24.0/19.0
Sex ratio (women per 100 men)	105
Life expectancy at birth 2000-2005 (women and men, years)	83/76
Infant mortality rate 2000-2005 (per 1,000 births)	4
Total fertility rate 2000-2005 (births per woman)	1.9
Contraceptive use (% of currently married women)	75[lm]
Urban population (%)	77
Urban population growth rate 2000-2005 (% per annum)	0.7
Rural population growth rate 2000-2005 (% per annum)	−0.4
Foreign born (%)	10.6[n]
Refugees and others of concern to UNHCR[o]	149851
Government education expenditure (% of GNP)	6.0
Primary-secondary gross enrolment ratio (w and m per 100)	108/108
Third-level students (women and men, % of total)	55/45
Newspaper circulation (per 1,000 inhabitants)	142
Television receivers (per 1,000 inhabitants)	633
Intentional homicides (per 100,000 inhabitants)	1
Parliamentary seats (women and men, % of total)	14/86

Environment	2000-2006
Threatened species	133
Forested area (% of land area)	28
CO2 emissions (000s Mt of carbon dioxide/per capita)[p]	408160/6.8[q]
Energy consumption per capita (kilograms oil equiv.)[j]	2811
Precipitation (mm)	650[r]
Average minimum and maximum temperatures (centigrade)[r]	8.5/15.5

a Prior to 1 January 1999, franc. b Sept. 2006. c July 2006. d August 2006. e Persons aged 15 years and over. f Excluding nationals residing abroad. g 2004. h Including French Guiana, Guadeloupe, Martinique and Reunion. i 1989. j Including Monaco. k Passenger vehicles only. l For women aged 20-49 in union or married. m 1994. n Estimated data. o Provisional. p Source: UNFCCC. q Including territories. r Paris.

French Guiana

Region	South America	
Largest urban agglom. (pop., 000s)	Cayenne (59)	
Currency	euro[a]	
Population in 2006 (proj., 000s)	191	
Surface area (square kms)	90000	
Population density (per square km)	2	

Economic indicators	2000	2005
Exchange rate (national currency per US$)	1.07	0.79[bc]
Consumer price index (2000=100)	100	111[d]
Unemployment (percentage of labour force)[e]	25.8	26.5
Tourist arrivals (000s)	65[f]	65[g]
Labour force participation, adult female pop. (%)	53.1[h]	...
Labour force participation, adult male pop. (%)	67.9[h]	...
Agricultural production index (1999-2001=100)	95	99[i]
Food production index (1999-2001=100)	95	99[i]
Motor vehicles (per 1,000 inhabitants)[j]	272.7	257.9[g]
Telephone lines (per 100 inhabitants)	30.7	30.2[kl]
Internet users, estimated (000s)	16.0	42.0

Social indicators	2000-2006
Population growth rate 2000-2005 (% per annum)	2.6
Population aged 0-14 years (%)	34.0
Population aged 60+ years (women and men, % of total)	7.0/6.0
Sex ratio (women per 100 men)	95
Life expectancy at birth 2000-2005 (women and men, years)	78/73
Infant mortality rate 2000-2005 (per 1,000 births)	14
Total fertility rate 2000-2005 (births per woman)	3.4
Urban population (%)	76
Urban population growth rate 2000-2005 (% per annum)	2.7
Rural population growth rate 2000-2005 (% per annum)	2.2
Foreign born (%)	45.0[m]
Television receivers (per 1,000 inhabitants)	243

Environment	2000-2006
Threatened species	53
Forested area (% of land area)	90
CO2 emissions (000s Mt of carbon dioxide/per capita)	1005/5.6
Energy consumption per capita (kilograms oil equiv.)	1683[m]
Precipitation (mm)	3674[n]
Average minimum and maximum temperatures (centigrade)[n]	22.5/30.1

a Prior to 1 January 1999, French franc. b November 2006. c Data refer to non-commercial rates derived from the Operational Rates of Exchange for United Nations Programmes. d August 2006. e Persons aged 15 years and over, June of each year. f 2001. g 2002. h 1999. i 2004. j Source: World Automotive Market Report, Auto and Truck International (Illinois). k Estimate. l 2003. m Estimated data. n Rochambeau.

French Polynesia

Region	Oceania-Polynesia	
Largest urban agglom. (pop., 000s)	Papeete (130)	
Currency	CFP franc	
Population in 2006 (proj., 000s)	260	
Surface area (square kms)	4000	
Population density (per square km)	65	

Economic indicators	2000	2005
Exchange rate (national currency per US$)[c]	121.12[ab]	93.79[d]
Consumer price index (2000=100)	100	109[e]
Tourist arrivals (000s)[f]	252[g]	212[hi]
GDP (million current US$)	3242	5388
GDP (per capita current US$)	13732	20998
Gross fixed capital formation (% of GDP)	13.0	13.0
Agricultural production index (1999-2001=100)	103	112[i]
Food production index (1999-2001=100)	103	112[i]
Primary energy production (000s Mt oil equiv.)	10	8[i]
Telephone lines (per 100 inhabitants)	22.7	20.9
Internet users, estimated (000s)	15.0	55.0

Total trade		Major trading partners			2005
	(million US$)	(% of exports)		(% of imports)	
Exports	210.3	Japan	28	France	35
Imports	1701.5	China, HK SAR	28	USA	10
		USA	14	Singapore	9

Social indicators	2000-2006
Population growth rate 2000-2005 (% per annum)	1.7
Population aged 0-14 years (%)	28.0
Population aged 60+ years (women and men, % of total)	8.0/8.0
Sex ratio (women per 100 men)	96
Life expectancy at birth 2000-2005 (women and men, years)	76/71
Infant mortality rate 2000-2005 (per 1,000 births)	9
Total fertility rate 2000-2005 (births per woman)	2.4
Urban population (%)	52
Urban population growth rate 2000-2005 (% per annum)	1.4
Rural population growth rate 2000-2005 (% per annum)	2.0
Foreign born (%)	13.2[j]
Government education expenditure (% of GNP)	8.9[k]
Third-level students (women and men, % of total)	50/50[l]
Television receivers (per 1,000 inhabitants)	226

Environment	2000-2006
Threatened species	127
Forested area (% of land area)	29
CO2 emissions (000s Mt of carbon dioxide/per capita)	694/2.8
Energy consumption per capita (kilograms oil equiv.)	864[j]
Precipitation (mm)	1761[m]
Average minimum and maximum temperatures (centigrade)[m]	22.3/29.5

a 2002. b September 2002. c Data refer to non-commercial rates derived from the Operational Rates of Exchange for United Nations Programmes. d November 2006. e August 2006. f Excluding nationals residing abroad. g Arrivals of non-resident tourists at national borders. Excluding nationals residing abroad. The figure of 252,200 has been estimated by the 'Institut de la Statistique (ISPF)'. Due to problems of E/D card distributions, the brekdown by country of origin could not be elaborated. h Arrivals of non-resident tourists at national borders (excluding same-day visitors). i 2004. j Estimated data. k 1984. l 1995. m Tahiti.

Gabon

Region	Middle Africa
Largest urban agglom. (pop., 000s)	Libreville (556)
Currency	CFA franc
Population in 2006 (proj., 000s)	1406
Surface area (square kms)	267668
Population density (per square km)	5
United Nations membership date	20 September 1960

Economic indicators	2000	2005
Exchange rate (national currency per US$)[a]	704.95	516.66[b]
Consumer price index (2000=100)[c]	100	110[d]
Unemployment (percentage of labour force)[e]	18.0[f]	...
Balance of payments, current account (million US$)	1001	575[g]
Tourist arrivals (000s)[h]	155	222[g]
GDP (million current US$)	5019	7919
GDP (per capita current US$)	3945	5723
Gross fixed capital formation (% of GDP)	21.0	22.0
Agricultural production index (1999-2001=100)	101	102[i]
Food production index (1999-2001=100)	101	102[i]
Primary energy production (000s Mt oil equiv.)	15652	10930[i]
Motor vehicles (per 1,000 inhabitants)	29.5[j]	...
Telephone lines (per 100 inhabitants)	3.2	2.8
Internet users, estimated (000s)	15.0	67.0

Total trade		Major trading partners			2005
	(million US$)[i]		(% of exports)[i]		(% of imports)[i]
Exports	2780.0	USA	49	France	41
Imports	964.9	France	9	Belgium	11
		China	6	USA	5

Social indicators	2000-2006
Population growth rate 2000-2005 (% per annum)	1.7
Population aged 0-14 years (%)	40.0
Population aged 60+ years (women and men, % of total)	7.0/6.0
Sex ratio (women per 100 men)	101
Life expectancy at birth 2000-2005 (women and men, years)	55/54
Infant mortality rate 2000-2005 (per 1,000 births)	58
Total fertility rate 2000-2005 (births per woman)	4.0
Contraceptive use (% of currently married women)	33
Urban population (%)	84
Urban population growth rate 2000-2005 (% per annum)	2.5
Rural population growth rate 2000-2005 (% per annum)	−2.1
Refugees and others of concern to UNHCR[k]	13388
Government education expenditure (% of GNP)	4.6
Primary-secondary gross enrolment ratio (w and m per 100)	86/90
Third-level students (women and men, % of total)	36/64[l]
Television receivers (per 1,000 inhabitants)	163
Parliamentary seats (women and men, % of total)	12/88

Environment	2000-2006
Threatened species	150
Forested area (% of land area)	85
CO2 emissions (000s Mt of carbon dioxide/per capita)	1225/0.9
Energy consumption per capita (kilograms oil equiv.)	450
Precipitation (mm)	2842[c]
Average minimum and maximum temperatures (centigrade)[c]	23.3/28.6

a Official rate. b October 2006. c Libreville. d June 2006. e Persons aged 10 years and over. f 1993. g 2003. h Arrivals of non-resident tourists at Libreville airport. i 2004. j 1995. k Provisional. l 1999.

Gambia

Region	Western Africa
Largest urban agglom. (pop., 000s)	Banjul (381)
Currency	dalasi
Population in 2006 (proj., 000s)	1556
Surface area (square kms)	11295
Population density (per square km)	138
United Nations membership date	21 September 1965

Economic indicators	2000	2005
Exchange rate (national currency per US$)	14.89	28.03[a]
Consumer price index (2000=100)[b]	100	152[c]
Tourist arrivals (000s)[d]	79	90[c]
GDP (million current US$)	421	480
GDP (per capita current US$)	320	316
Gross fixed capital formation (% of GDP)	17.0	24.0
Labour force participation, adult female pop. (%)	44.8[e]	...
Labour force participation, adult male pop. (%)	70.2[e]	...
Agricultural production index (1999-2001=100)	101	69[c]
Food production index (1999-2001=100)	101	69[c]
Motor vehicles (per 1,000 inhabitants)	8.9[f]	...
Telephone lines (per 100 inhabitants)[g]	2.7	2.9
Internet users, estimated (000s)[g]	12.0	49.0[c]

Total trade		Major trading partners			2005
	(million US$)	(% of exports)		(% of imports)	
Exports	5.1	Guinea	45	Côte d'Ivoire	13
Imports	259.6	UK	18	Germany	12
		Senegal	10	Denmark	11

Social indicators	2000-2006
Population growth rate 2000-2005 (% per annum)	2.8
Population aged 0-14 years (%)	40.0
Population aged 60+ years (women and men, % of total)	6.0/6.0
Sex ratio (women per 100 men)	102
Life expectancy at birth 2000-2005 (women and men, years)	57/54
Infant mortality rate 2000-2005 (per 1,000 births)	77
Total fertility rate 2000-2005 (births per woman)	4.8
Contraceptive use (% of currently married women)	10
Urban population (%)	54
Urban population growth rate 2000-2005 (% per annum)	4.7
Rural population growth rate 2000-2005 (% per annum)	0.9
Refugees and others of concern to UNHCR[h]	7932
Government education expenditure (% of GNP)	2.1
Primary-secondary gross enrolment ratio (w and m per 100)	65/66
Third-level students (women and men, % of total)	19/81
Television receivers (per 1,000 inhabitants)[g]	15
Parliamentary seats (women and men, % of total)	13/87

Environment	2000-2006
Threatened species	32
Forested area (% of land area)	48
CO2 emissions (000s Mt of carbon dioxide/per capita)	283/0.2
Energy consumption per capita (kilograms oil equiv.)	62[i]
Precipitation (mm)	977[i]
Average minimum and maximum temperatures (centigrade)[j]	19.9/32.0

a July 2006. b Banjul,Kombo,St.Mary. c 2004. d Charter tourist only. e 1993. f 1995.
g Year beginning 1 April. h Provisional. i Estimated data. j Banjul.

Georgia

Region	Western Asia
Largest urban agglom. (pop., 000s)	Tbilisi (1047)
Currency	lari
Population in 2006 (proj., 000s)	4434
Surface area (square kms)	69700
Population density (per square km)	64
United Nations membership date	31 July 1992

Economic indicators	2000	2005
Exchange rate (national currency per US$)[a]	1.98	1.74[b]
Consumer price index (2000=100)[c]	100	140[d]
Unemployment (percentage of labour force)[e]	10.8	13.8
Balance of payments, current account (million US$)	−269	−752
Tourist arrivals (000s)	387	368[f]
GDP (million current US$)	3044	6490
GDP (per capita current US$)	645	1450
Gross fixed capital formation (% of GDP)	26.0	29.0
Labour force participation, adult female pop. (%)	55.0	56.9[f]
Labour force participation, adult male pop. (%)	74.7	74.1[f]
Employment in industrial sector (%)	9.8	9.3
Employment in agricultural sector (%)	52.1	54.3
Agricultural production index (1999-2001=100)	92	99[f]
Food production index (1999-2001=100)	93	101[f]
Primary energy production (000s Mt oil equiv.)	680	634[f]
Motor vehicles (per 1,000 inhabitants)	66.0	102.0[f]
Telephone lines (per 100 inhabitants)	10.8	15.1
Internet users, estimated (000s)	23.0	175.6[f]

Total trade		Major trading partners			2005
	(million US$)	(% of exports)		(% of imports)	
Exports	866.2	Russian Fed.	18	Russian Fed.	15
Imports	2490.9	Turkey	14	Turkey	11
		Azerbaijan	10	Azerbaijan	9

Social indicators	2000-2006
Population growth rate 2000-2005 (% per annum)	−1.1
Population aged 0-14 years (%)	19.0
Population aged 60+ years (women and men, % of total)	20.0/15.0
Sex ratio (women per 100 men)	112
Life expectancy at birth 2000-2005 (women and men, years)	74/67
Infant mortality rate 2000-2005 (per 1,000 births)	40
Total fertility rate 2000-2005 (births per woman)	1.5
Contraceptive use (% of currently married women)	26[gh]
Urban population (%)	52
Urban population growth rate 2000-2005 (% per annum)	−1.3
Rural population growth rate 2000-2005 (% per annum)	−0.9
Refugees and others of concern to UNHCR[i]	238482
Government education expenditure (% of GNP)	3.0
Primary-secondary gross enrolment ratio (w and m per 100)	88/89
Third-level students (women and men, % of total)	51/49
Newspaper circulation (per 1,000 inhabitants)	5
Television receivers (per 1,000 inhabitants)	386
Intentional homicides (per 100,000 inhabitants)	4
Parliamentary seats (women and men, % of total)	9/81

Environment	2000-2006
Threatened species	49
Forested area (% of land area)	44
CO2 emissions (000s Mt of carbon dioxide/per capita)	3732/0.8
Energy consumption per capita (kilograms oil equiv.)	417
Precipitation (mm)	496[j]

a Official rate. b October 2006. c 5 cities. d February 2006. e Persons aged 15 years and over. f 2004. g 2005. h For women aged 15-44 in union or married. i Provisional. j Tbilisi.

Germany

Region	Western Europe
Largest urban agglom. (pop., 000s)	Berlin (3389)
Currency	euro[a]
Population in 2006 (proj., 000s)	82716
Surface area (square kms)	357022
Population density (per square km)	232
United Nations membership date	18 September 1973

Economic indicators	2000	2005
Exchange rate (national currency per US$)	1.07	0.79[b]
Consumer price index (2000=100)	100	111[c]
Industrial production index (1995=100)[d]	115	119[e]
Unemployment (percentage of labour force)[f]	7.9[g]	11.1
Balance of payments, current account (million US$)	−32	116
Tourist arrivals (000s)	18983	20137[h]
GDP (million current US$)	1900220	2794857
GDP (per capita current US$)	23076	33800
Gross fixed capital formation (% of GDP)	21.0	17.0
Labour force participation, adult female pop. (%)	49.0	50.3[h]
Labour force participation, adult male pop. (%)	66.6	65.0[h]
Employment in industrial sector (%)	33.1	29.7
Employment in agricultural sector (%)	2.7	2.4
Agricultural production index (1999-2001=100)	100	103[h]
Food production index (1999-2001=100)	100	103[h]
Primary energy production (000s Mt oil equiv.)	97749	95786[h]
Motor vehicles (per 1,000 inhabitants)	520.3	587.2[hi]
Telephone lines (per 100 inhabitants)	61.1	66.6
Internet users, estimated (000s)	24800.0	37500.0

Total trade		Major trading partners			2005
	(million US$)	(% of exports)		(% of imports)	
Exports	977028.1	France	10	France	9
Imports	776843.0	USA	9	Netherlands	8
		UK	8	USA	6

Social indicators	2000-2006
Population growth rate 2000-2005 (% per annum)	0.1
Population aged 0-14 years (%)	14.0
Population aged 60+ years (women and men, % of total)	28.0/22.0
Sex ratio (women per 100 men)	105
Life expectancy at birth 2000-2005 (women and men, years)	81/76
Infant mortality rate 2000-2005 (per 1,000 births)	4
Total fertility rate 2000-2005 (births per woman)	1.3
Contraceptive use (% of currently married women)	75[jk]
Urban population (%)	75
Urban population growth rate 2000-2005 (% per annum)	0.1
Rural population growth rate 2000-2005 (% per annum)	≺
Foreign born (%)	8.9[l]
Refugees and others of concern to UNHCR[m]	781116
Government education expenditure (% of GNP)	4.8
Primary-secondary gross enrolment ratio (w and m per 100)	100/101
Newspaper circulation (per 1,000 inhabitants)	291
Television receivers (per 1,000 inhabitants)	675
Intentional homicides (per 100,000 inhabitants)	1
Parliamentary seats (women and men, % of total)	30/70

Environment	2000-2006
Threatened species	83
Forested area (% of land area)	31
CO2 emissions (000s Mt of carbon dioxide/per capita)[n]	865370/10.5
Energy consumption per capita (kilograms oil equiv.)	3571
Precipitation (mm)	571[o]
Average minimum and maximum temperatures (centigrade)[o]	5.9/13.4

a Prior to 1 January 1999, deutsche mark. b September 2006. c July 2006. d Monthly indices are adjusted for differences in the number of working days. e August 2006. f Persons aged 15 years and over. g May. h 2004. i Beginning 2001, data refer to fiscal years ending 1 January. j For women aged 20-39 in union or married. k 1992. l Estimated data. m Provisional. n Source: UNFCCC. o Berlin.

Ghana

Region	Western Africa	
Largest urban agglom. (pop., 000s)	Accra (1981)	
Currency	cedi	
Population in 2006 (proj., 000s)	22556	
Surface area (square kms)	238533	
Population density (per square km)	95	
United Nations membership date	8 March 1957	

Economic indicators	2000	2005
Exchange rate (national currency per US$)[a]	7047.65	9224.44[b]
Consumer price index (2000=100)	100	251
Balance of payments, current account (million US$)	−387	−812
Tourist arrivals (000s)[c]	399	584[d]
GDP (million current US$)	4978	10393
GDP (per capita current US$)	251	470
Gross fixed capital formation (% of GDP)	24.0	23.0
Labour force participation, adult female pop. (%)	72.7	...
Labour force participation, adult male pop. (%)	76.7	...
Employment in industrial sector (%)	14.0	...
Employment in agricultural sector (%)	55.0	...
Agricultural production index (1999-2001=100)	99	121[d]
Food production index (1999-2001=100)	99	121[d]
Primary energy production (000s Mt oil equiv.)	568	454[d]
Motor vehicles (per 1,000 inhabitants)	10.7	10.2[e]
Telephone lines (per 100 inhabitants)	1.1	1.5
Internet users, estimated (000s)	30.0	401.3

Total trade	Major trading partners				2005
	(million US$)[d]	(% of exports)[d]			(% of imports)[d]
Exports	1779.1	Netherlands	25	China	9
Imports	4073.9	UK	13	USA	9
		Belgium	7	Germany	8

Social indicators	2000-2006
Population growth rate 2000-2005 (% per annum)	2.1
Population aged 0-14 years (%)	39.0
Population aged 60+ years (women and men, % of total)	6.0/5.0
Sex ratio (women per 100 men)	98
Life expectancy at birth 2000-2005 (women and men, years)	57/56
Infant mortality rate 2000-2005 (per 1,000 births)	62
Total fertility rate 2000-2005 (births per woman)	4.4
Contraceptive use (% of currently married women)	25
Urban population (%)	48
Urban population growth rate 2000-2005 (% per annum)	3.8
Rural population growth rate 2000-2005 (% per annum)	0.7
Foreign born (%)	7.8
Refugees and others of concern to UNHCR[f]	59034
Government education expenditure (% of GNP)	4.2[g]
Primary-secondary gross enrolment ratio (w and m per 100)	64/69
Third-level students (women and men, % of total)	32/68
Television receivers (per 1,000 inhabitants)	52
Parliamentary seats (women and men, % of total)	11/89

Environment	2000-2006
Threatened species	172
Forested area (% of land area)	28
CO2 emissions (000s Mt of carbon dioxide/per capita)	7745/0.4
Energy consumption per capita (kilograms oil equiv.)	112
Precipitation (mm)	807[h]
Average minimum and maximum temperatures (centigrade)[h]	23.4/30.8

a Principal rate. b October 2006. c Including nationals residing abroad. d 2004. e 2003. f Provisional. g 1999. h Accra.

Greece

Region	Southern Europe
Largest urban agglom. (pop., 000s)	Athens (3230)
Currency	euro[a]
Population in 2006 (proj., 000s)	11140
Surface area (square kms)	131957
Population density (per square km)	84
United Nations membership date	25 October 1945

Economic indicators	2000	2005
Exchange rate (national currency per US$)	365.62	0.79[b]
Consumer price index (2000=100)	100	122[c]
Industrial production index (1995=100)	123	114[d]
Unemployment (percentage of labour force)[e]	11.2	9.6
Balance of payments, current account (million US$)	−9820	−17879
Tourist arrivals (000s)[fg]	13096	13969[h]
GDP (million current US$)	115997	225201
GDP (per capita current US$)	10569	20252
Gross fixed capital formation (% of GDP)	23.0	24.0
Labour force participation, adult female pop. (%)	38.7	42.2[i]
Labour force participation, adult male pop. (%)	62.2	65.0[i]
Employment in industrial sector (%)	22.6	22.4
Employment in agricultural sector (%)	17.4	12.4
Agricultural production index (1999-2001=100)	102	94[i]
Food production index (1999-2001=100)	102	95[i]
Primary energy production (000s Mt oil equiv.)	9094	9891[i]
Motor vehicles (per 1,000 inhabitants)	389.9	496.0
Telephone lines (per 100 inhabitants)	53.6	56.7
Internet users, estimated (000s)	1000.0	2001.0

Total trade		Major trading partners			2005
	(million US$)	(% of exports)		(% of imports)	
Exports	17434.4	Germany	12	Germany	13
Imports	54893.9	Italy	11	Italy	12
		UK	7	Russian Fed.	8

Social indicators	2000-2006
Population growth rate 2000-2005 (% per annum)	0.3
Population aged 0-14 years (%)	14.0
Population aged 60+ years (women and men, % of total)	25.0/21.0
Sex ratio (women per 100 men)	102
Life expectancy at birth 2000-2005 (women and men, years)	81/76
Infant mortality rate 2000-2005 (per 1,000 births)	6
Total fertility rate 2000-2005 (births per woman)	1.3
Urban population (%)	59
Urban population growth rate 2000-2005 (% per annum)	0.3
Rural population growth rate 2000-2005 (% per annum)	0.2
Foreign born (%)	10.3
Refugees and others of concern to UNHCR[j]	14257
Government education expenditure (% of GNP)	4.0
Primary-secondary gross enrolment ratio (w and m per 100)	99/99
Third-level students (women and men, % of total)	52/48
Television receivers (per 1,000 inhabitants)	537
Intentional homicides (per 100,000 inhabitants)	1
Parliamentary seats (women and men, % of total)	13/87

Environment	2000-2006
Threatened species	111
Forested area (% of land area)	28
CO2 emissions (000s Mt of carbon dioxide/per capita)[k]	109980/9.9
Energy consumption per capita (kilograms oil equiv.)	2823
Precipitation (mm)	414[l]
Average minimum and maximum temperatures (centigrade)[l]	12.3/22.5

a Prior to 1 January 2001, drachma. b September 2006. c July 2006. d August 2006. e Persons aged 15 years and over, second quarter of each year. f Data based on surveys. g Arrivals of non-resident tourists at national borders (excluding same-day visitors). h 2003. i 2004. j Provisional. k Source: UNFCCC. l Athens.

Grenada

Region	Caribbean
Largest urban agglom. (pop., 000s)	St George's (32)
Currency	EC dollar
Population in 2006 (proj., 000s)	103[a]
Surface area (square kms)	344
Population density (per square km)	298
United Nations membership date	17 September 1974

Economic indicators	2000	2005
Exchange rate (national currency per US$)[b]	2.70	2.70[c]
Balance of payments, current account (million US$)	−84	−116[d]
Tourist arrivals (000s)	129	134[e]
GDP (million current US$)	410	454
GDP (per capita current US$)	4031	4415
Gross fixed capital formation (% of GDP)	44.0	38.0
Labour force participation, adult female pop. (%)	55.0[f]	...
Labour force participation, adult male pop. (%)	75.7[f]	...
Employment in industrial sector (%)	23.9[f]	...
Employment in agricultural sector (%)	13.8[f]	...
Agricultural production index (1999-2001=100)	103	99[e]
Food production index (1999-2001=100)	103	99[e]
Motor vehicles (per 1,000 inhabitants)	182.1	196.6[g]
Telephone lines (per 100 inhabitants)	30.9	32.0
Internet users, estimated (000s)	4.1	19.0[h]

Total trade		Major trading partners			2005
	(million US$)[e]	(% of exports)[e]		(% of imports)[e]	
Exports	31.6	USA	23	USA	43
Imports	250.5	Netherlands	16	Trinidad Tbg	19
		Belgium	8	UK	7

Social indicators	2000-2006
Population growth rate 2000-2005 (% per annum)	0.3
Population aged 0-14 years (%)	35.0
Population aged 60+ years (women and men, % of total)	12.0/8.0
Sex ratio (women per 100 men)	103[gi]
Infant mortality rate 2000-2005 (per 1,000 births)	17[j]
Contraceptive use (% of currently married women)	54[kl]
Urban population (%)	31
Urban population growth rate 2000-2005 (% per annum)	<
Rural population growth rate 2000-2005 (% per annum)	0.4
Foreign born (%)	9.8[m]
Government education expenditure (% of GNP)	6.0
Primary-secondary gross enrolment ratio (w and m per 100)	96/95
Third-level students (women and men, % of total)	55/45[n]
Television receivers (per 1,000 inhabitants)	383
Parliamentary seats (women and men, % of total)	32/68

Environment	2000-2006
Threatened species	25
Forested area (% of land area)	15
CO2 emissions (000s Mt of carbon dioxide/per capita)	221/2.2
Energy consumption per capita (kilograms oil equiv.)	925

a 2005. b Official rate. c October 2006. d 2002. e 2004. f 1998. g 2001. h 2003. i Refers to de facto population count. j Data tabulated by date of registration rather than occurrence. k For women aged 15-44 in union or married. l 1990. m Estimated data. n 1995.

Guadeloupe

Region	Caribbean
Largest urban agglom. (pop., 000s)	Pointe-à-Pitre (19)
Currency	euro[a]
Population in 2006 (proj., 000s)	452
Surface area (square kms)	1705
Population density (per square km)	265

Economic indicators	2000	2005
Exchange rate (national currency per US$)	1.07	0.79[bc]
Consumer price index (2000=100)	100	114[d]
Unemployment (percentage of labour force)[e]	25.7	26.2
Tourist arrivals (000s)	603[f]	456[gh]
Labour force participation, adult male pop. (%)	63.8[i]	...
Agricultural production index (1999-2001=100)	98	103[h]
Food production index (1999-2001=100)	98	103[h]
Motor vehicles (per 1,000 inhabitants)[j]	347.0	340.7[k]
Telephone lines (per 100 inhabitants)	48.0	48.7[lm]
Internet users, estimated (000s)	25.0	85.0

Social indicators	2000-2006
Population growth rate 2000-2005 (% per annum)	0.9
Population aged 0-14 years (%)	25.0
Population aged 60+ years (women and men, % of total)	15.0/13.0
Sex ratio (women per 100 men)	107
Life expectancy at birth 2000-2005 (women and men, years)	82/75
Infant mortality rate 2000-2005 (per 1,000 births)	7
Total fertility rate 2000-2005 (births per woman)	2.1
Urban population (%)	100
Urban population growth rate 2000-2005 (% per annum)	0.9
Rural population growth rate 2000-2005 (% per annum)	−10.3
Foreign born (%)	19.4[n]
Television receivers (per 1,000 inhabitants)	296

Environment	2000-2006
Threatened species	39
Forested area (% of land area)	49
CO_2 emissions (000s Mt of carbon dioxide/per capita)	1713/3.9
Energy consumption per capita (kilograms oil equiv.)	1211[n]
Precipitation (mm)	1779[o]
Average minimum and maximum temperatures (centigrade)[o]	22.1/30.5

a Prior to 1 January 1999, French franc. b November 2006. c Data refer to non-commercial rates derived from the Operational Rates of Exchange for United Nations Programmes. d June 2006. e Persons aged 15 years and over, June of each year. f Arrivals of non-resident tourists in all types of accommodation establishments. Air arrivals. Excluding the north islands (Saint Martin and Saint Barthelemy). g Arrivals of non-resident tourists in all types of accommodation establishments. Air arrivals. Air arrivals in hotels only. Excluding the north islands (Saint Martin and Saint Barthelemy). h 2004. i 1999. j Source: World Automotive Market Report, Auto and Truck International (Illinois). k 2002. l Estimate. m 2003. n Estimated data. o Le Raizet.

Guam

Region	Oceania-Micronesia
Largest urban agglom. (pop., 000s)	Hagåtña (144)
Currency	US dollar
Population in 2006 (proj., 000s)	172
Surface area (square kms)	549
Population density (per square km)	314

Economic indicators	2000	2005
Consumer price index (2000=100)	100	101[a]
Unemployment (percentage of labour force)[b]	5.5[c]	...
Tourist arrivals (000s)	1287	1157[d]
Agricultural production index (1999-2001=100)	100	107[d]
Food production index (1999-2001=100)	100	107[d]
Motor vehicles (per 1,000 inhabitants)	586.3	464.0[e]
Telephone lines (per 100 inhabitants)	48.0	50.9[af]
Internet users, estimated (000s)	25.0	65.0

Social indicators	2000-2006
Population growth rate 2000-2005 (% per annum)	1.8
Population aged 0-14 years (%)	30.0
Population aged 60+ years (women and men, % of total)	10.0/8.0
Sex ratio (women per 100 men)	96
Life expectancy at birth 2000-2005 (women and men, years)	77/72
Infant mortality rate 2000-2005 (per 1,000 births)	10
Total fertility rate 2000-2005 (births per woman)	3.0
Urban population (%)	94
Urban population growth rate 2000-2005 (% per annum)	1.9
Rural population growth rate 2000-2005 (% per annum)	−1.0
Foreign born (%)	62.2[g]
Government education expenditure (% of GNP)	8.3[h]
Third-level students (women and men, % of total)	59/41[i]
Television receivers (per 1,000 inhabitants)	272

Environment	2000-2006
Threatened species	32
Forested area (% of land area)	38
CO2 emissions (000s Mt of carbon dioxide/per capita)	4087/25.0
Energy consumption per capita (kilograms oil equiv.)	8493[e]
Precipitation (mm)	2313
Average minimum and maximum temperatures (centigrade)	23.7/30.9

a 2003. b Persons aged 16 years and over. c 1993. d 2004. e 2002. f Estimate.
g Estimated data. h 1985. i 1995.

Guatemala

Region	Central America
Largest urban agglom. (pop., 000s)	Guatemala City (984)
Currency	quetzal
Population in 2006 (proj., 000s)	12911
Surface area (square kms)	108889
Population density (per square km)	119
United Nations membership date	21 November 1945

Economic indicators	2000	2005
Exchange rate (national currency per US$)	7.73	7.58[a]
Consumer price index (2000=100)[b]	100	155[c]
Unemployment (percentage of labour force)[d]	1.4	3.4[e]
Balance of payments, current account (million US$)	−1050	−1188[f]
Tourist arrivals (000s)	826	1182[f]
GDP (million current US$)	19289	31923
GDP (per capita current US$)	1727	2534
Gross fixed capital formation (% of GDP)	16.0	15.0
Labour force participation, adult female pop. (%)	23.4[g]	...
Labour force participation, adult male pop. (%)	74.1[g]	...
Employment in industrial sector (%)	20.4	20.0[h]
Employment in agricultural sector (%)	36.4	38.7[h]
Agricultural production index (1999-2001=100)	100	101[f]
Food production index (1999-2001=100)	99	104[f]
Primary energy production (000s Mt oil equiv.)	1327	1216[f]
Motor vehicles (per 1,000 inhabitants)	91.9	108.0[f]
Telephone lines (per 100 inhabitants)	5.9	8.9
Internet users, estimated (000s)	80.0	1000.0

Total trade		Major trading partners		2005	
	(million US$)	(% of exports)		(% of imports)	
Exports	5380.8	USA	50	USA	34
Imports	10499.5	El Salvador	12	Mexico	9
		Honduras	7	China	7

Social indicators	2000-2006
Population growth rate 2000-2005 (% per annum)	2.4
Population aged 0-14 years (%)	43.0
Population aged 60+ years (women and men, % of total)	6.0/6.0
Sex ratio (women per 100 men)	105
Life expectancy at birth 2000-2005 (women and men, years)	71/63
Infant mortality rate 2000-2005 (per 1,000 births)	39
Total fertility rate 2000-2005 (births per woman)	4.6
Contraceptive use (% of currently married women)	43
Urban population (%)	47
Urban population growth rate 2000-2005 (% per annum)	3.3
Rural population growth rate 2000-2005 (% per annum)	1.6
Foreign born (%)	0.4[i]
Refugees and others of concern to UNHCR[j]	394
Government education expenditure (% of GNP)	1.7
Primary-secondary gross enrolment ratio (w and m per 100)	82/90
Third-level students (women and men, % of total)	43/57
Television receivers (per 1,000 inhabitants)	158
Parliamentary seats (women and men, % of total)	8/92

Environment	2000-2006
Threatened species	219
Forested area (% of land area)	26
CO2 emissions (000s Mt of carbon dioxide/per capita)	10711/0.9
Energy consumption per capita (kilograms oil equiv.)	290
Precipitation (mm)	1186[k]
Average minimum and maximum temperatures (centigrade)[k]	14.7/25.0

a October 2006. b Guatemala. c August 2006. d Persons aged 10 years and over.
e 2003. f 2004. g 1999. h 2002. i Estimated data. j Provisional. k Guatemala City.

Guinea

Region	Western Africa
Largest urban agglom. (pop., 000s)	Conakry (1425)
Currency	franc
Population in 2006 (proj., 000s)	9603
Surface area (square kms)	245857
Population density (per square km)	39
United Nations membership date	12 December 1958

Economic indicators	2000	2005
Exchange rate (national currency per US$)	1882.27	5089.00[a]
Consumer price index (2000=100)[b]	100	223[c]
Balance of payments, current account (million US$)	−155	−175[d]
Tourist arrivals (000s)	33[e]	45[d]
GDP (million current US$)	3134	3058
GDP (per capita current US$)	372	325
Gross fixed capital formation (% of GDP)	20.0	14.0
Agricultural production index (1999-2001=100)	99	111[d]
Food production index (1999-2001=100)	98	114[d]
Primary energy production (000s Mt oil equiv.)	35	38[d]
Motor vehicles (per 1,000 inhabitants)	4.8[fg]	...
Telephone lines (per 100 inhabitants)	0.3	0.3
Internet users, estimated (000s)	8.0	50.0

Total trade		Major trading partners			2005
	(million US$)[h]	(% of exports)[h]		(% of imports)[h]	
Exports	525.4	France	24	France	16
Imports	666.5	Ireland	10	Côte d'Ivoire	15
		Spain	10	USA	8

Social indicators	2000-2006
Population growth rate 2000-2005 (% per annum)	2.2
Population aged 0-14 years (%)	44.0
Population aged 60+ years (women and men, % of total)	6.0/5.0
Sex ratio (women per 100 men)	95
Life expectancy at birth 2000-2005 (women and men, years)	54/53
Infant mortality rate 2000-2005 (per 1,000 births)	106
Total fertility rate 2000-2005 (births per woman)	5.9
Urban population (%)	33
Urban population growth rate 2000-2005 (% per annum)	3.4
Rural population growth rate 2000-2005 (% per annum)	1.6
Foreign born (%)	9.1[i]
Refugees and others of concern to UNHCR[j]	67336
Government education expenditure (% of GNP)	1.9
Primary-secondary gross enrolment ratio (w and m per 100)	45/62
Third-level students (women and men, % of total)	16/84
Television receivers (per 1,000 inhabitants)	18
Parliamentary seats (women and men, % of total)	19/81

Environment	2000-2006
Threatened species	83
Forested area (% of land area)	28
CO2 emissions (000s Mt of carbon dioxide/per capita)	1341/0.1
Energy consumption per capita (kilograms oil equiv.)	46[i]
Precipitation (mm)	3776[b]
Average minimum and maximum temperatures (centigrade)[k]	22.9/29.9

a June 2006. b Conakry. c April 2006. d 2004. e Air arrivals at Conakry Airport. f Source: AAMA Motor Vehicle Facts and Figures, American Automobile Manufacturers Association (Michigan). g 1995. h 2002. i Estimated data. j Provisional. k Conakry/Gbessia.

Guinea-Bissau

Region	Western Africa
Largest urban agglom. (pop., 000s)	Bissau (367)
Currency	CFA franc
Population in 2006 (proj., 000s)	1634
Surface area (square kms)	36125
Population density (per square km)	45
United Nations membership date	17 September 1974

Economic indicators	2000	2005
Exchange rate (national currency per US$)[a]	704.95	516.66[b]
Consumer price index (2000=100)[c]	...	107[de]
Balance of payments, current account (million US$)	−27[f]	−6[g]
Tourist arrivals (000s)	8[f]	...
GDP (million current US$)	215	298
GDP (per capita current US$)	158	188
Gross fixed capital formation (% of GDP)	15.0	20.0
Agricultural production index (1999-2001=100)	100	110[h]
Food production index (1999-2001=100)	100	110[h]
Motor vehicles (per 1,000 inhabitants)	6.0[ij]	
Telephone lines (per 100 inhabitants)	0.9	0.8
Internet users, estimated (000s)	3.0	31.0

Social indicators	2000-2006
Population growth rate 2000-2005 (% per annum)	3.0
Population aged 0-14 years (%)	48.0
Population aged 60+ years (women and men, % of total)	5.0/4.0
Sex ratio (women per 100 men)	102
Life expectancy at birth 2000-2005 (women and men, years)	46/43
Infant mortality rate 2000-2005 (per 1,000 births)	120
Total fertility rate 2000-2005 (births per woman)	7.1
Contraceptive use (% of currently married women)	8
Urban population (%)	30
Urban population growth rate 2000-2005 (% per annum)	2.9
Rural population growth rate 2000-2005 (% per annum)	3.0
Foreign born (%)	1.4[k]
Refugees and others of concern to UNHCR[l]	7782
Government education expenditure (% of GNP)	5.6[m]
Primary-secondary gross enrolment ratio (w and m per 100)	38/59
Third-level students (women and men, % of total)	16/84
Television receivers (per 1,000 inhabitants)	45
Parliamentary seats (women and men, % of total)	14/86

Environment	2000-2006
Threatened species	32
Forested area (% of land area)	61
CO2 emissions (000s Mt of carbon dioxide/per capita)	270/0.2
Energy consumption per capita (kilograms oil equiv.)	65[k]
Precipitation (mm)[n]	1756[e]
Average minimum and maximum temperatures (centigrade)[e]	19.7/32.1

a Official rate. b October 2006. c Base: 2003=100. d June 2006. e Bissau. f 2001. g 2003. h 2004. i Source: AAMA Motor Vehicle Facts and Figures, American Automobile Manufacturers Association (Michigan). j 1992. k Estimated data. l Provisional. m 1999. n January to November only.

Guyana

Region	South America	
Largest urban agglom. (pop., 000s)	Georgetown (134)	
Currency	dollar	
Population in 2006 (proj., 000s)	752	
Surface area (square kms)	214969	
Population density (per square km)	3	
United Nations membership date	20 September 1966	

Economic indicators	2000	2005
Exchange rate (national currency per US$)[a]	184.75	200.00[b]
Consumer price index (2000=100)[c]	100	137[d]
Balance of payments, current account (million US$)	−82	−30[e]
Tourist arrivals (000s)	105	122[ef]
GDP (million current US$)	713	786
GDP (per capita current US$)	958	1046
Gross fixed capital formation (% of GDP)	39.0	34.0
Labour force participation, adult female pop. (%)	39.3[g]	...
Labour force participation, adult male pop. (%)	81.2[g]	...
Employment in industrial sector (%)	22.6[h]	...
Employment in agricultural sector (%)	27.8[h]	...
Agricultural production index (1999-2001=100)	102	105[e]
Food production index (1999-2001=100)	102	105[e]
Motor vehicles (per 1,000 inhabitants)	103.3	102.7[ij]
Telephone lines (per 100 inhabitants)	9.2	14.7
Internet users, estimated (000s)	50.0	160.0

Total trade		Major trading partners			2005
	(million US$)		(% of exports)		(% of imports)
Exports	538.7	UK	20	Trinidad Tbg	33
Imports	778.1	Canada	16	USA	31
		USA	15	UK	4

Social indicators	2000-2006
Population growth rate 2000-2005 (% per annum)	0.2
Population aged 0-14 years (%)	29.0
Population aged 60+ years (women and men, % of total)	8.0/6.0
Sex ratio (women per 100 men)	106
Life expectancy at birth 2000-2005 (women and men, years)	66/60
Infant mortality rate 2000-2005 (per 1,000 births)	49
Total fertility rate 2000-2005 (births per woman)	2.3
Contraceptive use (% of currently married women)	37
Urban population (%)	28
Urban population growth rate 2000-2005 (% per annum)	−0.1
Rural population growth rate 2000-2005 (% per annum)	0.3
Foreign born (%)	0.2[k]
Government education expenditure (% of GNP)	5.8
Primary-secondary gross enrolment ratio (w and m per 100)	110/110
Third-level students (women and men, % of total)	65/35
Newspaper circulation (per 1,000 inhabitants)	76
Television receivers (per 1,000 inhabitants)	167
Parliamentary seats (women and men, % of total)	29/71

Environment	2000-2006
Threatened species	71
Forested area (% of land area)	79
CO2 emissions (000s Mt of carbon dioxide/per capita)	1632/2.2
Energy consumption per capita (kilograms oil equiv.)	626
Precipitation (mm)	2260[c]
Average minimum and maximum temperatures (centigrade)[c]	24.0/29.6

a Principal rate. b September 2006. c Georgetown. d March 2006. e 2004. f Arrivals to Timehri airport only. g 1993. h 1997. i Source: World Automotive Market Report, Auto and Truck International (Illinois). j 2002. k Estimated data.

Haiti

Region	Caribbean
Largest urban agglom. (pop., 000s)	Port-au-Prince (2129)
Currency	gourde
Population in 2006 (proj., 000s)	8650
Surface area (square kms)	27750
Population density (per square km)	312
United Nations membership date	24 October 1945

Economic indicators	2000	2005
Exchange rate (national currency per US$)[a]	22.52	39.86[b]
Consumer price index (2000=100)	100	281[c]
Industrial production index (1995=100)[d]	118	138
Unemployment (percentage of labour force)[e]	23.9[f]	...
Balance of payments, current account (million US$)	−85	−13[g]
Tourist arrivals (000s)	140	96[h]
GDP (million current US$)	3515	3884
GDP (per capita current US$)	443	455
Gross fixed capital formation (% of GDP)	13.0	13.0
Labour force participation, adult female pop. (%)	50.6[i]	...
Labour force participation, adult male pop. (%)	58.8[i]	...
Employment in industrial sector (%)	10.7[i]	...
Employment in agricultural sector (%)	50.6[i]	...
Agricultural production index (1999-2001=100)	103	101[h]
Food production index (1999-2001=100)	103	101[h]
Primary energy production (000s Mt oil equiv.)	24	24[h]
Motor vehicles (per 1,000 inhabitants)	19.8[i]	...
Telephone lines (per 100 inhabitants)	0.9	1.7
Internet users, estimated (000s)	20.0	600.0

Social indicators	2000-2006
Population growth rate 2000-2005 (% per annum)	1.4
Population aged 0-14 years (%)	38.0
Population aged 60+ years (women and men, % of total)	7.0/5.0
Sex ratio (women per 100 men)	103
Life expectancy at birth 2000-2005 (women and men, years)	52/51
Infant mortality rate 2000-2005 (per 1,000 births)	62
Total fertility rate 2000-2005 (births per woman)	4.0
Contraceptive use (% of currently married women)	28
Urban population (%)	39
Urban population growth rate 2000-2005 (% per annum)	3.1
Rural population growth rate 2000-2005 (% per annum)	0.4
Foreign born (%)	0.3[i]
Refugees and others of concern to UNHCR	2
Government education expenditure (% of GNP)	1.5[k]
Television receivers (per 1,000 inhabitants)	63
Parliamentary seats (women and men, % of total)	6/94

Environment	2000-2006
Threatened species	119
Forested area (% of land area)	3
CO2 emissions (000s Mt of carbon dioxide/per capita)	1741/0.2
Energy consumption per capita (kilograms oil equiv.)	62

a Principal rate. b June 2006. c May 2006. d Manufacturing. e Persons aged 10 years and over. f 1998. g 2003. h 2004. i 1999. j Estimated data. k 1990.

Honduras

Region	Central America
Largest urban agglom. (pop., 000s)	Tegucigalpa (927)
Currency	lempira
Population in 2006 (proj., 000s)	7362
Surface area (square kms)	112088
Population density (per square km)	66
United Nations membership date	17 December 1945

Economic indicators	2000	2005
Exchange rate (national currency per US$)[a]	15.14	18.90[b]
Consumer price index (2000=100)	100	158[c]
Unemployment (percentage of labour force)[d]	3.7[e]	4.1
Balance of payments, current account (million US$)	−262	−86
Tourist arrivals (000s)	471	672[f]
GDP (million current US$)	6025	8374
GDP (per capita current US$)	938	1162
Gross fixed capital formation (% of GDP)	26.0	23.0
Labour force participation, adult female pop. (%)	44.2	38.6[f]
Labour force participation, adult male pop. (%)	87.1	82.8[f]
Employment in industrial sector (%)	22.0[e]	20.9
Employment in agricultural sector (%)	35.1[e]	39.2
Agricultural production index (1999-2001=100)	102	109[f]
Food production index (1999-2001=100)	102	111[f]
Primary energy production (000s Mt oil equiv.)	194	202[f]
Motor vehicles (per 1,000 inhabitants)	13.6[e]	...
Telephone lines (per 100 inhabitants)	4.8	6.9
Internet users, estimated (000s)	75.0	260.0

Total trade		Major trading partners			2005
	(million US$)		(% of exports)		(% of imports)
Exports	1883.2	USA	52	USA	37
Imports	4565.2	El Salvador	7	Guatemala	8
		Germany	6	Mexico	6

Social indicators	2000-2006
Population growth rate 2000-2005 (% per annum)	2.3
Population aged 0-14 years (%)	39.0
Population aged 60+ years (women and men, % of total)	6.0/5.0
Sex ratio (women per 100 men)	98
Life expectancy at birth 2000-2005 (women and men, years)	70/66
Infant mortality rate 2000-2005 (per 1,000 births)	32
Total fertility rate 2000-2005 (births per woman)	3.7
Contraceptive use (% of currently married women)	62[g]
Urban population (%)	46
Urban population growth rate 2000-2005 (% per annum)	3.2
Rural population growth rate 2000-2005 (% per annum)	1.5
Foreign born (%)	0.7[h]
Refugees and others of concern to UNHCR[i]	72
Primary-secondary gross enrolment ratio (w and m per 100)	96/89
Third-level students (women and men, % of total)	59/41
Television receivers (per 1,000 inhabitants)	143
Parliamentary seats (women and men, % of total)	23/77

Environment	2000-2006
Threatened species	212
Forested area (% of land area)	48
CO2 emissions (000s Mt of carbon dioxide/per capita)	6507/0.9
Energy consumption per capita (kilograms oil equiv.)	337
Precipitation (mm)	872[j]
Average minimum and maximum temperatures (centigrade)[j]	16.7/28.0

a Principal rate. b October 2006. c June 2006. d Persons aged 10 years and over, March. e 1999. f 2004. g For women aged 15-44 in union or married. h Estimated data. i Provisional. j Tegucigalpa.

Hungary

Region	Eastern Europe
Largest urban agglom. (pop., 000s)	Budapest (1693)
Currency	forint
Population in 2006 (proj., 000s)	10071
Surface area (square kms)	93032
Population density (per square km)	108
United Nations membership date	14 December 1955

Economic indicators	2000	2005
Exchange rate (national currency per US$)[a]	284.73	206.42[b]
Consumer price index (2000=100)	100	141[c]
Industrial production index (1995=100)	168	221[d]
Unemployment (percentage of labour force)[e]	6.4	7.2
Balance of payments, current account (million US$)	−4004	−8106
Tourist arrivals (000s)	2992	3270[f]
GDP (million current US$)	47035	109239
GDP (per capita current US$)	4600	10818
Gross fixed capital formation (% of GDP)	23.0	23.0
Labour force participation, adult female pop. (%)	45.5	47.0[f]
Labour force participation, adult male pop. (%)	61.1	61.2[f]
Employment in industrial sector (%)	33.7	32.4
Employment in agricultural sector (%)	6.5	5.0
Agricultural production index (1999-2001=100)	94	112[f]
Food production index (1999-2001=100)	94	112[f]
Primary energy production (000s Mt oil equiv.)	8590	7953[f]
Motor vehicles (per 1,000 inhabitants)	268.8	328.4
Telephone lines (per 100 inhabitants)	37.3	33.2
Internet users, estimated (000s)	715.0	3000.0

Total trade		Major trading partners			2005
	(million US$)	(% of exports)		(% of imports)	
Exports	63240.6	Germany	29	Germany	27
Imports	66741.3	Austria	6	Russian Fed.	7
		Italy	5	China	7

Social indicators	2000-2006
Population growth rate 2000-2005 (% per annum)	−0.3
Population aged 0-14 years (%)	16.0
Population aged 60+ years (women and men, % of total)	25.0/17.0
Sex ratio (women per 100 men)	110
Life expectancy at birth 2000-2005 (women and men, years)	77/68
Infant mortality rate 2000-2005 (per 1,000 births)	8
Total fertility rate 2000-2005 (births per woman)	1.3
Contraceptive use (% of currently married women)	77[gh]
Urban population (%)	66
Urban population growth rate 2000-2005 (% per annum)	0.3
Rural population growth rate 2000-2005 (% per annum)	−1.3
Foreign born (%)	2.8
Refugees and others of concern to UNHCR[i]	8779
Government education expenditure (% of GNP)	6.3
Primary-secondary gross enrolment ratio (w and m per 100)	97/97
Third-level students (women and men, % of total)	57/43
Newspaper circulation (per 1,000 inhabitants)	159
Television receivers (per 1,000 inhabitants)	530
Intentional homicides (per 100,000 inhabitants)	2
Parliamentary seats (women and men, % of total)	10/90

Environment	2000-2006
Threatened species	60
Forested area (% of land area)	20
CO2 emissions (000s Mt of carbon dioxide/per capita)[j]	60460/6.0
Energy consumption per capita (kilograms oil equiv.)	2374
Precipitation (mm)	516[k]
Average minimum and maximum temperatures (centigrade)[k]	6.3/15.0

a Official rate. b October 2006. c September 2006. d August 2006. e Persons aged 15 to 74 years. f 2004. g For women aged 18-41 in union or married. h 1992. i Provisional. j Source: UNFCCC. k Budapest.

Iceland

Region	Northern Europe
Largest urban agglom. (pop., 000s)	Reykjavik (185)
Currency	króna
Population in 2006 (proj., 000s)	297
Surface area (square kms)	103000
Population density (per square km)	3
United Nations membership date	19 November 1946

Economic indicators	2000	2005
Exchange rate (national currency per US$)[a]	84.70	68.08[b]
Consumer price index (2000=100)	100	133[c]
Unemployment (percentage of labour force)[d]	2.3	2.6
Balance of payments, current account (million US$)	−847	−2627
Tourist arrivals (000s)	634	836[e]
GDP (million current US$)	8628	15814
GDP (per capita current US$)	30675	53687
Gross fixed capital formation (% of GDP)	23.0	29.0
Labour force participation, adult female pop. (%)	71.3	69.8[e]
Labour force participation, adult male pop. (%)	81.3	79.7[e]
Employment in industrial sector (%)	23.0	23.1[f]
Employment in agricultural sector (%)	8.3	7.2[f]
Agricultural production index (1999-2001=100)	101	103[e]
Food production index (1999-2001=100)	101	104[e]
Primary energy production (000s Mt oil equiv.)[g]	660	741[e]
Motor vehicles (per 1,000 inhabitants)[g]	639.9	726.5
Telephone lines (per 100 inhabitants)[h]	69.9	65.9
Internet users, estimated (000s)	168.0	258.0

Total trade		Major trading partners			2005
	(million US$)	(% of exports)		(% of imports)	
Exports	3090.5	UK	18	Germany	14
Imports	4978.9	Germany	16	USA	9
		Netherlands	13	Sweden	8

Social indicators	2000-2006
Population growth rate 2000-2005 (% per annum)	0.9
Population aged 0-14 years (%)	22.0
Population aged 60+ years (women and men, % of total)	17.0/15.0
Sex ratio (women per 100 men)	100
Life expectancy at birth 2000-2005 (women and men, years)	83/79
Infant mortality rate 2000-2005 (per 1,000 births)	3
Total fertility rate 2000-2005 (births per woman)	2.0
Urban population (%)	93
Urban population growth rate 2000-2005 (% per annum)	1.0
Rural population growth rate 2000-2005 (% per annum)	−0.3
Foreign born (%)	5.5[i]
Refugees and others of concern to UNHCR[j]	375
Government education expenditure (% of GNP)	8.2
Primary-secondary gross enrolment ratio (w and m per 100)	104/104
Third-level students (women and men, % of total)	65/35
Newspaper circulation (per 1,000 inhabitants)	324
Television receivers (per 1,000 inhabitants)	345
Intentional homicides (per 100,000 inhabitants)	1
Parliamentary seats (women and men, % of total)	33/67

Environment	2000-2006
Threatened species	20
Forested area (% of land area)	<
CO2 emissions (000s Mt of carbon dioxide/per capita)[k]	2180/7.5
Energy consumption per capita (kilograms oil equiv.)	4918
Precipitation (mm)	798[l]
Average minimum and maximum temperatures (centigrade)[l]	1.9/7.0

a Official rate. b October 2006. c September 2006. d Persons aged 16 to 74 years, April and Nov. of each year. e 2004. f 2002. g Excluding tractors and semi-trailer combinations. h Year ending 30 September. i Estimated data. j Provisional. k Source: UNFCCC. l Reykjavik.

India

Region	South-central Asia
Largest urban agglom. (pop., 000s)	Bombay (18196)
Currency	rupee
Population in 2006 (proj., 000s)	1119539
Surface area (square kms)	3287263
Population density (per square km)	341
United Nations membership date	30 October 1945

Economic indicators	2000	2005
Exchange rate (national currency per US$)	46.75	45.03[a]
Consumer price index (2000=100)[b]	100	129[c]
Industrial production index (1995=100)	132	189[d]
Unemployment (percentage of labour force)	4.3[e]	...
Balance of payments, current account (million US$)	−4601	6853[f]
Tourist arrivals (000s)[g]	2649	3457[h]
GDP (million current US$)	464937	800783
GDP (per capita current US$)	455	726
Gross fixed capital formation (% of GDP)	22.0	23.0
Labour force participation, adult female pop. (%)	34.3	33.7[h]
Labour force participation, adult male pop. (%)	82.4	82.0[h]
Employment in industrial sector (%)	12.9[i]	...
Employment in agricultural sector (%)	66.7[i]	...
Agricultural production index (1999-2001=100)	99	106[h]
Food production index (1999-2001=100)	99	105[h]
Primary energy production (000s Mt oil equiv.)	253283	300205[h]
Motor vehicles (per 1,000 inhabitants)	14.2	18.2[h]
Telephone lines (per 100 inhabitants)[j]	3.2	4.5
Internet users, estimated (000s)[j]	5500.0	60000.0

Total trade		Major trading partners	2005
	(million US$)	(% of exports)	(% of imports)
Exports	103404.2	USA 17	...
Imports	149750.0	Untd Arab Em 8	...
		China 7	...

Social indicators	2000-2006
Population growth rate 2000-2005 (% per annum)	1.6
Population aged 0-14 years (%)	32.0
Population aged 60+ years (women and men, % of total)	9.0/7.0
Sex ratio (women per 100 men)	95
Life expectancy at birth 2000-2005 (women and men, years)	65/62
Infant mortality rate 2000-2005 (per 1,000 births)	68
Total fertility rate 2000-2005 (births per woman)	3.1
Urban population (%)	29
Urban population growth rate 2000-2005 (% per annum)	2.3
Rural population growth rate 2000-2005 (% per annum)	1.3
Foreign born (%)	0.6[k]
Refugees and others of concern to UNHCR[l]	139586
Government education expenditure (% of GNP)	3.3
Primary-secondary gross enrolment ratio (w and m per 100)	75/85
Third-level students (women and men, % of total)	38/62
Television receivers (per 1,000 inhabitants)[j]	84
Parliamentary seats (women and men, % of total)	9/91

Environment	2000-2006
Threatened species	569
Forested area (% of land area)	22
CO2 emissions (000s Mt of carbon dioxide/per capita)	1275608/1.2
Energy consumption per capita (kilograms oil equiv.)	338
Precipitation (mm)	797[m]
Average minimum and maximum temperatures (centigrade)[m]	18.8/31.4

a October 2006. b Industrial workers. c July 2006. d August 2006. e January. f 2003.
g Excluding nationals residing abroad. h 2004. i 1995. j Year beginning 1 April.
k Estimated data. l Provisional. m New Delhi.

Indonesia

Region	South-eastern Asia
Largest urban agglom. (pop., 000s)	Jakarta (13215)
Currency	rupiah
Population in 2006 (proj., 000s)	225465
Surface area (square kms)	1904569
Population density (per square km)	118
United Nations membership date	28 September 1950

Economic indicators	2000	2005
Exchange rate (national currency per US$)	9595.00	9110.00[a]
Consumer price index (2000=100)	100	176[b]
Industrial production index (1995=100)[c]	92	105[d]
Unemployment (percentage of labour force)[e]	6.1	9.1[f]
Balance of payments, current account (million US$)	7992	929
Tourist arrivals (000s)	5064	5321[g]
GDP (million current US$)	165021	281276
GDP (per capita current US$)	789	1263
Gross fixed capital formation (% of GDP)	20.0	22.0
Labour force participation, adult female pop. (%)	51.5[h]	49.2[g]
Labour force participation, adult male pop. (%)	84.6[h]	86.0[g]
Employment in industrial sector (%)	17.5	18.0
Employment in agricultural sector (%)	45.1	44.0
Agricultural production index (1999-2001=100)	101	119[g]
Food production index (1999-2001=100)	101	117[g]
Primary energy production (000s Mt oil equiv.)	213239	234670[g]
Motor vehicles (per 1,000 inhabitants)	25.9	31.0[i]
Telephone lines (per 100 inhabitants)	3.2	5.7
Internet users, estimated (000s)	1900.0	16000.0

Total trade	(million US$)	Major trading partners			2005
		(% of exports)		(% of imports)	
Exports	85659.9	Japan	21	Singapore	16
Imports	57700.9	USA	12	Japan	12
		Singapore	9	China	10

Social indicators	2000-2006
Population growth rate 2000-2005 (% per annum)	1.3
Population aged 0-14 years (%)	28.0
Population aged 60+ years (women and men, % of total)	9.0/8.0
Sex ratio (women per 100 men)	100
Life expectancy at birth 2000-2005 (women and men, years)	69/65
Infant mortality rate 2000-2005 (per 1,000 births)	43
Total fertility rate 2000-2005 (births per woman)	2.4
Contraceptive use (% of currently married women)	60[j]
Urban population (%)	48
Urban population growth rate 2000-2005 (% per annum)	4.0
Rural population growth rate 2000-2005 (% per annum)	−1.0
Foreign born (%)	0.2[k]
Refugees and others of concern to UNHCR[l]	528
Government education expenditure (% of GNP)	1.0
Primary-secondary gross enrolment ratio (w and m per 100)	90/91
Third-level students (women and men, % of total)	44/56
Television receivers (per 1,000 inhabitants)	152
Parliamentary seats (women and men, % of total)	11/89

Environment	2000-2006
Threatened species	857
Forested area (% of land area)	58
CO2 emissions (000s Mt of carbon dioxide/per capita)	295596/1.4
Energy consumption per capita (kilograms oil equiv.)	467
Average minimum and maximum temperatures (centigrade)[m]	25.0/31.8

a October 2006. b July 2006. c Manufacturing. d 1st quarter 2006. e Persons aged 15 years and over, May of each year. f 2002. g 2004. h 1999. i 2003. j 2002/03. k Estimated data. l Provisional. m Jakarta.

Iran, Islamic Republic of

Region	South-central Asia
Largest urban agglom. (pop., 000s)	Teheran (7314)
Currency	rial
Population in 2006 (proj., 000s)	70324
Surface area (square kms)	1648195
Population density (per square km)	43
United Nations membership date	24 October 1945

Economic indicators	2000	2005
Exchange rate (national currency per US$)[a]	2262.93	9216.00[b]
Consumer price index (2000=100)	100	210[c]
Unemployment (percentage of labour force)[d]	12.8[e]	11.5
Balance of payments, current account (million US$)	12481	...
Tourist arrivals (000s)	1342	1659[f]
GDP (million current US$)	112695	216713
GDP (per capita current US$)	1698	3117
Gross fixed capital formation (% of GDP)	24.0	21.0
Labour force participation, adult female pop. (%)	10.6[g]	...
Labour force participation, adult male pop. (%)	74.8[g]	...
Employment in industrial sector (%)	30.7[g]	30.4
Employment in agricultural sector (%)	23.0[g]	24.9
Agricultural production index (1999-2001=100)	100	115[f]
Food production index (1999-2001=100)	100	115[f]
Primary energy production (000s Mt oil equiv.)	252785	283254[f]
Motor vehicles (per 1,000 inhabitants)[h]	23.0	29.2[e]
Telephone lines (per 100 inhabitants)[i]	14.9	27.3
Internet users, estimated (000s)[i]	625.0	7000.0

Total trade		Major trading partners		2005
	(million US$)	(% of exports)		(% of imports)
Exports	60012.0	...	Untd Arab Em	19
Imports	38674.7	...	Germany	13
		...	France	7

Social indicators	2000-2005
Population growth rate 2000-2005 (% per annum)	0.9
Population aged 0-14 years (%)	29.0
Population aged 60+ years (women and men, % of total)	7.0/6.0
Sex ratio (women per 100 men)	97
Life expectancy at birth 2000-2005 (women and men, years)	72/69
Infant mortality rate 2000-2005 (per 1,000 births)	34
Total fertility rate 2000-2005 (births per woman)	2.1
Urban population (%)	67
Urban population growth rate 2000-2005 (% per annum)	1.8
Rural population growth rate 2000-2005 (% per annum)	−0.7
Foreign born (%)	1.7[j]
Refugees and others of concern to UNHCR[k]	716611
Government education expenditure (% of GNP)	4.8
Primary-secondary gross enrolment ratio (w and m per 100)	90/89
Third-level students (women and men, % of total)	51/49
Television receivers (per 1,000 inhabitants)[i]	174
Parliamentary seats (women and men, % of total)	4/96

Environment	2000-2006
Threatened species	78
Forested area (% of land area)	5
CO2 emissions (000s Mt of carbon dioxide/per capita)	382092/5.6
Energy consumption per capita (kilograms oil equiv.)	2323
Precipitation (mm)	230[l]
Average minimum and maximum temperatures (centigrade)[l]	11.7/22.5

a Official rate. b October 2006. c April 2006. d Persons aged 10 years and over.
e 2002. f 2004. g 1996. h Source: World Automotive Market Report, Auto and Truck
International (Illinois). i Year beginning 22 March. j 1986. k Provisional. l Tehran.

Iraq

Region	Western Asia
Largest urban agglom. (pop., 000s)	Baghdad (5904)
Currency	dinar
Population in 2006 (proj., 000s)	29551
Surface area (square kms)	438317
Population density (per square km)	67
United Nations membership date	21 December 1945

Economic indicators	2000	2005
Exchange rate (national currency per US$)[a]	0.31	1488.00[b]
Unemployment (percentage of labour force)[c]	...	26.8[d]
Tourist arrivals (000s)	78	127[e]
GDP (million current US$)	20969	33379
GDP (per capita current US$)	836	1159
Gross fixed capital formation (% of GDP)	7.0	7.0
Labour force participation, adult female pop. (%)	...	6.0[d]
Labour force participation, adult male pop. (%)	...	69.1[d]
Primary energy production (000s Mt oil equiv.)	129901	101482[d]
Motor vehicles (per 1,000 inhabitants)	43.9[f]	41.4[g]
Telephone lines (per 100 inhabitants)	2.9	4.0
Internet users, estimated (000s)	12.5[e]	36.0[d]

Social indicators	2000-2006
Population growth rate 2000-2005 (% per annum)	2.8
Population aged 0-14 years (%)	41.0
Population aged 60+ years (women and men, % of total)	5.0/4.0
Sex ratio (women per 100 men)	97
Life expectancy at birth 2000-2005 (women and men, years)	60/57
Infant mortality rate 2000-2005 (per 1,000 births)	94
Total fertility rate 2000-2005 (births per woman)	4.8
Contraceptive use (% of currently married women)	14[hi]
Urban population (%)	67
Urban population growth rate 2000-2005 (% per annum)	2.5
Rural population growth rate 2000-2005 (% per annum)	3.3
Refugees and others of concern to UNHCR[j]	1634280
Government education expenditure (% of GNP)	4.6[k]
Primary-secondary gross enrolment ratio (w and m per 100)	64/83
Third-level students (women and men, % of total)	36/64
Television receivers (per 1,000 inhabitants)	83
Parliamentary seats (women and men, % of total)	24/76

Environment	2000-2006
Threatened species	41
Forested area (% of land area)	2
CO2 emissions (000s Mt of carbon dioxide/per capita)	73007/2.7
Energy consumption per capita (kilograms oil equiv.)	988

a Principal rate. b July 2006. c Persons aged 15 years and over. d 2004. e 2001.
f Source: United Nations Economic and Social Commission for Western Asia (ESCWA).
g 2002. h Households of nationals of the country. i 1989. j Provisional. k 1988.

Ireland

Region	Northern Europe
Largest urban agglom. (pop., 000s)	Dublin (1037)
Currency	euro[a]
Population in 2006 (proj., 000s)	4210
Surface area (square kms)	70273
Population density (per square km)	60
United Nations membership date	14 December 1955

Economic indicators	2000	2005
Exchange rate (national currency per US$)	1.07	0.79[b]
Consumer price index (2000=100)	100	125[b]
Industrial production index (1995=100)	202	258[c]
Unemployment (percentage of labour force)[d]	4.3	4.2
Balance of payments, current account (million US$)	–516	–3945
Tourist arrivals (000s)[e]	6646	6982[f]
GDP (million current US$)	96166	201763
GDP (per capita current US$)	25298	48642
Gross fixed capital formation (% of GDP)	24.0	27.0
Labour force participation, adult female pop. (%)	47.2	49.4[f]
Labour force participation, adult male pop. (%)	71.0	70.9[f]
Employment in industrial sector (%)	28.5	27.8
Employment in agricultural sector (%)	7.8	5.9
Agricultural production index (1999-2001=100)	99	98[f]
Food production index (1999-2001=100)	99	98[f]
Primary energy production (000s Mt oil equiv.)	2282	1885[f]
Motor vehicles (per 1,000 inhabitants)	408.1	447.3[gh]
Telephone lines (per 100 inhabitants)[i]	48.4	49.0
Internet users, estimated (000s)[i]	679.0	1400.0

Total trade		Major trading partners			2005
	(million US$)	(% of exports)		(% of imports)	
Exports	109993.7	USA	19	UK	31
Imports	70292.4	UK	17	USA	14
		Belgium	15	Germany	8

Social indicators	2000-2006
Population growth rate 2000-2005 (% per annum)	1.7
Population aged 0-14 years (%)	20.0
Population aged 60+ years (women and men, % of total)	16.0/14.0
Sex ratio (women per 100 men)	101
Life expectancy at birth 2000-2005 (women and men, years)	80/75
Infant mortality rate 2000-2005 (per 1,000 births)	5
Total fertility rate 2000-2005 (births per woman)	1.9
Urban population (%)	60
Urban population growth rate 2000-2005 (% per annum)	2.2
Rural population growth rate 2000-2005 (% per annum)	1.1
Refugees and others of concern to UNHCR[j]	9527
Government education expenditure (% of GNP)	5.5
Primary-secondary gross enrolment ratio (w and m per 100)	110/107
Third-level students (women and men, % of total)	55/45
Newspaper circulation (per 1,000 inhabitants)	148
Television receivers (per 1,000 inhabitants)[i]	694
Intentional homicides (per 100,000 inhabitants)	1
Parliamentary seats (women and men, % of total)	14/86

Environment	2000-2006
Threatened species	24
Forested area (% of land area)	10
CO2 emissions (000s Mt of carbon dioxide/per capita)[k]	44450/11.1
Energy consumption per capita (kilograms oil equiv.)	3618
Precipitation (mm)	733[l]
Average minimum and maximum temperatures (centigrade)[l]	6.3/12.8

a Prior 1 January 1999, pound. b September 2006. c August 2006. d Persons aged 15 years and over, March-May of each year. e Including tourists from North Ireland. f 2004. g 2003. h Data refer to fiscal years ending 30 September. i Year beginning 1 April. j Provisional. k Source: UNFCCC. l Dublin.

Israel

Region	Western Asia
Largest urban agglom. (pop., 000s)	Tel Aviv-Jaffa (3012)
Currency	new sheqel
Population in 2006 (proj., 000s)	6847
Surface area (square kms)	22145
Population density (per square km)	309
United Nations membership date	11 May 1949

Economic indicators	2000	2005
Exchange rate (national currency per US$)	4.04	4.29[a]
Consumer price index (2000=100)	100	112[b]
Industrial production index (1995=100)	123	119[c]
Unemployment (percentage of labour force)[d]	8.8	9.0
Balance of payments, current account (million US$)	−1285	2385
Tourist arrivals (000s)[e]	2417	1506[f]
GDP (million current US$)	121024	129648
GDP (per capita current US$)	19891	19280
Gross fixed capital formation (% of GDP)	19.0	16.0
Labour force participation, adult female pop. (%)	48.2	49.1[g]
Labour force participation, adult male pop. (%)	60.7	60.1[g]
Employment in industrial sector (%)	24.0	21.7
Employment in agricultural sector (%)	2.2	2.0
Agricultural production index (1999-2001=100)	105	108[f]
Food production index (1999-2001=100)	106	108[f]
Primary energy production (000s Mt oil equiv.)	100	114[f]
Motor vehicles (per 1,000 inhabitants)	287.6	296.5[f]
Telephone lines (per 100 inhabitants)	47.4	42.6
Internet users, estimated (000s)	1270.0	1685.9

Total trade		Major trading partners			2005
	(million US$)	(% of exports)		(% of imports)	
Exports	42770.7	USA	36	USA	13
Imports	45032.3	Belgium	9	Belgium	10
		China, HK SAR	6	Germany	7

Social indicators	2000-2006
Population growth rate 2000-2005 (% per annum)	2.0
Population aged 0-14 years (%)	28.0
Population aged 60+ years (women and men, % of total)	15.0/12.0
Sex ratio (women per 100 men)	102
Life expectancy at birth 2000-2005 (women and men, years)	82/78
Infant mortality rate 2000-2005 (per 1,000 births)	5
Total fertility rate 2000-2005 (births per woman)	2.9
Contraceptive use (% of currently married women)	68[hij]
Urban population (%)	92
Urban population growth rate 2000-2005 (% per annum)	2.0
Rural population growth rate 2000-2005 (% per annum)	1.6
Refugees and others of concern to UNHCR[k]	1548
Government education expenditure (% of GNP)	7.5
Primary-secondary gross enrolment ratio (w and m per 100)	102/102
Third-level students (women and men, % of total)	56/44
Television receivers (per 1,000 inhabitants)[l]	354
Parliamentary seats (women and men, % of total)	14/86

Environment	2000-2006
Threatened species	82
Forested area (% of land area)	6
CO2 emissions (000s Mt of carbon dioxide/per capita)	68427/10.6
Energy consumption per capita (kilograms oil equiv.)	3050
Average minimum and maximum temperatures (centigrade)[m]	16.4/24.0

a October 2006. b June 2006. c April 2006. d Persons aged 15 years and over.
e Excluding nationals residing abroad. f 2004. g 2003. h For women aged 18-39 in union
or married. i 1987/88. j Data refers to the Jewish population. k Provisional. l Data
prior to 1986 refer to year beginning 1 April. m Tel Aviv.

Italy

Region	Southern Europe
Largest urban agglom. (pop., 000s)	Rome (3348)
Currency	euro[a]
Population in 2006 (proj., 000s)	58140
Surface area (square kms)	301318
Population density (per square km)	193
United Nations membership date	14 December 1955

Economic indicators	2000	2005
Exchange rate (national currency per US$)	1.07	0.79[b]
Consumer price index (2000=100)[c]	100	115[d]
Industrial production index (1995=100)	108	62[e]
Unemployment (percentage of labour force)[f]	10.5	7.7
Balance of payments, current account (million US$)	−5781	−27724
Tourist arrivals (000s)[g]	41181	37071[h]
GDP (million current US$)	1097343	1762475
GDP (per capita current US$)	19013	30339
Gross fixed capital formation (% of GDP)	20.0	21.0
Labour force participation, adult female pop. (%)	35.8	38.3[h]
Labour force participation, adult male pop. (%)	62.2	61.3[h]
Employment in industrial sector (%)	31.9	30.7
Employment in agricultural sector (%)	5.3	4.2
Agricultural production index (1999-2001=100)	100	98[h]
Food production index (1999-2001=100)	100	98[h]
Primary energy production (000s Mt oil equiv.)[i]	25134	22222[h]
Motor vehicles (per 1,000 inhabitants)	623.1	654.6[h]
Telephone lines (per 100 inhabitants)	47.4	43.1
Internet users, estimated (000s)	13200.0	28000.0

Total trade		Major trading partners			2005
	(million US$)[i]		(% of exports)[i]		(% of imports)[i]
Exports	367866.5	Germany	13	Germany	17
Imports	380560.8	France	12	France	10
		USA	8	Netherlands	5

Social indicators	2000-2006
Population growth rate 2000-2005 (% per annum)	0.1
Population aged 0-14 years (%)	14.0
Population aged 60+ years (women and men, % of total)	28.0/23.0
Sex ratio (women per 100 men)	106
Life expectancy at birth 2000-2005 (women and men, years)	83/77
Infant mortality rate 2000-2005 (per 1,000 births)	5
Total fertility rate 2000-2005 (births per woman)	1.3
Urban population (%)	68
Urban population growth rate 2000-2005 (% per annum)	0.2
Rural population growth rate 2000-2005 (% per annum)	−0.1
Foreign born (%)	2.8[j]
Refugees and others of concern to UNHCR[k]	21561
Government education expenditure (% of GNP)	4.9
Primary-secondary gross enrolment ratio (w and m per 100)	99/100
Third-level students (women and men, % of total)	56/44
Newspaper circulation (per 1,000 inhabitants)	109
Television receivers (per 1,000 inhabitants)	494
Intentional homicides (per 100,000 inhabitants)	1
Parliamentary seats (women and men, % of total)	16/84

Environment	2000-2006
Threatened species	147
Forested area (% of land area)	34
CO2 emissions (000s Mt of carbon dioxide/per capita)[l]	487280/8.4[i]
Energy consumption per capita (kilograms oil equiv.)[i]	3108
Precipitation (mm)	943[m]
Average minimum and maximum temperatures (centigrade)[n]	10.6/20.3

a Prior to 1 January 1999, lire. b September 2006. c Excluding tobacco. d May 2006. e August 2006. f Persons aged 15 years and over. g Excluding seasonal and border workers. h 2004. i Including San Marino. j Estimated data. k Provisional. l Source: UNFCCC. m Milan. n Rome.

Jamaica

Region	Caribbean
Largest urban agglom. (pop., 000s)	Kingston (576)
Currency	dollar
Population in 2006 (proj., 000s)	2662
Surface area (square kms)	10991
Population density (per square km)	242
United Nations membership date	18 September 1962

Economic indicators	2000	2005
Exchange rate (national currency per US$)	45.41	66.41[a]
Consumer price index (2000=100)	100	177[b]
Unemployment (percentage of labour force)[c]	15.5	10.9
Balance of payments, current account (million US$)	−367	−1079
Tourist arrivals (000s)[d]	1323	1415[e]
GDP (million current US$)	7889	10063
GDP (per capita current US$)	3052	3796
Gross fixed capital formation (% of GDP)	27.0	33.0
Labour force participation, adult female pop. (%)	56.5[f]	55.3[e]
Labour force participation, adult male pop. (%)	73.0	73.7[e]
Employment in industrial sector (%)	16.9	17.7
Employment in agricultural sector (%)	20.8	18.0
Agricultural production index (1999-2001=100)	95	99[e]
Food production index (1999-2001=100)	95	99[e]
Primary energy production (000s Mt oil equiv.)	10	14[e]
Motor vehicles (per 1,000 inhabitants)[g]	72.4	74.8[h]
Telephone lines (per 100 inhabitants)	19.0	12.9
Internet users, estimated (000s)	80.0	1067.0[e]

Total trade		Major trading partners			2005
	(million US$)[e]	(% of exports)[e]		(% of imports)[e]	
Exports	1411.7	USA	21	USA	40
Imports	3817.2	Canada	19	Trinidad Tbg	13
		China	12	Japan	5

Social indicators	2000-2006
Population growth rate 2000-2005 (% per annum)	0.5
Population aged 0-14 years (%)	31.0
Population aged 60+ years (women and men, % of total)	11.0/10.0
Sex ratio (women per 100 men)	102
Life expectancy at birth 2000-2005 (women and men, years)	73/69
Infant mortality rate 2000-2005 (per 1,000 births)	15
Total fertility rate 2000-2005 (births per woman)	2.4
Urban population (%)	53
Urban population growth rate 2000-2005 (% per annum)	1.0
Rural population growth rate 2000-2005 (% per annum)	−<
Government education expenditure (% of GNP)	5.3
Primary-secondary gross enrolment ratio (w and m per 100)	93/92
Third-level students (women and men, % of total)	70/30
Television receivers (per 1,000 inhabitants)	380
Parliamentary seats (women and men, % of total)	14/86

Environment	2000-2006
Threatened species	269
Forested area (% of land area)	30
CO2 emissions (000s Mt of carbon dioxide/per capita)	10737/4.1
Energy consumption per capita (kilograms oil equiv.)	1252
Precipitation (mm)	813[i]
Average minimum and maximum temperatures (centigrade)[i]	22.9/31.4

a October 2006. b May 2006. c Persons aged 14 years and over. d Arrivals of non-resident tourists by air. Including nationals residing abroad. E/D cards. e 2004. f 1999. g Source: World Automotive Market Report, Auto and Truck International (Illinois). h 2001. i Kingston.

Japan

Region	Eastern Asia
Largest urban agglom. (pop., 000s)	Tokyo (35197)
Currency	yen
Population in 2006 (proj., 000s)	128219
Surface area (square kms)	377873
Population density (per square km)	339
United Nations membership date	18 December 1956

Economic indicators	2000	2005
Exchange rate (national currency per US$)	114.90	117.65[a]
Consumer price index (2000=100)	100	99[b]
Industrial production index (1995=100)	106	112[c]
Unemployment (percentage of labour force)[d]	4.7	4.4
Balance of payments, current account (million US$)	120	166
Tourist arrivals (000s)[e]	4757	6138[f]
GDP (million current US$)	4649615	4558950
GDP (per capita current US$)	36601	35593
Gross fixed capital formation (% of GDP)	25.0	23.0
Labour force participation, adult female pop. (%)	49.3	48.2[f]
Labour force participation, adult male pop. (%)	76.4	73.4[f]
Employment in industrial sector (%)	31.2	27.9
Employment in agricultural sector (%)	5.1	4.4
Agricultural production index (1999-2001=100)	101	98[f]
Food production index (1999-2001=100)	101	98[f]
Primary energy production (000s Mt oil equiv.)	44423	38953[f]
Motor vehicles (per 1,000 inhabitants)	560.5	570.7[f]
Telephone lines (per 100 inhabitants)	48.8	45.9
Internet users, estimated (000s)	38000.0	66010.0

Total trade		Major trading partners			2005
	(million US$)		(% of exports)		(% of imports)
Exports	594940.9	USA	23	China	21
Imports	515866.4	China	13	USA	13
		Korea Rep.	8	Saudi Arabia	6

Social indicators	2000-2006
Population growth rate 2000-2005 (% per annum)	0.2
Population aged 0-14 years (%)	14.0
Population aged 60+ years (women and men, % of total)	29.0/24.0
Sex ratio (women per 100 men)	105
Life expectancy at birth 2000-2005 (women and men, years)	85/78
Infant mortality rate 2000-2005 (per 1,000 births)	3
Total fertility rate 2000-2005 (births per woman)	1.3
Contraceptive use (% of currently married women)	56
Urban population (%)	66
Urban population growth rate 2000-2005 (% per annum)	0.4
Rural population growth rate 2000-2005 (% per annum)	−0.2
Foreign born (%)	1.3[g]
Refugees and others of concern to UNHCR[h]	2474
Government education expenditure (% of GNP)	3.6
Primary-secondary gross enrolment ratio (w and m per 100)	101/101
Third-level students (women and men, % of total)	46/54
Newspaper circulation (per 1,000 inhabitants)	566
Television receivers (per 1,000 inhabitants)	843
Intentional homicides (per 100,000 inhabitants)	1
Parliamentary seats (women and men, % of total)	11/89

Environment	2000-2006
Threatened species	215
Forested area (% of land area)	64
CO2 emissions (000s Mt of carbon dioxide/per capita)[i]	1259430/9.9
Energy consumption per capita (kilograms oil equiv.)	3460
Precipitation (mm)	1467[j]
Average minimum and maximum temperatures (centigrade)[j]	12.6/19.7

a October 2006. b August 2006. c July 2006. d Persons aged 15 years and over.
e Excluding nationals residing abroad. f 2004. g Estimated data. h Provisional. i Source:
UNFCCC. j Tokyo.

Jordan

Region	Western Asia
Largest urban agglom. (pop., 000s)	Amman (1292)
Currency	dinar
Population in 2006 (proj., 000s)	5837
Surface area (square kms)	89342
Population density (per square km)	65
United Nations membership date	14 December 1955

Economic indicators	2000	2005
Exchange rate (national currency per US$)[a]	0.71	0.71[b]
Consumer price index (2000=100)	100	120[c]
Industrial production index (1995=100)	110	159[d]
Balance of payments, current account (million US$)	59	−2311
Tourist arrivals (000s)	1580	2853[e]
GDP (million current US$)	8461	12535
GDP (per capita current US$)	1702	2198
Gross fixed capital formation (% of GDP)	21.0	24.0
Labour force participation, adult female pop. (%)	11.6	10.4[e]
Labour force participation, adult male pop. (%)	63.6	63.7[e]
Employment in industrial sector (%)	21.8	21.8[f]
Employment in agricultural sector (%)	4.9	3.6[f]
Agricultural production index (1999-2001=100)	111	118[e]
Food production index (1999-2001=100)	111	118[e]
Primary energy production (000s Mt oil equiv.)	221	254[e]
Motor vehicles (per 1,000 inhabitants)	72.5	102.9[e]
Telephone lines (per 100 inhabitants)	12.3	11.4
Internet users, estimated (000s)	127.3	629.5[e]

Total trade		Major trading partners			2005
	(million US$)	(% of exports)		(% of imports)	
Exports	4278.7	USA	26	Saudi Arabia	24
Imports	10454.6	Iraq	17	China	9
		India	8	Germany	8

Social indicators	2000-2006
Population growth rate 2000-2005 (% per annum)	2.7
Population aged 0-14 years (%)	37.0
Population aged 60+ years (women and men, % of total)	5.0/5.0
Sex ratio (women per 100 men)	93
Life expectancy at birth 2000-2005 (women and men, years)	73/70
Infant mortality rate 2000-2005 (per 1,000 births)	23
Total fertility rate 2000-2005 (births per woman)	3.5
Contraceptive use (% of currently married women)	56
Urban population (%)	82
Urban population growth rate 2000-2005 (% per annum)	3.2
Rural population growth rate 2000-2005 (% per annum)	0.8
Foreign born (%)	38.6[g]
Refugees and others of concern to UNHCR[h]	17544
Government education expenditure (% of GNP)	5.0[i]
Primary-secondary gross enrolment ratio (w and m per 100)	94/93
Third-level students (women and men, % of total)	51/49
Television receivers (per 1,000 inhabitants)	190
Parliamentary seats (women and men, % of total)	8/92

Environment	2000-2006
Threatened species	47
Forested area (% of land area)	1
CO2 emissions (000s Mt of carbon dioxide/per capita)	17117/3.2
Energy consumption per capita (kilograms oil equiv.)	857
Precipitation (mm)	269[j]
Average minimum and maximum temperatures (centigrade)[j]	11.3/23.5

a Official rate. b October 2006. c June 2006. d August 2006. e 2004. f 2003. g Estimated data. h Provisional. i 1999. j Amman.

Kazakhstan

Region	South-central Asia
Largest urban agglom. (pop., 000s)	Almaty (1156)
Currency	tenge
Population in 2006 (proj., 000s)	14812
Surface area (square kms)	2724900
Population density (per square km)	5
United Nations membership date	2 March 1992

Economic indicators	2000	2005
Exchange rate (national currency per US$)[a]	144.50	127.82[b]
Consumer price index (2000=100)	100	149[c]
Unemployment (percentage of labour force)[d]	10.4[e]	8.4[f]
Balance of payments, current account (million US$)	366	−486
Tourist arrivals (000s)	1471	3073[f]
GDP (million current US$)	18292	56088
GDP (per capita current US$)	1217	3783
Gross fixed capital formation (% of GDP)	17.0	26.0
Labour force participation, adult female pop. (%)	64.9[g]	64.8[f]
Labour force participation, adult male pop. (%)	76.0[g]	75.6[f]
Employment in industrial sector (%)	16.3[e]	17.4[f]
Employment in agricultural sector (%)	35.5[e]	33.5[f]
Agricultural production index (1999-2001=100)	91	105[f]
Food production index (1999-2001=100)	90	103[f]
Primary energy production (000s Mt oil equiv.)	79687	121026[f]
Motor vehicles (per 1,000 inhabitants)	83.6	101.8[f]
Telephone lines (per 100 inhabitants)	12.2	16.9
Internet users, estimated (000s)	100.0	400.0[f]

Total trade	Major trading partners		2005
(million US$)[f]	(% of exports)[f]	(% of imports)[f]	
Exports 19938.6	Switzerland 19	Russian Fed. 37	
Imports 12635.8	Italy 16	Germany 8	
	Russian Fed. 14	China 6	

Social indicators	2000-2006
Population growth rate 2000-2005 (% per annum)	−0.3
Population aged 0-14 years (%)	23.0
Population aged 60+ years (women and men, % of total)	14.0/9.0
Sex ratio (women per 100 men)	109
Life expectancy at birth 2000-2005 (women and men, years)	69/58
Infant mortality rate 2000-2005 (per 1,000 births)	61
Total fertility rate 2000-2005 (births per woman)	2.0
Urban population (%)	57
Urban population growth rate 2000-2005 (% per annum)	0.1
Rural population growth rate 2000-2005 (% per annum)	−0.8
Foreign born (%)	22.3[h]
Refugees and others of concern to UNHCR[i]	57906
Government education expenditure (% of GNP)	2.6
Primary-secondary gross enrolment ratio (w and m per 100)	101/103
Third-level students (women and men, % of total)	57/43
Television receivers (per 1,000 inhabitants)	497
Intentional homicides (per 100,000 inhabitants)	13
Parliamentary seats (women and men, % of total)	9/91

Environment	2000-2006
Threatened species	274
Forested area (% of land area)	5
CO2 emissions (000s Mt of carbon dioxide/per capita)	159494/10.7
Energy consumption per capita (kilograms oil equiv.)	3621
Precipitation (mm)	641[j]
Average minimum and maximum temperatures (centigrade)[j]	3.8/14.6

a Official rate. b October 2006. c February 2006. d Persons aged 15 years and over.
e 2001. f 2004. g 2002. h 1989. i Provisional. j Almaty.

Kenya

Region	Eastern Africa
Largest urban agglom. (pop., 000s)	Nairobi (2773)
Currency	shilling
Population in 2006 (proj., 000s)	35106
Surface area (square kms)	580367
Population density (per square km)	60
United Nations membership date	16 December 1963

Economic indicators	2000	2005
Exchange rate (national currency per US$)[a]	78.04	72.02[b]
Consumer price index (2000=100)[c]	100	188[d]
Balance of payments, current account (million US$)	−199	−378[e]
Tourist arrivals (000s)	899	1199[e]
GDP (million current US$)	12705	19184
GDP (per capita current US$)	414	560
Gross fixed capital formation (% of GDP)	17.0	20.0
Employment in industrial sector (%)	19.5[f]	...
Employment in agricultural sector (%)	18.6[f]	...
Agricultural production index (1999-2001=100)	96	105[e]
Food production index (1999-2001=100)	96	104[e]
Primary energy production (000s Mt oil equiv.)	166	336[e]
Motor vehicles (per 1,000 inhabitants)	16.3	18.1[e]
Telephone lines (per 100 inhabitants)[g]	1.0	0.8
Internet users, estimated (000s)[h]	100.0	1111.0

Total trade		Major trading partners			2005
	(million US$)[e]	(% of exports)[e]		(% of imports)[e]	
Exports	2683.2	Uganda	18	Untd Arab Em	11
Imports	4563.5	UK	11	South Africa	10
		Netherlands	8	Saudi Arabia	9

Social indicators	2000-2006
Population growth rate 2000-2005 (% per annum)	2.2
Population aged 0-14 years (%)	43.0
Population aged 60+ years (women and men, % of total)	4.0/4.0
Sex ratio (women per 100 men)	100
Life expectancy at birth 2000-2005 (women and men, years)	46/48
Infant mortality rate 2000-2005 (per 1,000 births)	68
Total fertility rate 2000-2005 (births per woman)	5.0
Contraceptive use (% of currently married women)	39
Urban population (%)	21
Urban population growth rate 2000-2005 (% per annum)	3.2
Rural population growth rate 2000-2005 (% per annum)	1.9
Foreign born (%)	1.1[i]
Refugees and others of concern to UNHCR[j]	267731
Government education expenditure (% of GNP)	7.1
Primary-secondary gross enrolment ratio (w and m per 100)	78/83
Third-level students (women and men, % of total)	37/63
Television receivers (per 1,000 inhabitants)[h]	48
Parliamentary seats (women and men, % of total)	7/93

Environment	2000-2006
Threatened species	13
Forested area (% of land area)	30
CO2 emissions (000s Mt of carbon dioxide/per capita)	8790/0.3
Energy consumption per capita (kilograms oil equiv.)	95
Precipitation (mm)	1024[c]
Average minimum and maximum temperatures (centigrade)[k]	12.0/23.4

a Official rate. b October 2006. c Nairobi. d February 2006. e 2004. f 1999. g Year beginning 30 June. h Year ending 30 June. i Estimated data. j Provisional. k Nairobi/ Dagoretti.

Kiribati

Region	Oceania-Micronesia
Largest urban agglom. (pop., 000s)	Tarawa (47)
Currency	Australian dollar
Population in 2006 (proj., 000s)	84
Surface area (square kms)	726
Population density (per square km)	116
United Nations membership date	14 September 1999

Economic indicators	2000	2005
Exchange rate (national currency per US$)[a]	1.81	1.30[b]
Tourist arrivals (000s)[c]	5	3[d]
GDP (million current US$)	47	72
GDP (per capita current US$)	526	721
Gross fixed capital formation (% of GDP)	43.0	43.0
Agricultural production index (1999-2001=100)	97	107[d]
Food production index (1999-2001=100)	97	107[d]
Telephone lines (per 100 inhabitants)	4.0	5.1[de]
Internet users, estimated (000s)	1.5	2.0

Total trade		Major trading partners		2005
	(million US$)	(% of exports)		(% of imports)
Exports	3.6	...	Australia	36
Imports	74.0	...	Fiji	21
		...	Japan	17

Social indicators	2000-2006
Population growth rate 2000-2005 (% per annum)	2.1
Population aged 60+ years (women and men, % of total)	6.0/102.9
Life expectancy at birth 2000-2005 (women and men, years)	67/58[f]
Infant mortality rate 2000-2005 (per 1,000 births)	43
Total fertility rate 2000-2005 (births per woman)	4.3
Urban population (%)	47
Urban population growth rate 2000-2005 (% per annum)	4.0
Rural population growth rate 2000-2005 (% per annum)	0.4
Foreign born (%)	2.9[g]
Government education expenditure (% of GNP)	9.3
Primary-secondary gross enrolment ratio (w and m per 100)	109/98
Television receivers (per 1,000 inhabitants)	44
Parliamentary seats (women and men, % of total)	5/95

Environment	2000-2006
Threatened species	50
Forested area (% of land area)	38
CO2 emissions (000s Mt of carbon dioxide/per capita)	31/0.3
Energy consumption per capita (kilograms oil equiv.)	92[g]
Average minimum and maximum temperatures (centigrade)	27.6/28.1

a Official rate. b October 2006. c Air arrivals, Tarawa and Christmas Island. d 2004.
e Estimate. f 2000. g Estimated data.

Korea, Democratic People's Republic of

Region Eastern Asia
Largest urban agglom. (pop., 000s) Pyongyang (3351)
Currency won
Population in 2006 (proj., 000s) 22583
Surface area (square kms) 120538
Population density (per square km) 187
United Nations membership date 17 September 1991

Economic indicators	2000	2005
Exchange rate (national currency per US$)[ab]	2.15	140.95[c]
Tourist arrivals (000s)	128[d]	...
GDP (million current US$)	10608	12260
GDP (per capita current US$)	462	517
Agricultural production index (1999-2001=100)	96	109[e]
Food production index (1999-2001=100)	96	110[e]
Primary energy production (000s Mt oil equiv.)	19669	20118[e]
Telephone lines (per 100 inhabitants)	2.3	4.4

Social indicators	2000-2006
Population growth rate 2000-2005 (% per annum)	0.6
Population aged 0-14 years (%)	25.0
Population aged 60+ years (women and men, % of total)	13.0/10.0
Sex ratio (women per 100 men)	100
Life expectancy at birth 2000-2005 (women and men, years)	66/60
Infant mortality rate 2000-2005 (per 1,000 births)	46
Total fertility rate 2000-2005 (births per woman)	2.0
Contraceptive use (% of currently married women)	62[f]
Urban population (%)	62
Urban population growth rate 2000-2005 (% per annum)	1.0
Rural population growth rate 2000-2005 (% per annum)	−0.2
Foreign born (%)	0.2[g]
Third-level students (women and men, % of total)	51/49[h]
Television receivers (per 1,000 inhabitants)	172
Parliamentary seats (women and men, % of total)	20/80

Environment	2000-2006
Threatened species	60
Forested area (% of land area)	68
CO2 emissions (000s Mt of carbon dioxide/per capita)	77601/3.5
Energy consumption per capita (kilograms oil equiv.)	937
Precipitation (mm)	940[i]
Average minimum and maximum temperatures (centigrade)[i]	5.6/15.7

a Introduced a new exchange rate system in 2002. b Data refer to non-commercial rates derived from the Operational Rates of Exchange for United Nations Programmes. c October 2006. d 1997. e 2004. f 1990/92. g Estimated data. h 1998. i Pyongyang.

Korea, Republic of

Region	Eastern Asia
Largest urban agglom. (pop., 000s)	Seoul (9645)
Currency	won
Population in 2006 (proj., 000s)	47983
Surface area (square kms)	99538
Population density (per square km)	482
United Nations membership date	17 September 1991

Economic indicators	2000	2005
Exchange rate (national currency per US$)	1264.50	944.20[a]
Consumer price index (2000=100)	100	122[b]
Industrial production index (1995=100)	154	218[c]
Unemployment (percentage of labour force)[d]	4.4[e]	3.7
Balance of payments, current account (million US$)	12251	16559
Tourist arrivals (000s)[f]	5322	5818[g]
GDP (million current US$)	511659	787627
GDP (per capita current US$)	10938	16472
Gross fixed capital formation (% of GDP)	31.0	29.0
Labour force participation, adult female pop. (%)	48.6	49.8[g]
Labour force participation, adult male pop. (%)	74.2	74.8[g]
Employment in industrial sector (%)	28.1	26.8
Employment in agricultural sector (%)	10.6	7.9
Agricultural production index (1999-2001=100)	100	92[g]
Food production index (1999-2001=100)	100	92[g]
Primary energy production (000s Mt oil equiv.)	11719	13209[g]
Motor vehicles (per 1,000 inhabitants)	257.0	312.5[g]
Telephone lines (per 100 inhabitants)[h]	56.2	49.2
Internet users, estimated (000s)	19040.0	33010.0

Total trade		Major trading partners			2005
	(million US$)	(% of exports)			(% of imports)
Exports	284418.2	China	22	Japan	19
Imports	261235.6	USA	15	China	15
		Japan	8	USA	12

Social indicators	2000-2006
Population growth rate 2000-2005 (% per annum)	0.4
Population aged 0-14 years (%)	20.0
Population aged 60+ years (women and men, % of total)	16.0/12.0
Sex ratio (women per 100 men)	100
Life expectancy at birth 2000-2005 (women and men, years)	80/73
Infant mortality rate 2000-2005 (per 1,000 births)	4
Total fertility rate 2000-2005 (births per woman)	1.2
Urban population (%)	81
Urban population growth rate 2000-2005 (% per annum)	0.7
Rural population growth rate 2000-2005 (% per annum)	−0.7
Foreign born (%)	1.2
Refugees and others of concern to UNHCR[i]	588
Government education expenditure (% of GNP)	4.6
Primary-secondary gross enrolment ratio (w and m per 100)	99/99
Third-level students (women and men, % of total)	37/63
Television receivers (per 1,000 inhabitants)	477
Intentional homicides (per 100,000 inhabitants)	2
Parliamentary seats (women and men, % of total)	13/87

Environment	2000-2006
Threatened species	28
Forested area (% of land area)	63
CO2 emissions (000s Mt of carbon dioxide/per capita)	456751/9.6
Energy consumption per capita (kilograms oil equiv.)	3225
Precipitation (mm)	1344[j]
Average minimum and maximum temperatures (centigrade)[j]	8.2/16.9

a October 2006. b September 2006. c August 2006. d Persons aged 15 years and over. e Estimates based on the 2000 Population Census results. f Including nationals residing abroad and crew members. g 2004. h Telephone subscribers . i Provisional. j Seoul.

Kuwait

Region	Western Asia
Largest urban agglom. (pop., 000s)	Kuwait City (1810)
Currency	dinar
Population in 2006 (proj., 000s)	2765
Surface area (square kms)	17818
Population density (per square km)	155
United Nations membership date	14 May 1963

Economic indicators	2000	2005
Exchange rate (national currency per US$)[a]	0.31	0.29[b]
Consumer price index (2000=100)	100	111[c]
Unemployment (percentage of labour force)[d]	0.8	1.1[e]
Balance of payments, current account (million US$)	14672	18884[f]
Tourist arrivals (000s)	78	91[f]
GDP (million current US$)	37718	74214
GDP (per capita current US$)	16916	27621
Gross fixed capital formation (% of GDP)	10.0	12.0
Labour force participation, adult female pop. (%)	43.5[g]	...
Labour force participation, adult male pop. (%)	83.6[g]	...
Agricultural production index (1999-2001=100)	92	126[f]
Food production index (1999-2001=100)	92	126[f]
Primary energy production (000s Mt oil equiv.)[h]	116771	135111[f]
Motor vehicles (per 1,000 inhabitants)	359.2	391.7[f]
Telephone lines (per 100 inhabitants)	21.3	19.0
Internet users, estimated (000s)	150.0	700.0

Total trade	(million US$)[i]	Major trading partners (% of exports)		2005 (% of imports)[i]
Exports	16164.5	...	USA	11
Imports	7869.0	...	Germany	10
		...	Japan	10

Social indicators	2000-2006
Population growth rate 2000-2005 (% per annum)	3.7
Population aged 0-14 years (%)	24.0
Population aged 60+ years (women and men, % of total)	3.0/3.0
Sex ratio (women per 100 men)	67
Life expectancy at birth 2000-2005 (women and men, years)	79/75
Infant mortality rate 2000-2005 (per 1,000 births)	10
Total fertility rate 2000-2005 (births per woman)	2.4
Urban population (%)	98
Urban population growth rate 2000-2005 (% per annum)	3.7
Rural population growth rate 2000-2005 (% per annum)	2.7
Foreign born (%)	59.9[j]
Refugees and others of concern to UNHCR[k]	102726
Government education expenditure (% of GNP)	7.6
Primary-secondary gross enrolment ratio (w and m per 100)	94/90
Third-level students (women and men, % of total)	71/29
Television receivers (per 1,000 inhabitants)	407
Intentional homicides (per 100,000 inhabitants)	1
Parliamentary seats (women and men, % of total)	1/99

Environment	2000-2006
Threatened species	28
Forested area (% of land area)	<
CO2 emissions (000s Mt of carbon dioxide/per capita)	78602/31.1
Energy consumption per capita (kilograms oil equiv.)[h]	10559
Precipitation (mm)	107[l]
Average minimum and maximum temperatures (centigrade)[l]	18.7/32.4

a Official rate. b October 2006. c May 2006. d 31st December of each year. e 2002.
f 2004. g 1995. h Part Neutral Zone. i 2001. j 1985. k Provisional. l Kuwait.

Kyrgyzstan

Region	South-central Asia	
Largest urban agglom. (pop., 000s)	Bishkek (798)	
Currency	som	
Population in 2006 (proj., 000s)	5325	
Surface area (square kms)	199900	
Population density (per square km)	27	
United Nations membership date	2 March 1992	

Economic indicators	2000	2005
Exchange rate (national currency per US$)[a]	48.30	39.08[b]
Consumer price index (2000=100)	100	127[c]
Unemployment (percentage of labour force)[d]	12.5[e]	8.5[f]
Balance of payments, current account (million US$)	−124	−101[f]
Tourist arrivals (000s)	59	398[fg]
GDP (million current US$)	1370	2441
GDP (per capita current US$)	277	464
Gross fixed capital formation (% of GDP)	18.0	15.0
Labour force participation, adult female pop. (%)	...	55.3[e]
Labour force participation, adult male pop. (%)	...	74.0[e]
Employment in industrial sector (%)	10.5	10.3[e]
Employment in agricultural sector (%)	53.1	52.7[e]
Agricultural production index (1999-2001=100)	100	98[f]
Food production index (1999-2001=100)	100	98[f]
Primary energy production (000s Mt oil equiv.)	1416	1449[f]
Motor vehicles (per 1,000 inhabitants)[h]	38.3	38.2
Telephone lines (per 100 inhabitants)	7.7	8.3
Internet users, estimated (000s)	51.6	280.0

Total trade		Major trading partners			2005
	(million US$)	(% of exports)		(% of imports)	
Exports	672.0	Untd Arab Em	26	Russian Fed.	34
Imports	1107.8	Russian Fed.	20	Kazakhstan	16
		Kazakhstan	17	China	9

Social indicators	2000-2006
Population growth rate 2000-2005 (% per annum)	1.2
Population aged 0-14 years (%)	31.0
Population aged 60+ years (women and men, % of total)	9.0/6.0
Sex ratio (women per 100 men)	103
Life expectancy at birth 2000-2005 (women and men, years)	71/63
Infant mortality rate 2000-2005 (per 1,000 births)	55
Total fertility rate 2000-2005 (births per woman)	2.7
Urban population (%)	36
Urban population growth rate 2000-2005 (% per annum)	1.4
Rural population growth rate 2000-2005 (% per annum)	1.1
Refugees and others of concern to UNHCR[i]	103096
Government education expenditure (% of GNP)	4.6
Primary-secondary gross enrolment ratio (w and m per 100)	92/91
Third-level students (women and men, % of total)	54/46
Television receivers (per 1,000 inhabitants)	188
Intentional homicides (per 100,000 inhabitants)	7
Parliamentary seats (women and men, % of total)	1/99

Environment	2000-2006
Threatened species	21
Forested area (% of land area)	5
CO2 emissions (000s Mt of carbon dioxide/per capita)	5328/1.0
Energy consumption per capita (kilograms oil equiv.)	547
Precipitation (mm)	442[j]
Average minimum and maximum temperatures (centigrade)[j]	4.8/17.0

a Official rate. b October 2006. c August 2006. d Persons aged 15 years and over, Nov. e 2002. f 2004. g New data source: Department of Customs Control. h Passenger-cars only. i Provisional. j Bishkek.

Lao People's Democratic Republic

Region	South-eastern Asia
Largest urban agglom. (pop., 000s)	Vientiane (702)
Currency	kip
Population in 2006 (proj., 000s)	6058
Surface area (square kms)	236800
Population density (per square km)	26
United Nations membership date	14 December 1955

Economic indicators	2000	2005
Exchange rate (national currency per US$)	8218.00	10187.00[a]
Balance of payments, current account (million US$)	–8	–82[b]
Tourist arrivals (000s)	191	236[c]
GDP (million current US$)	1733	2872
GDP (per capita current US$)	328	485
Gross fixed capital formation (% of GDP)	10.0	17.0
Employment in industrial sector (%)	3.5[d]	...
Employment in agricultural sector (%)	85.4[d]	...
Agricultural production index (1999-2001=100)	105	115[c]
Food production index (1999-2001=100)	103	117[c]
Primary energy production (000s Mt oil equiv.)	260	310[c]
Telephone lines (per 100 inhabitants)	0.8	1.3
Internet users, estimated (000s)	6.0	25.0

Social indicators	2000-2006
Population growth rate 2000-2005 (% per annum)	2.3
Population aged 0-14 years (%)	41.0
Population aged 60+ years (women and men, % of total)	6.0/5.0
Sex ratio (women per 100 men)	100
Life expectancy at birth 2000-2005 (women and men, years)	56/53
Infant mortality rate 2000-2005 (per 1,000 births)	88
Total fertility rate 2000-2005 (births per woman)	4.8
Contraceptive use (% of currently married women)	32
Urban population (%)	21
Urban population growth rate 2000-2005 (% per annum)	4.1
Rural population growth rate 2000-2005 (% per annum)	1.9
Government education expenditure (% of GNP)	2.5
Primary-secondary gross enrolment ratio (w and m per 100)	73/86
Third-level students (women and men, % of total)	38/62
Television receivers (per 1,000 inhabitants)	57
Parliamentary seats (women and men, % of total)	25/75

Environment	2000-2006
Threatened species	106
Forested area (% of land area)	54
CO2 emissions (000s Mt of carbon dioxide/per capita)	1254/0.2
Energy consumption per capita (kilograms oil equiv.)	65[e]
Precipitation (mm)	1661[f]
Average minimum and maximum temperatures (centigrade)[f]	21.8/31.1

a June 2006. b 2001. c 2004. d 1995. e Estimated data. f Vientiane.

Latvia

Region	Northern Europe
Largest urban agglom. (pop., 000s)	Riga (729)
Currency	lats
Population in 2006 (proj., 000s)	2295
Surface area (square kms)	64600
Population density (per square km)	36
United Nations membership date	17 September 1991

Economic indicators	2000	2005
Exchange rate (national currency per US$)[a]	0.61	0.55[b]
Consumer price index (2000=100)	100	131[c]
Industrial production index (1995=100)	105	148[d]
Unemployment (percentage of labour force)	14.4[e]	8.7[f]
Balance of payments, current account (million US$)	−371	−1959
Tourist arrivals (000s)[g]	509	1079[h]
GDP (million current US$)	7726	15244
GDP (per capita current US$)	3256	6608
Gross fixed capital formation (% of GDP)	25.0	30.0
Labour force participation, adult female pop. (%)	50.3	57.2[h]
Labour force participation, adult male pop. (%)	64.7	68.7[h]
Employment in industrial sector (%)	26.3	25.8
Employment in agricultural sector (%)	14.5	12.1
Agricultural production index (1999-2001=100)	99	117[h]
Food production index (1999-2001=100)	99	117[h]
Primary energy production (000s Mt oil equiv.)	258	274[h]
Motor vehicles (per 1,000 inhabitants)	280.4	375.4
Telephone lines (per 100 inhabitants)	30.3	31.7
Internet users, estimated (000s)	150.0	1030.0

Total trade		Major trading partners			2005
	(million US$)	(% of exports)		(% of imports)	
Exports	5302.7	Lithuania	11	Germany	14
Imports	8770.5	Estonia	10	Lithuania	13
		Germany	10	Russian Fed.	9

Social indicators	2000-2006
Population growth rate 2000-2005 (% per annum)	−0.6
Population aged 0-14 years (%)	15.0
Population aged 60+ years (women and men, % of total)	27.0/17.0
Sex ratio (women per 100 men)	119
Life expectancy at birth 2000-2005 (women and men, years)	77/66
Infant mortality rate 2000-2005 (per 1,000 births)	10
Total fertility rate 2000-2005 (births per woman)	1.3
Urban population (%)	68
Urban population growth rate 2000-2005 (% per annum)	−0.6
Rural population growth rate 2000-2005 (% per annum)	−0.4
Foreign born (%)	18.3
Refugees and others of concern to UNHCR[i]	418658
Government education expenditure (% of GNP)	5.4
Primary-secondary gross enrolment ratio (w and m per 100)	95/96
Third-level students (women and men, % of total)	62/38
Newspaper circulation (per 1,000 inhabitants)	138
Television receivers (per 1,000 inhabitants)	859
Intentional homicides (per 100,000 inhabitants)	11
Parliamentary seats (women and men, % of total)	19/81

Environment	2000-2006
Threatened species	27
Forested area (% of land area)	47
CO_2 emissions (000s Mt of carbon dioxide/per capita)[j]	7430/3.2
Energy consumption per capita (kilograms oil equiv.)	1402
Precipitation (mm)	633[k]
Average minimum and maximum temperatures (centigrade)[k]	2.3/9.9

a Official rate. b October 2006. c September 2006. d August 2006. e Persons aged 15 years and over. f Persons aged 15 to 74 years. g Non-resident departures. Survey of persons crossing the state border. h 2004. i Provisional. j Source: UNFCCC. k Riga.

Lebanon

Region	Western Asia
Largest urban agglom. (pop., 000s)	Beirut (1777)
Currency	pound
Population in 2006 (proj., 000s)	3614
Surface area (square kms)	10400
Population density (per square km)	348
United Nations membership date	24 October 1945

Economic indicators	2000	2005
Exchange rate (national currency per US$)	1507.50	1507.50[a]
Balance of payments, current account (million US$)	−4385[b]	−1881
Tourist arrivals (000s)[c]	742	1278[d]
GDP (million current US$)	16679	21184
GDP (per capita current US$)	4909	5923
Gross fixed capital formation (% of GDP)	21.0	17.0
Agricultural production index (1999-2001=100)	105	100[d]
Food production index (1999-2001=100)	106	101[d]
Primary energy production (000s Mt oil equiv.)	76	96[d]
Motor vehicles (per 1,000 inhabitants)[e]	432.9	429.1[f]
Telephone lines (per 100 inhabitants)	17.5	27.7
Internet users, estimated (000s)	300.0	700.0

Total trade		Major trading partners			2005
	(million US$)[d]	(% of exports)[d]		(% of imports)[d]	
Exports	1745.6	Iraq	15	Italy	10
Imports	9396.5	Switzerland	11	France	8
		Syria	8	Germany	8

Social indicators	2000-2006
Population growth rate 2000-2005 (% per annum)	1.0
Population aged 0-14 years (%)	29.0
Population aged 60+ years (women and men, % of total)	11.0/10.0
Sex ratio (women per 100 men)	104
Life expectancy at birth 2000-2005 (women and men, years)	74/70
Infant mortality rate 2000-2005 (per 1,000 births)	22
Total fertility rate 2000-2005 (births per woman)	2.3
Urban population (%)	87
Urban population growth rate 2000-2005 (% per annum)	1.2
Rural population growth rate 2000-2005 (% per annum)	0.1
Foreign born (%)	18.2[g]
Refugees and others of concern to UNHCR[h]	2547
Government education expenditure (% of GNP)	2.5
Primary-secondary gross enrolment ratio (w and m per 100)	99/97
Third-level students (women and men, % of total)[i]	52/48
Newspaper circulation (per 1,000 inhabitants)	65
Television receivers (per 1,000 inhabitants)	357
Parliamentary seats (women and men, % of total)	5/95

Environment	2000-2006
Threatened species	39
Forested area (% of land area)	4
CO2 emissions (000s Mt of carbon dioxide/per capita)	18998/5.4
Energy consumption per capita (kilograms oil equiv.)	1286
Precipitation (mm)	826[j]

a October 2006. b 2002. c Excluding Syrian nationals. d 2004. e Source: United Nations Economic and Social Commission for Western Asia (ESCWA). f 2001. g Estimated data. h Provisional. i East Bank only. j Beirut.

Lesotho

Region	Southern Africa
Largest urban agglom. (pop., 000s)	Maseru (172)
Currency	loti
Population in 2006 (proj., 000s)	1791
Surface area (square kms)	30355
Population density (per square km)	59
United Nations membership date	17 October 1966

Economic indicators	2000	2005
Exchange rate (national currency per US$)[a]	7.57	7.45[b]
Consumer price index (2000=100)	100	147[c]
Unemployment (percentage of labour force)[d]	39.3[e]	...
Balance of payments, current account (million US$)	−151	−76[f]
Tourist arrivals (000s)	302	304[f]
GDP (million current US$)	859	1335
GDP (per capita current US$)	481	744
Gross fixed capital formation (% of GDP)	45.0	39.0
Labour force participation, adult female pop. (%)	56.2[e]	...
Labour force participation, adult male pop. (%)	69.2[e]	...
Employment in industrial sector (%)	15.2[e]	...
Employment in agricultural sector (%)	56.5[e]	...
Agricultural production index (1999-2001=100)	100	106[f]
Food production index (1999-2001=100)	100	106[f]
Motor vehicles (per 1,000 inhabitants)	11.4[g]	...
Telephone lines (per 100 inhabitants)	1.2	2.7
Internet users, estimated (000s)	4.0	43.0[f]

Total trade		Major trading partners			2005
	(million US$)[h]	(% of exports)[h]		(% of imports)[h]	
Exports	358.0	USA	44	South Africa	77
Imports	799.6	South Africa	42	China, HK SAR	5
		Canada	7	China	4

Social indicators	2000-2006
Population growth rate 2000-2005 (% per annum)	0.1
Population aged 0-14 years (%)	39.0
Population aged 60+ years (women and men, % of total)	8.0/7.0
Sex ratio (women per 100 men)	115
Life expectancy at birth 2000-2005 (women and men, years)	38/35
Infant mortality rate 2000-2005 (per 1,000 births)	67
Total fertility rate 2000-2005 (births per woman)	3.7
Contraceptive use (% of currently married women)	30
Urban population (%)	19
Urban population growth rate 2000-2005 (% per annum)	1.0
Rural population growth rate 2000-2005 (% per annum)	−0.1
Foreign born (%)	0.3[i]
Government education expenditure (% of GNP)	7.3
Primary-secondary gross enrolment ratio (w and m per 100)	92/89
Third-level students (women and men, % of total)	61/39
Television receivers (per 1,000 inhabitants)	44
Parliamentary seats (women and men, % of total)	17/83

Environment	2000-2006
Threatened species	15
Forested area (% of land area)	1
CO2 emissions (000s Mt of carbon dioxide/per capita)[j]	636/0.4[k]

a Principal rate. b October 2006. c June 2006. d Persons aged 15 years and over.
e 1997. f 2004. g 1987. h 2002. i Estimated data. j Source: UNFCCC. k 1994.

Liberia

Region	Western Africa
Largest urban agglom. (pop., 000s)	Monrovia (936)
Currency	dollar
Population in 2006 (proj., 000s)	3356
Surface area (square kms)	111369
Population density (per square km)	30
United Nations membership date	2 November 1945

Economic indicators	2000	2005
Exchange rate (national currency per US$)[a]	42.75	59.50[b]
GDP (million current US$)	561	561
GDP (per capita current US$)	183	171
Gross fixed capital formation (% of GDP)	7.0	9.0
Agricultural production index (1999-2001=100)	102	101[c]
Food production index (1999-2001=100)	103	97[c]
Primary energy production (000s Mt oil equiv.)	17[d]	17[e]
Motor vehicles (per 1,000 inhabitants)[f]	9.8	9.3[e]
Telephone lines (per 100 inhabitants)	0.2	0.2[cg]
Internet users, estimated (000s)	0.5	1.0[h]

Social indicators	2000-2006
Population growth rate 2000-2005 (% per annum)	1.4
Population aged 0-14 years (%)	47.0
Population aged 60+ years (women and men, % of total)	4.0/3.0
Sex ratio (women per 100 men)	100
Life expectancy at birth 2000-2005 (women and men, years)	44/41
Infant mortality rate 2000-2005 (per 1,000 births)	142
Total fertility rate 2000-2005 (births per woman)	6.8
Contraceptive use (% of currently married women)	6[i]
Urban population (%)	58
Urban population growth rate 2000-2005 (% per annum)	2.7
Rural population growth rate 2000-2005 (% per annum)	−0.4
Foreign born (%)	5.4[j]
Refugees and others of concern to UNHCR[k]	579085
Government education expenditure (% of GNP)	4.9[l]
Primary-secondary gross enrolment ratio (w and m per 100)	58/79
Third-level students (women and men, % of total)	43/57
Television receivers (per 1,000 inhabitants)	25
Parliamentary seats (women and men, % of total)	14/86

Environment	2000-2006
Threatened species	109
Forested area (% of land area)	31
CO2 emissions (000s Mt of carbon dioxide/per capita)	464/0.1
Energy consumption per capita (kilograms oil equiv.)	53[j]

a Principal rate. b September 2006. c 2004. d 1998. e 2002. f Source: World Automotive Market Report, Auto and Truck International (Illinois). g Estimate. h 2001. i 1986. j Estimated data. k Provisional. l 1980.

Libyan Arab Jamahiriya

Region	Northern Africa
Largest urban agglom. (pop., 000s)	Tripoli (2098)
Currency	dinar
Population in 2006 (proj., 000s)	5968
Surface area (square kms)	1759540
Population density (per square km)	3
United Nations membership date	14 December 1955

Economic indicators	2000	2005
Exchange rate (national currency per US$)[a]	0.54	1.31[b]
Balance of payments, current account (million US$)	7740	3705[c]
Tourist arrivals (000s)	174	149[c]
GDP (million current US$)	34265	37173
GDP (per capita current US$)	6457	6351
Gross fixed capital formation (% of GDP)	13.0	20.0
Agricultural production index (1999-2001=100)	97	104[c]
Food production index (1999-2001=100)	97	104[c]
Primary energy production (000s Mt oil equiv.)	73783	83943[c]
Motor vehicles (per 1,000 inhabitants)[d]	137.0	138.3[e]
Telephone lines (per 100 inhabitants)	10.8	13.6
Internet users, estimated (000s)	10.0	205.0[c]

Total trade		Major trading partners			2005
	(million US$)	(% of exports)		(% of imports)	
Exports	10194.9[f]	Italy	43	Italy	18
Imports	6317.6[c]	Germany	15	Monaco	13
		Spain	15	Germany	12

Social indicators	2000-2006
Population growth rate 2000-2005 (% per annum)	2.0
Population aged 0-14 years (%)	30.0
Population aged 60+ years (women and men, % of total)	6.0/7.0
Sex ratio (women per 100 men)	94
Life expectancy at birth 2000-2005 (women and men, years)	76/71
Infant mortality rate 2000-2005 (per 1,000 births)	19
Total fertility rate 2000-2005 (births per woman)	3.0
Urban population (%)	85
Urban population growth rate 2000-2005 (% per annum)	2.4
Rural population growth rate 2000-2005 (% per annum)	−0.2
Foreign born (%)	10.9[g]
Refugees and others of concern to UNHCR[h]	12366
Government education expenditure (% of GNP)	7.7[i]
Primary-secondary gross enrolment ratio (w and m per 100)	110/106
Third-level students (women and men, % of total)	51/49
Television receivers (per 1,000 inhabitants)	149
Parliamentary seats (women and men, % of total)	8/92

Environment	2000-2006
Threatened species	31
Forested area (% of land area)	<
CO2 emissions (000s Mt of carbon dioxide/per capita)	50274/8.9
Energy consumption per capita (kilograms oil equiv.)	2616
Precipitation (mm)	334[j]
Average minimum and maximum temperatures (centigrade)[j]	15.6/25.4

a Official rate. b October 2006. c 2004. d Beginning 2000, excluding government passenger-cars. e 2001. f 2000. g Estimated data. h Provisional. i 1986. j Tripoli.

Liechtenstein

Region	Western Europe
Largest urban agglom. (pop., 000s)	Vaduz (5)
Currency	Swiss franc
Population in 2006 (proj., 000s)	35[a]
Surface area (square kms)	160
Population density (per square km)	208
United Nations membership date	18 September 1990

Economic indicators	2000	2005[c]
Exchange rate (national currency per US$)[b]	1.07	1.24[c]
Tourist arrivals (000s)	62	49[d]
GDP (million current US$)	2484	3482
GDP (per capita current US$)	75583	100860
Gross fixed capital formation (% of GDP)	23.0	21.0
Agricultural production index (1999-2001=100)	100	100[e]
Food production index (1999-2001=100)	100	100[e]
Telephone lines (per 100 inhabitants)	...	58.8[f]
Internet users, estimated (000s)	12.0	22.0

Social indicators	2000-2006
Population growth rate 2000-2005 (% per annum)	1.0
Population aged 0-14 years (%)	18.1[g]
Population aged 60+ years (women and men, % of total)	17.0/14.0
Sex ratio (women per 100 men)	103[hi]
Total fertility rate 2000-2005 (births per woman)	1.4
Urban population (%)	15
Urban population growth rate 2000-2005 (% per annum)	0.2
Rural population growth rate 2000-2005 (% per annum)	1.1
Foreign born (%)	35.9[j]
Refugees and others of concern to UNHCR[k]	210
Primary-secondary gross enrolment ratio (w and m per 100)	105/114
Third-level students (women and men, % of total)	27/73
Television receivers (per 1,000 inhabitants)	535
Parliamentary seats (women and men, % of total)	24/76

Environment	2000-2006
Threatened species	7
Forested area (% of land area)	47
CO2 emissions (000s Mt of carbon dioxide/per capita)[l]	240/7.1

a 2005. b Data refer to non-commercial rates derived from the Operational Rates of Exchange for United Nations Programmes. c October 2006. d 2004. e 2003. f 2002. g De facto estimate. h Refers to de facto population count. i 2000. j Estimated data. k Provisional. l Source: UNFCCC.

Lithuania

Region	Northern Europe
Largest urban agglom. (pop., 000s)	Vilnius (553)
Currency	litas
Population in 2006 (proj., 000s)	3417
Surface area (square kms)	65300
Population density (per square km)	52
United Nations membership date	17 September 1991

Economic indicators	2000	2005
Exchange rate (national currency per US$)[a]	4.00	2.71[b]
Consumer price index (2000=100)	100	108[c]
Industrial production index (1995=100)	110	186[d]
Unemployment (percentage of labour force)[e]	16.4	8.3
Balance of payments, current account (million US$)	−675	−1771
Tourist arrivals (000s)	1083	1800[f]
GDP (million current US$)	11462	24864
GDP (per capita current US$)	3275	7247
Gross fixed capital formation (% of GDP)	19.0	22.0
Labour force participation, adult female pop. (%)	54.8	51.7[f]
Labour force participation, adult male pop. (%)	67.1	63.6[f]
Employment in industrial sector (%)	26.8	29.1
Employment in agricultural sector (%)	18.7	14.0
Agricultural production index (1999-2001=100)	109	112[f]
Food production index (1999-2001=100)	109	112[f]
Primary energy production (000s Mt oil equiv.)	1156	1743[f]
Motor vehicles (per 1,000 inhabitants)	367.5	419.9[f]
Telephone lines (per 100 inhabitants)[g]	32.2	23.4
Internet users, estimated (000s)	225.0	1221.7

Total trade		Major trading partners			2005
	(million US$)	(% of exports)		(% of imports)	
Exports	12070.4	Russian Fed.	11	Russian Fed.	28
Imports	15704.4	Latvia	10	Germany	15
		Germany	9	Poland	8

Social indicators	2000-2006
Population growth rate 2000-2005 (% per annum)	−0.4
Population aged 0-14 years (%)	17.0
Population aged 60+ years (women and men, % of total)	25.0/16.0
Sex ratio (women per 100 men)	115
Life expectancy at birth 2000-2005 (women and men, years)	78/66
Infant mortality rate 2000-2005 (per 1,000 births)	9
Total fertility rate 2000-2005 (births per woman)	1.3
Urban population (%)	67
Urban population growth rate 2000-2005 (% per annum)	−0.5
Rural population growth rate 2000-2005 (% per annum)	−0.1
Foreign born (%)	5.9
Refugees and others of concern to UNHCR[h]	9294
Government education expenditure (% of GNP)	5.4
Primary-secondary gross enrolment ratio (w and m per 100)	100/101
Third-level students (women and men, % of total)	60/40
Newspaper circulation (per 1,000 inhabitants)	31
Television receivers (per 1,000 inhabitants)	518
Intentional homicides (per 100,000 inhabitants)	7
Parliamentary seats (women and men, % of total)	22/78

Environment	2000-2006
Threatened species	21
Forested area (% of land area)	32
CO2 emissions (000s Mt of carbon dioxide/per capita)[i]	12290/3.6
Energy consumption per capita (kilograms oil equiv.)	1714
Precipitation (mm)	683[j]
Average minimum and maximum temperatures (centigrade)[j]	2.4/10.1

a Official rate. b October 2006. c September 2006. d August 2006. e Persons aged 15 years and over. f 2004. g Excluding public call offices. h Provisional. i Source: UNFCCC. j Vilnius.

Luxembourg

Region	Western Europe
Largest urban agglom. (pop., 000s)	Luxembourg (77)
Currency	euro[a]
Population in 2006 (proj., 000s)	471
Surface area (square kms)	2586
Population density (per square km)	182
United Nations membership date	24 October 1945

Economic indicators	2000	2005
Exchange rate (national currency per US$)	1.07	0.79[b]
Consumer price index (2000=100)[c]	100	116[b]
Industrial production index (1995=100)	122	142[d]
Unemployment (percentage of labour force)[e]	2.7	4.7
Balance of payments, current account (million US$)	2562	3756
Tourist arrivals (000s)	852	874[f]
GDP (million current US$)	20270	36468
GDP (per capita current US$)	46573	78442
Gross fixed capital formation (% of GDP)	21.0	20.0
Labour force participation, adult female pop. (%)	41.7	39.9[f]
Labour force participation, adult male pop. (%)	66.1	64.2[f]
Employment in industrial sector (%)	36.2	20.9
Employment in agricultural sector (%)	1.5	1.2
Agricultural production index (1999-2001=100)	...	94[f]
Food production index (1999-2001=100)	...	94[f]
Primary energy production (000s Mt oil equiv.)	76	78[f]
Motor vehicles (per 1,000 inhabitants)	736.9	769.0[f]
Telephone lines (per 100 inhabitants)	56.8	52.6
Internet users, estimated (000s)	100.0	315.0

Total trade		Major trading partners			2005
	(million US$)	(% of exports)		(% of imports)	
Exports	12714.9	Germany	25	Belgium	34
Imports	17585.5	France	17	Germany	25
		Belgium	11	France	12

Social indicators	2000-2006
Population growth rate 2000-2005 (% per annum)	1.3
Population aged 0-14 years (%)	19.0
Population aged 60+ years (women and men, % of total)	21.0/16.0
Sex ratio (women per 100 men)	103
Life expectancy at birth 2000-2005 (women and men, years)	81/75
Infant mortality rate 2000-2005 (per 1,000 births)	5
Total fertility rate 2000-2005 (births per woman)	1.7
Urban population (%)	83
Urban population growth rate 2000-2005 (% per annum)	1.1
Rural population growth rate 2000-2005 (% per annum)	2.4
Foreign born (%)	32.5
Refugees and others of concern to UNHCR[g]	1822[f]
Government education expenditure (% of GNP)	3.6[h]
Primary-secondary gross enrolment ratio (w and m per 100)	98/96
Third-level students (women and men, % of total)	53/47
Newspaper circulation (per 1,000 inhabitants)	276
Television receivers (per 1,000 inhabitants)	598
Intentional homicides (per 100,000 inhabitants)	2
Parliamentary seats (women and men, % of total)	23/77

Environment	2000-2006
Threatened species	10
CO2 emissions (000s Mt of carbon dioxide/per capita)[i]	10690/23.6
Energy consumption per capita (kilograms oil equiv.)	9711
Precipitation (mm)	875[j]
Average minimum and maximum temperatures (centigrade)[j]	4.7/12.3

a Prior to 1 January 1999, franc. b September 2006. c Excluding tobacco. d June 2006. e Persons aged 16 to 64 years. f 2004. g Provisional. h 1999. i Source: UNFCCC. j Luxembourg.

Madagascar

Region	Eastern Africa
Largest urban agglom. (pop., 000s)	Antananarivo (1585)
Currency	ariary
Population in 2006 (proj., 000s)	19105
Surface area (square kms)	587041
Population density (per square km)	33
United Nations membership date	20 September 1960

Economic indicators	2000	2005
Exchange rate (national currency per US$)[a]	1310.09	2120.13[b]
Consumer price index (2000=100)	100	184[c]
Unemployment (percentage of labour force)	...	4.5[d]
Balance of payments, current account (million US$)	−283	−439[e]
Tourist arrivals (000s)[f]	160	229[g]
GDP (million current US$)	3878	4950
GDP (per capita current US$)	239	266
Gross fixed capital formation (% of GDP)	15.0	25.0
Labour force participation, adult female pop. (%)	...	79.2[e]
Labour force participation, adult male pop. (%)	...	86.3[e]
Employment in industrial sector (%)	...	6.7[d]
Employment in agricultural sector (%)	...	78.0[d]
Agricultural production index (1999-2001=100)	100	107[g]
Food production index (1999-2001=100)	100	108[g]
Primary energy production (000s Mt oil equiv.)	46	55[g]
Motor vehicles (per 1,000 inhabitants)	4.8[h]	...
Telephone lines (per 100 inhabitants)	0.3	0.4
Internet users, estimated (000s)	30.0	100.0

Total trade		Major trading partners			2005
	(million US$)[g]	(% of exports)[g]		(% of imports)[g]	
Exports	426.6	France	36	France	15
Imports	1204.2	USA	21	China	10
		Mauritius	6	Bahrain	9

Social indicators	2000-2006
Population growth rate 2000-2005 (% per annum)	2.8
Population aged 0-14 years (%)	44.0
Population aged 60+ years (women and men, % of total)	5.0/4.0
Sex ratio (women per 100 men)	101
Life expectancy at birth 2000-2005 (women and men, years)	57/54
Infant mortality rate 2000-2005 (per 1,000 births)	79
Total fertility rate 2000-2005 (births per woman)	5.4
Contraceptive use (% of currently married women)	27[i]
Urban population (%)	27
Urban population growth rate 2000-2005 (% per annum)	3.4
Rural population growth rate 2000-2005 (% per annum)	2.6
Foreign born (%)	0.4[j]
Government education expenditure (% of GNP)	3.4
Third-level students (women and men, % of total)	47/53
Television receivers (per 1,000 inhabitants)	18
Parliamentary seats (women and men, % of total)	8/92

Environment	2000-2006
Threatened species	538
Forested area (% of land area)	20
CO2 emissions (000s Mt of carbon dioxide/per capita)	2345/0.1
Energy consumption per capita (kilograms oil equiv.)	38[j]
Precipitation (mm)	1365[k]
Average minimum and maximum temperatures (centigrade)[l]	13.8/24.0

a Introduced a new currency, Ariary (ISO Code MGA). 1.00 MGA is equivalent to 5.00 Madagascar Franc. Use of operatonal rate system began in 1 Jan 2005. b October 2006. c August 2006. d 2002. e 2003. f Arrivals of non-resident tourists by air. g 2004. h 1998. i 2003/04. j Estimated data. k Antananarivo. l Antananarivo/Ivato.

Malawi

Region	Eastern Africa
Largest urban agglom. (pop., 000s)	Blantyre-Limbe (676)
Currency	kwacha
Population in 2006 (proj., 000s)	13166
Surface area (square kms)	118484
Population density (per square km)	111
United Nations membership date	1 December 1964

Economic indicators	2000	2005
Exchange rate (national currency per US$)[a]	80.08	138.25[b]
Consumer price index (2000=100)	100	230[c]
Balance of payments, current account (million US$)	−73	−201[d]
Tourist arrivals (000s)[e]	228	471[f]
GDP (million current US$)	1744	2140
GDP (per capita current US$)	151	166
Gross fixed capital formation (% of GDP)	12.0	9.0
Labour force participation, adult female pop. (%)	75.5[g]	...
Labour force participation, adult male pop. (%)	79.3[g]	...
Agricultural production index (1999-2001=100)	104	94[f]
Food production index (1999-2001=100)	103	96[f]
Primary energy production (000s Mt oil equiv.)	147	158[f]
Telephone lines (per 100 inhabitants)	0.5	0.8
Internet users, estimated (000s)	15.0	52.5

Total trade		Major trading partners			2005
	(million US$)	(% of exports)		(% of imports)	
Exports	495.5	South Africa	19	South Africa	33
Imports	1165.2	UK	12	Mozambique	13
		USA	11	Zimbabwe	8

Social indicators	2000-2006
Population growth rate 2000-2005 (% per annum)	2.3
Population aged 0-14 years (%)	47.0
Population aged 60+ years (women and men, % of total)	5.0/4.0
Sex ratio (women per 100 men)	101
Life expectancy at birth 2000-2005 (women and men, years)	40/40
Infant mortality rate 2000-2005 (per 1,000 births)	111
Total fertility rate 2000-2005 (births per woman)	6.1
Contraceptive use (% of currently married women)	31
Urban population (%)	17
Urban population growth rate 2000-2005 (% per annum)	4.8
Rural population growth rate 2000-2005 (% per annum)	1.8
Foreign born (%)	2.5[h]
Refugees and others of concern to UNHCR[i]	9571
Government education expenditure (% of GNP)	6.2
Primary-secondary gross enrolment ratio (w and m per 100)	83/84
Third-level students (women and men, % of total)	35/65
Television receivers (per 1,000 inhabitants)	6
Parliamentary seats (women and men, % of total)	14/86

Environment	2000-2006
Threatened species	157
Forested area (% of land area)	27
CO2 emissions (000s Mt of carbon dioxide/per capita)	885/0.1
Energy consumption per capita (kilograms oil equiv.)	34[h]
Precipitation (mm)	1289[j]
Average minimum and maximum temperatures (centigrade)[j]	12.2/24.1

a Official rate. b September 2006. c April 2006. d 2002. e Departures. f 2004. g 1998. h Estimated data. i Provisional. j Mzuzu.

Malaysia

Region	South-eastern Asia
Largest urban agglom. (pop., 000s)	Kuala Lumpur (1405)
Currency	ringgit
Population in 2006 (proj., 000s)	25796
Surface area (square kms)	329847
Population density (per square km)	78
United Nations membership date	17 September 1957

Economic indicators	2000	2005
Exchange rate (national currency per US$)[a]	3.80	3.65[b]
Consumer price index (2000=100)	100	114[b]
Industrial production index (1995=100)	148	203[c]
Unemployment (percentage of labour force)[d]	3.0	3.6[e]
Balance of payments, current account (million US$)	8488	14872[f]
Tourist arrivals (000s)[g]	10222	15703[f]
GDP (million current US$)	90320	130770
GDP (per capita current US$)	3927	5159
Gross fixed capital formation (% of GDP)	26.0	20.0
Labour force participation, adult female pop. (%)	46.7	46.7[h]
Labour force participation, adult male pop. (%)	83.3	...
Employment in industrial sector (%)	32.2	30.1[f]
Employment in agricultural sector (%)	18.4	14.8[f]
Agricultural production index (1999-2001=100)	99	121[f]
Food production index (1999-2001=100)	98	120[f]
Primary energy production (000s Mt oil equiv.)	76795	90405[f]
Motor vehicles (per 1,000 inhabitants)	16.8	17.5[f]
Telephone lines (per 100 inhabitants)	19.9	16.8
Internet users, estimated (000s)	4977.0	11016.0

Total trade		Major trading partners			2005
	(million US$)	(% of exports)		(% of imports)	
Exports	140962.9	USA	20	Japan	15
Imports	114583.6	Singapore	16	USA	13
		Japan	9	Singapore	12

Social indicators	2000-2006
Population growth rate 2000-2005 (% per annum)	1.9
Population aged 0-14 years (%)	32.0
Population aged 60+ years (women and men, % of total)	7.0/7.0
Sex ratio (women per 100 men)	97
Life expectancy at birth 2000-2005 (women and men, years)	75/71
Infant mortality rate 2000-2005 (per 1,000 births)	10
Total fertility rate 2000-2005 (births per woman)	2.9
Contraceptive use (% of currently married women)[i]	55[j]
Urban population (%)	67
Urban population growth rate 2000-2005 (% per annum)	3.7
Rural population growth rate 2000-2005 (% per annum)	−1.2
Refugees and others of concern to UNHCR[k]	106083
Government education expenditure (% of GNP)	8.5
Primary-secondary gross enrolment ratio (w and m per 100)	87/82
Third-level students (women and men, % of total)	57/43
Newspaper circulation (per 1,000 inhabitants)	95
Television receivers (per 1,000 inhabitants)	219
Parliamentary seats (women and men, % of total)	13/87

Environment	2000-2006
Threatened species	917
Forested area (% of land area)	59
CO2 emissions (000s Mt of carbon dioxide/per capita)	156680/6.4
Energy consumption per capita (kilograms oil equiv.)	2531
Precipitation (mm)	2427[l]
Average minimum and maximum temperatures (centigrade)[l]	23.0/32.4

a Official rate. b September 2006. c June 2006. d Persons aged 15 to 64 years. e 2003. f 2004. g Including Singapore residents crossing the frontier by road through Johore Causeway. h 2000. i The data refer only to Peninsular Malaysia. j 1994. k Provisional. l Kuala Lumpur.

Maldives

Region	South-central Asia	
Largest urban agglom. (pop., 000s)	Male (89)	
Currency	rufiyaa	
Population in 2006 (proj., 000s)	337	
Surface area (square kms)	298	
Population density (per square km)	1132	
United Nations membership date	21 September 1965	

Economic indicators	2000	2005
Exchange rate (national currency per US$)	11.77	12.80[a]
Consumer price index (2000=100)[b]	100	108[c]
Balance of payments, current account (million US$)	−51	−134[d]
Tourist arrivals (000s)[e]	467	617[d]
GDP (million current US$)	624	770
GDP (per capita current US$)	2151	2338
Gross fixed capital formation (% of GDP)	26.0	27.0
Labour force participation, adult female pop. (%)	37.4	...
Labour force participation, adult male pop. (%)	71.7	...
Employment in industrial sector (%)	19.0	...
Employment in agricultural sector (%)	13.7	...
Agricultural production index (1999-2001=100)	103	115[d]
Food production index (1999-2001=100)	103	115[d]
Motor vehicles (per 1,000 inhabitants)	2.4	13.6
Telephone lines (per 100 inhabitants)	9.1	9.6[d]
Internet users, estimated (000s)	6.0	19.0[d]

Total trade	(million US$)	Major trading partners	(% of exports)		2005 (% of imports)
Exports	154.2	Untd Arab Em	24	Singapore	24
Imports	744.9	Thailand	15	Untd Arab Em	16
		Japan	15	India	11

Social indicators	2000-2006
Population growth rate 2000-2005 (% per annum)	2.5
Population aged 0-14 years (%)	41.0
Population aged 60+ years (women and men, % of total)	5.0/5.0
Sex ratio (women per 100 men)	95
Life expectancy at birth 2000-2005 (women and men, years)	66/67
Infant mortality rate 2000-2005 (per 1,000 births)	43
Total fertility rate 2000-2005 (births per woman)	4.3
Urban population (%)	30
Urban population growth rate 2000-2005 (% per annum)	4.0
Rural population growth rate 2000-2005 (% per annum)	1.9
Foreign born (%)	1.1[f]
Government education expenditure (% of GNP)	8.6
Primary-secondary gross enrolment ratio (w and m per 100)	92/91
Third-level students (women and men, % of total)	70/30
Television receivers (per 1,000 inhabitants)	143
Parliamentary seats (women and men, % of total)	12/88

Environment	2000-2006
Threatened species	15
CO2 emissions (000s Mt of carbon dioxide/per capita)	442/1.4
Energy consumption per capita (kilograms oil equiv.)	809
Precipitation (mm)	1901[b]
Average minimum and maximum temperatures (centigrade)[b]	25.8/30.6

a October 2006. b Male. c January 2006. d 2004. e Arrivals by air. f Estimated data.

Mali

Region	Western Africa
Largest urban agglom. (pop., 000s)	Bamako (1368)
Currency	CFA franc
Population in 2006 (proj., 000s)	13918
Surface area (square kms)	1240192
Population density (per square km)	11
United Nations membership date	28 September 1960

Economic indicators	2000	2005
Exchange rate (national currency per US$)[a]	704.95	516.66[b]
Consumer price index (2000=100)[c]	100	115[d]
Unemployment (percentage of labour force)[e]	3.3[f]	8.8[g]
Balance of payments, current account (million US$)	−255	−409[g]
Tourist arrivals (000s)[h]	86	113[g]
GDP (million current US$)	2670	5181
GDP (per capita current US$)	229	383
Gross fixed capital formation (% of GDP)	19.0	23.0
Agricultural production index (1999-2001=100)	88	114[g]
Food production index (1999-2001=100)	95	110[g]
Primary energy production (000s Mt oil equiv.)	20	21[g]
Motor vehicles (per 1,000 inhabitants)[i]	4.3	4.2[j]
Telephone lines (per 100 inhabitants)	0.4	0.7
Internet users, estimated (000s)	15.0	60.0

Total trade		Major trading partners			2005
	(million US$)[g]	(% of exports)[g]		(% of imports)[g]	
Exports	982.7	South Africa	31	France	16
Imports	1360.1	Switzerland	20	Senegal	12
		Senegal	6	Côte d'Ivoire	9

Social indicators	2000-2006
Population growth rate 2000-2005 (% per annum)	3.0
Population aged 0-14 years (%)	48.0
Population aged 60+ years (women and men, % of total)	5.0/4.0
Sex ratio (women per 100 men)	101
Life expectancy at birth 2000-2005 (women and men, years)	48/47
Infant mortality rate 2000-2005 (per 1,000 births)	133
Total fertility rate 2000-2005 (births per woman)	6.9
Contraceptive use (% of currently married women)	8
Urban population (%)	30
Urban population growth rate 2000-2005 (% per annum)	4.8
Rural population growth rate 2000-2005 (% per annum)	2.2
Foreign born (%)	0.4[k]
Refugees and others of concern to UNHCR[l]	13066
Government education expenditure (% of GNP)	3.0[m]
Primary-secondary gross enrolment ratio (w and m per 100)	38/52
Third-level students (women and men, % of total)	31/69
Television receivers (per 1,000 inhabitants)	36
Parliamentary seats (women and men, % of total)	10/90

Environment	2000-2006
Threatened species	30
Forested area (% of land area)	11
CO2 emissions (000s Mt of carbon dioxide/per capita)	553/0.0
Energy consumption per capita (kilograms oil equiv.)	19[k]
Precipitation (mm)	991[c]
Average minimum and maximum temperatures (centigrade)[c]	21.3/35.0

a Official rate. b October 2006. c Bamako. d June 2006. e Persons aged 14 years and over. f 1997. g 2004. h Arrivals by air. i Source: World Automotive Market Report, Auto and Truck International (Illinois). j 2001. k Estimated data. l Provisional. m 1999.

Malta

Region	Southern Europe
Largest urban agglom. (pop., 000s)	Valletta (210)
Currency	lira
Population in 2006 (proj., 000s)	403
Surface area (square kms)	316
Population density (per square km)	1277
United Nations membership date	1 December 1964

Economic indicators	2000	2005
Exchange rate (national currency per US$)[a]	0.44	0.34[b]
Consumer price index (2000=100)	100	116[c]
Unemployment (percentage of labour force)[d]	6.7	7.5
Balance of payments, current account (million US$)	−498	−613
Tourist arrivals (000s)	1216	1158[e]
GDP (million current US$)	3793	5573
GDP (per capita current US$)	9682	13877
Gross fixed capital formation (% of GDP)	22.0	22.0
Labour force participation, adult female pop. (%)	28.3[f]	29.6[e]
Labour force participation, adult male pop. (%)	80.8[g]	69.7[e]
Employment in industrial sector (%)	33.0	29.4[e]
Employment in agricultural sector (%)	1.7	2.1[e]
Agricultural production index (1999-2001=100)	99	107[e]
Food production index (1999-2001=100)	99	107[e]
Motor vehicles (per 1,000 inhabitants)	669.5	742.8[e]
Telephone lines (per 100 inhabitants)	52.4	50.4
Internet users, estimated (000s)	51.0	127.2

Total trade		Major trading partners			2005
	(million US$)	(% of exports)		(% of imports)	
Exports	2275.0	France	15	Italy	32
Imports	3660.5	USA	14	UK	11
		Singapore	12	France	9

Social indicators	2000-2006
Population growth rate 2000-2005 (% per annum)	0.5
Population aged 0-14 years (%)	18.0
Population aged 60+ years (women and men, % of total)	21.0/16.0
Sex ratio (women per 100 men)	102
Life expectancy at birth 2000-2005 (women and men, years)	81/76
Infant mortality rate 2000-2005 (per 1,000 births)	7
Total fertility rate 2000-2005 (births per woman)	1.5
Urban population (%)	95
Urban population growth rate 2000-2005 (% per annum)	0.9
Rural population growth rate 2000-2005 (% per annum)	−6.3
Foreign born (%)	2.2[h]
Refugees and others of concern to UNHCR[i]	2088
Government education expenditure (% of GNP)	4.6
Primary-secondary gross enrolment ratio (w and m per 100)	102/106
Third-level students (women and men, % of total)	56/44
Television receivers (per 1,000 inhabitants)[j]	554
Intentional homicides (per 100,000 inhabitants)	1
Parliamentary seats (women and men, % of total)	9/91

Environment	2000-2006
Threatened species	28
Forested area (% of land area)	–
CO2 emissions (000s Mt of carbon dioxide/per capita)	2467/6.2
Energy consumption per capita (kilograms oil equiv.)	2004
Precipitation (mm)	553[k]
Average minimum and maximum temperatures (centigrade)[k]	14.9/22.3

a Official rate. b October 2006. c July 2006. d Persons aged 15 years and over. e 2004.
f 2001. g 1998. h Estimated data. i Provisional. j Year ending 30 September. k Luqa.

Marshall Islands

Region	Oceania-Micronesia
Largest urban agglom. (pop., 000s)	Majuro (27)
Currency	US dollar
Population in 2006 (proj., 000s)	62[a]
Surface area (square kms)	181
Population density (per square km)	281
United Nations membership date	17 September 1991

Economic indicators	2000	2005
Exchange rate (national currency per US$)	1.00	1.00
Consumer price index (2000=100)[b]	100	107
Unemployment (percentage of labour force)	30.9[c]	25.4[d]
Tourist arrivals (000s)[e]	5	9[fg]
GDP (million current US$)	99	111
GDP (per capita current US$)	1896	1791
Gross fixed capital formation (% of GDP)	57.0	57.0
Agricultural production index (1999-2001=100)	93	93[h]
Food production index (1999-2001=100)	93	93[h]
Telephone lines (per 100 inhabitants)	7.8	8.3
Internet users, estimated (000s)	0.8	2.2

Total trade	Major trading partners		2005
(million US$)	(% of exports)		(% of imports)[f]
Exports	...	... USA	61
Imports	68.2[i]	... Japan	5
		... Australia	2

Social indicators	2000-2006
Population growth rate 2000-2005 (% per annum)	3.5
Population aged 0-14 years (%)	42.0
Population aged 60+ years (women and men, % of total)	3.0/3.0
Sex ratio (women per 100 men)	95[jk]
Life expectancy at birth 2000-2005 (women and men, years)	69/59[kl]
Infant mortality rate 2000-2005 (per 1,000 births)	37
Total fertility rate 2000-2005 (births per woman)	5.7
Urban population (%)	67
Foreign born (%)	3.2[m]
Government education expenditure (% of GNP)	11.9
Primary-secondary gross enrolment ratio (w and m per 100)	99/101
Third-level students (women and men, % of total)	56/44
Parliamentary seats (women and men, % of total)	3/97

Environment	2000-2006
Threatened species	14
Precipitation (mm)	2407[n]
Average minimum and maximum temperatures (centigrade)[n]	26.7/27.7

a 2005. b Majuro. c Persons aged 15 years and over, 1999. d Persons aged 16 years and over, Labor force survey. e Arrivals by air. f Air and sea arrivals. g 2004. h 2003. i 2000. j Refers to de facto population count. k 1999. l Published by the Secretariat of the Pacific Community. m Estimated data. n Kwajalein.

Martinique

Region	Caribbean	
Largest urban agglom. (pop., 000s)	Fort-de-France (91)	
Currency	euro[a]	
Population in 2006 (proj., 000s)	397	
Surface area (square kms)	1102	
Population density (per square km)	361	

Economic indicators	2000	2005
Exchange rate (national currency per US$)[b]	1.07	0.79[c]
Consumer price index (2000=100)	100	114[d]
Unemployment (percentage of labour force)[e]	26.3	21.7
Tourist arrivals (000s)	526	471[f]
Agricultural production index (1999-2001=100)	104	99[f]
Food production index (1999-2001=100)	104	99[f]
Motor vehicles (per 1,000 inhabitants)	310.3[g]	...
Telephone lines (per 100 inhabitants)	44.7	44.5[hi]
Internet users, estimated (000s)	30.0	130.0

Social indicators	2000-2006
Population growth rate 2000-2005 (% per annum)	0.5
Population aged 0-14 years (%)	22.0
Population aged 60+ years (women and men, % of total)	18.0/15.0
Sex ratio (women per 100 men)	111
Life expectancy at birth 2000-2005 (women and men, years)	82/75
Infant mortality rate 2000-2005 (per 1,000 births)	7
Total fertility rate 2000-2005 (births per woman)	2.0
Urban population (%)	98
Urban population growth rate 2000-2005 (% per annum)	0.5
Rural population growth rate 2000-2005 (% per annum)	−0.5
Television receivers (per 1,000 inhabitants)	178

Environment	2000-2006
Threatened species	33
Forested area (% of land area)	44
CO2 emissions (000s Mt of carbon dioxide/per capita)	1341/3.4
Energy consumption per capita (kilograms oil equiv.)	1434[j]
Precipitation (mm)	2030[k]
Average minimum and maximum temperatures (centigrade)[k]	22.8/29.7

a Prior to 1 January 1999, French franc. b Data refer to non-commercial rates derived from the Operational Rates of Exchange for United Nations Programmes. c November 2006. d June 2006. e Persons aged 15 years and over, June of each year. f 2004. g 1995. h Estimate. i 2003. j Estimated data. k Le Lamentin.

Mauritania

Region	Western Africa	
Largest urban agglom. (pop., 000s)	Nouakchott (637)	
Currency	ouguiya	
Population in 2006 (proj., 000s)	3158	
Surface area (square kms)	1025520	
Population density (per square km)	3	
United Nations membership date	27 October 1961	

Economic indicators	2000	2005
Exchange rate (national currency per US$)	252.30	270.61[a]
Consumer price index (2000=100)	100	147[b]
Tourist arrivals (000s)	30	...
GDP (million current US$)	928	1672
GDP (per capita current US$)	351	545
Gross fixed capital formation (% of GDP)	13.0	13.0
Agricultural production index (1999-2001=100)	102	109[c]
Food production index (1999-2001=100)	102	109[c]
Primary energy production (000s Mt oil equiv.)	3	3[c]
Motor vehicles (per 1,000 inhabitants)[d]	11.5	10.8[e]
Telephone lines (per 100 inhabitants)	0.7	1.3
Internet users, estimated (000s)	5.0	20.0

Total trade		Major trading partners		2005
	(million US$)	(% of exports)		(% of imports)
Exports	175.0	Japan	44	...
Imports	1343.3	Spain	26	...
		Nigeria	6	...

Social indicators	2000-2006
Population growth rate 2000-2005 (% per annum)	3.0
Population aged 0-14 years (%)	43.0
Population aged 60+ years (women and men, % of total)	6.0/5.0
Sex ratio (women per 100 men)	102
Life expectancy at birth 2000-2005 (women and men, years)	54/51
Infant mortality rate 2000-2005 (per 1,000 births)	97
Total fertility rate 2000-2005 (births per woman)	5.8
Contraceptive use (% of currently married women)	8[f]
Urban population (%)	40
Urban population growth rate 2000-2005 (% per annum)	3.2
Rural population growth rate 2000-2005 (% per annum)	2.8
Foreign born (%)	2.4[g]
Refugees and others of concern to UNHCR[h]	30224
Government education expenditure (% of GNP)	3.7
Primary-secondary gross enrolment ratio (w and m per 100)	57/59
Third-level students (women and men, % of total)	24/76
Television receivers (per 1,000 inhabitants)	41
Parliamentary seats (women and men, % of total)	4/96

Environment	2000-2006
Threatened species	39
Forested area (% of land area)	<
CO2 emissions (000s Mt of carbon dioxide/per capita)	2503/0.9
Energy consumption per capita (kilograms oil equiv.)	269[g]

a June 2006. b May 2006. c 2004. d Source: World Automotive Market Report, Auto and Truck International (Illinois). e 2002. f 2000/01. g Estimated data. h Provisional.

Mauritius

Region	Eastern Africa
Largest urban agglom. (pop., 000s)	Port Louis (146)
Currency	rupee
Population in 2006 (proj., 000s)	1256[a]
Surface area (square kms)	2040
Population density (per square km)	616
United Nations membership date	24 April 1968

Economic indicators	2000	2005
Exchange rate (national currency per US$)	27.88	32.96[b]
Consumer price index (2000=100)	100	135[c]
Unemployment (percentage of labour force)	9.8[d]	9.6[e]
Balance of payments, current account (million US$)	−37	−340
Tourist arrivals (000s)	656	719[f]
GDP (million current US$)	4552	6288
GDP (per capita current US$)	3839	5052
Gross fixed capital formation (% of GDP)	23.0	22.0
Labour force participation, adult female pop. (%)	38.5	39.7[f]
Labour force participation, adult male pop. (%)	78.3	77.4[f]
Employment in industrial sector (%)	38.8	32.4
Employment in agricultural sector (%)	12.1	10.0
Agricultural production index (1999-2001=100)	101	106[f]
Food production index (1999-2001=100)	101	106[f]
Primary energy production (000s Mt oil equiv.)	8	10[f]
Motor vehicles (per 1,000 inhabitants)	102.6	132.6
Telephone lines (per 100 inhabitants)	23.5	28.8
Internet users, estimated (000s)	87.0	300.0

Total trade		Major trading partners			2005
	(million US$)	(% of exports)		(% of imports)	
Exports	2004.4	UK	32	China	10
Imports	3160.1	France	17	South Africa	9
		USA	10	France	8

Social indicators	2000-2006
Population growth rate 2000-2005 (% per annum)	1.0[a]
Population aged 0-14 years (%)	25.0
Population aged 60+ years (women and men, % of total)	11.0/8.0
Sex ratio (women per 100 men)	101[a]
Life expectancy at birth 2000-2005 (women and men, years)	76/69[a]
Infant mortality rate 2000-2005 (per 1,000 births)	15
Total fertility rate 2000-2005 (births per woman)	2.0[a]
Contraceptive use (% of currently married women)	75[g]
Urban population (%)[a]	42
Urban population growth rate 2000-2005 (% per annum)[a]	0.8
Rural population growth rate 2000-2005 (% per annum)[a]	1.1
Foreign born (%)	1.3
Government education expenditure (% of GNP)	4.7
Primary-secondary gross enrolment ratio (w and m per 100)	94/95
Third-level students (women and men, % of total)	58/42
Newspaper circulation (per 1,000 inhabitants)	116
Television receivers (per 1,000 inhabitants)	370
Intentional homicides (per 100,000 inhabitants)	3
Parliamentary seats (women and men, % of total)	17/83

Environment	2000-2006
Threatened species	157
Forested area (% of land area)	8
CO2 emissions (000s Mt of carbon dioxide/per capita)	3150/2.6
Energy consumption per capita (kilograms oil equiv.)	835
Precipitation (mm)	711[h]
Average minimum and maximum temperatures (centigrade)[h]	21.6/19.4

a Including Agalega, Rodrigues and Saint Brandon. b October 2006. c April 2006.
d Persons aged 12 years and over, 1995. e Persons aged 15 years and over. f 2004.
g For women aged 15-44 in union or married. h Port Louis.

Mexico

Region	Central America
Largest urban agglom. (pop., 000s)	Mexico City (19411)
Currency	peso
Population in 2006 (proj., 000s)	108327
Surface area (square kms)	1958201
Population density (per square km)	55
United Nations membership date	7 November 1945

Economic indicators	2000	2005
Exchange rate (national currency per US$)[a]	9.57	10.71[b]
Consumer price index (2000=100)	100	130[c]
Industrial production index (1995=100)[d]	142	151[e]
Unemployment (percentage of labour force)[f]	2.6	3.5
Balance of payments, current account (million US$)	–18707	–4804
Tourist arrivals (000s)[g]	20641	20618[h]
GDP (million current US$)	580792	768438
GDP (per capita current US$)	5803	7180
Gross fixed capital formation (% of GDP)	21.0	19.0
Labour force participation, adult female pop. (%)	39.0	40.3[h]
Labour force participation, adult male pop. (%)	82.9	81.9[h]
Employment in industrial sector (%)	26.9	25.7
Employment in agricultural sector (%)	17.6	15.1
Agricultural production index (1999-2001=100)	99	107[h]
Food production index (1999-2001=100)	99	108[h]
Primary energy production (000s Mt oil equiv.)	212599	237911[h]
Motor vehicles (per 1,000 inhabitants)	162.3	191.0[i]
Telephone lines (per 100 inhabitants)[j]	12.5	18.2
Internet users, estimated (000s)	5058.0	18091.8

Total trade		Major trading partners			2005
	(million US$)	(% of exports)		(% of imports)	
Exports	214207.3	USA	86	USA	54
Imports	221819.0	Canada	2	China	8
		Spain	1	Japan	6

Social indicators	2000-2006
Population growth rate 2000-2005 (% per annum)	1.3
Population aged 0-14 years (%)	31.0
Population aged 60+ years (women and men, % of total)	8.0/7.0
Sex ratio (women per 100 men)	105
Life expectancy at birth 2000-2005 (women and men, years)	77/72
Infant mortality rate 2000-2005 (per 1,000 births)	21
Total fertility rate 2000-2005 (births per woman)	2.4
Urban population (%)	76
Urban population growth rate 2000-2005 (% per annum)	1.7
Rural population growth rate 2000-2005 (% per annum)	0.3
Foreign born (%)	0.5
Refugees and others of concern to UNHCR[k]	3390
Government education expenditure (% of GNP)	5.9
Primary-secondary gross enrolment ratio (w and m per 100)	95/94
Third-level students (women and men, % of total)	50/50
Newspaper circulation (per 1,000 inhabitants)	92
Television receivers (per 1,000 inhabitants)	276
Intentional homicides (per 100,000 inhabitants)	10
Parliamentary seats (women and men, % of total)	24/76

Environment	2000-2006
Threatened species	771
Forested area (% of land area)	29
CO$_2$ emissions (000s Mt of carbon dioxide/per capita)	416698/4.0
Energy consumption per capita (kilograms oil equiv.)	1365
Precipitation (mm)	816[l]
Average minimum and maximum temperatures (centigrade)[l]	9.6/23.4

a Principal rate. b October 2006. c May 2006. d Including construction. e June 2006.
f Persons aged 14 years and over, second quarter of each year. g Including nationals residing abroad. h 2004. i 2002. j Lines in service. k Provisional. l Mexico City.

Micronesia, Federated States of

Region	Oceania-Micronesia
Largest urban agglom. (pop., 000s)	Palikir (7)
Currency	US dollar
Population in 2006 (proj., 000s)	111
Surface area (square kms)	702
Population density (per square km)	158
United Nations membership date	17 September 1991

Economic indicators	2000	2005
Exchange rate (national currency per US$)	1.00	1.00
Tourist arrivals (000s)[a]	21	19[b]
GDP (million current US$)	214	239
GDP (per capita current US$)	1998	2168
Gross fixed capital formation (% of GDP)	35.0	35.0
Agricultural production index (1999-2001=100)	100	100[b]
Food production index (1999-2001=100)	100	100[b]
Telephone lines (per 100 inhabitants)	9.0	11.2
Internet users, estimated (000s)	4.0	14.0

Social indicators	2000-2006
Population growth rate 2000-2005 (% per annum)	0.6
Population aged 0-14 years (%)	39.0
Population aged 60+ years (women and men, % of total)	5.0/4.0
Sex ratio (women per 100 men)	99
Life expectancy at birth 2000-2005 (women and men, years)	68/67
Infant mortality rate 2000-2005 (per 1,000 births)	38
Total fertility rate 2000-2005 (births per woman)	4.4
Urban population (%)	22
Urban population growth rate 2000-2005 (% per annum)	0.6
Rural population growth rate 2000-2005 (% per annum)	0.6
Government education expenditure (% of GNP)	6.7
Television receivers (per 1,000 inhabitants)	25
Parliamentary seats (women and men, % of total)	1/99

Environment	2000-2006
Threatened species	33
CO2 emissions (000s Mt of carbon dioxide/per capita)[c]	141/1.3

a Arrivals in the States of Kosrae, Chuuk, Pohnpei and Yap; excluding FSM citizens.
b 2004. c Source: UNFCCC.

Moldova, Republic of

Region	Eastern Europe
Largest urban agglom. (pop., 000s)	Chisinau (598)
Currency	leu
Population in 2006 (proj., 000s)	4195
Surface area (square kms)	33851
Population density (per square km)	124
United Nations membership date	2 March 1992

Economic indicators	2000	2005
Exchange rate (national currency per US$)[a]	12.38	13.23[b]
Consumer price index (2000=100)[c]	100	181[d]
Unemployment (percentage of labour force)[e]	8.5	7.3
Balance of payments, current account (million US$)	−108	−307
Tourist arrivals (000s)[f]	18	24[g]
GDP (million current US$)	1288	2917
GDP (per capita current US$)	301	694
Gross fixed capital formation (% of GDP)	15.0	24.0
Labour force participation, adult female pop. (%)	54.2[h]	47.9[g]
Labour force participation, adult male pop. (%)	59.3[h]	51.7[g]
Employment in industrial sector (%)	13.9	16.0
Employment in agricultural sector (%)	50.9	40.6
Agricultural production index (1999-2001=100)	103	113[g]
Food production index (1999-2001=100)	102	116[g]
Primary energy production (000s Mt oil equiv.)	5	5[g]
Motor vehicles (per 1,000 inhabitants)[ij]	57.4	70.9
Telephone lines (per 100 inhabitants)	13.7	22.1
Internet users, estimated (000s)	52.6	550.0

Total trade		Major trading partners			2005
	(million US$)	(% of exports)		(% of imports)	
Exports	1091.3	Russian Fed.	32	Ukraine	21
Imports	2293.0	Italy	12	Russian Fed.	12
		Romania	10	Romania	11

Social indicators	2000-2006
Population growth rate 2000-2005 (% per annum)	−0.3
Population aged 0-14 years (%)	18.0
Population aged 60+ years (women and men, % of total)	16.0/11.0
Sex ratio (women per 100 men)	109
Life expectancy at birth 2000-2005 (women and men, years)	71/64
Infant mortality rate 2000-2005 (per 1,000 births)	26
Total fertility rate 2000-2005 (births per woman)	1.2
Contraceptive use (% of currently married women)	62
Urban population (%)	47
Urban population growth rate 2000-2005 (% per annum)	−0.1
Rural population growth rate 2000-2005 (% per annum)	−0.5
Foreign born (%)	13.6[k]
Refugees and others of concern to UNHCR[l]	1762
Government education expenditure (% of GNP)	4.2
Primary-secondary gross enrolment ratio (w and m per 100)	87/85
Third-level students (women and men, % of total)	57/43
Television receivers (per 1,000 inhabitants)	307
Intentional homicides (per 100,000 inhabitants)	10
Parliamentary seats (women and men, % of total)	22/78

Environment	2000-2006
Threatened species	27
Forested area (% of land area)	10
CO2 emissions (000s Mt of carbon dioxide/per capita)	7240/1.7
Energy consumption per capita (kilograms oil equiv.)	836
Precipitation (mm)	547[m]
Average minimum and maximum temperatures (centigrade)[m]	5.6/14.3

a Official rate. b October 2006. c June of each year. d April 2006. e Persons aged 15 years and over. f Excluding the regions of the left bank of the Dniestr and the munic- ipality of Bender. g 2004. h 2002. i Including motor vehicles owned by enterprises with main activity as road transport enterprises. j Excluding the data from the left side of river Nistru and municipality Bender. k 1989. l Provisional. m Chisinau.

Monaco

Region	Western Europe		
Largest urban agglom. (pop., 000s)	Monaco (...)		
Currency	euro[a]		
Population in 2006 (proj., 000s)	33		
Surface area (square kms)	1		
Population density (per square km)	21477		
United Nations membership date	28 May 1993		

Economic indicators	2000	2005
Exchange rate (national currency per US$)[b]	7.58	0.79[cd]
Tourist arrivals (000s)	300	250[e]
GDP (million current US$)	727	1203
GDP (per capita current US$)	21776	34128
Gross fixed capital formation (% of GDP)	19.0	20.0
Telephone lines (per 100 inhabitants)	...	102.8[e]
Internet users, estimated (000s)	13.5	18.0

Social indicators	2000-2006
Population growth rate 2000-2005 (% per annum)	1.1
Population aged 0-14 years (%)	13.0
Population aged 60+ years (women and men, % of total)	32.0/26.0[f]
Urban population (%)	100
Urban population growth rate 2000-2005 (% per annum)	1.1
Rural population growth rate 2000-2005 (% per annum)	–
Foreign born (%)	68.8[g]
Television receivers (per 1,000 inhabitants)	762
Parliamentary seats (women and men, % of total)	21/79

Environment	2000-2006
Threatened species	6
CO2 emissions (000s Mt of carbon dioxide/per capita)[h]	130/3.8

a Prior to 1 January 2002, French franc. b Data refer to non-commercial rates derived from the Operational Rates of Exchange for United Nations Programmes. c October 2006. d Provisional. e 2004. f De jure. g Estimated data. h Source: UNFCCC.

Mongolia

Region	Eastern Asia
Largest urban agglom. (pop., 000s)	Ulan Bator (863)
Currency	tugrik
Population in 2006 (proj., 000s)	2679
Surface area (square kms)	1566500
Population density (per square km)	2
United Nations membership date	27 October 1961

Economic indicators	2000	2005
Exchange rate (national currency per US$)	1097.00	1168.00[a]
Consumer price index (2000=100)[b]	100	147[c]
Unemployment (percentage of labour force)[d]	4.6	3.3
Balance of payments, current account (million US$)	−156	−25[e]
Tourist arrivals (000s)[f]	137	301[e]
GDP (million current US$)	946	1867
GDP (per capita current US$)	379	706
Gross fixed capital formation (% of GDP)	32.0	32.0
Labour force participation, adult female pop. (%)	55.5	57.5[e]
Labour force participation, adult male pop. (%)	68.6	59.7[e]
Employment in industrial sector (%)	14.1	16.1[e]
Employment in agricultural sector (%)	48.6	40.2[e]
Agricultural production index (1999-2001=100)	109	94[e]
Food production index (1999-2001=100)	110	94[e]
Primary energy production (000s Mt oil equiv.)	1548	2111[e]
Telephone lines (per 100 inhabitants)	5.0	5.9
Internet users, estimated (000s)	30.0	268.3

Total trade		Major trading partners			2005
	(million US$)		(% of exports)		(% of imports)
Exports	1064.4	China	48	Russian Fed.	35
Imports	1182.6	USA	14	China	25
		Canada	11	Japan	6

Social indicators	2000-2006
Population growth rate 2000-2005 (% per annum)	1.2
Population aged 0-14 years (%)	30.0
Population aged 60+ years (women and men, % of total)	6.0/5.0
Sex ratio (women per 100 men)	100
Life expectancy at birth 2000-2005 (women and men, years)	66/62
Infant mortality rate 2000-2005 (per 1,000 births)	58
Total fertility rate 2000-2005 (births per woman)	2.5
Contraceptive use (% of currently married women)	67
Urban population (%)	57
Urban population growth rate 2000-2005 (% per annum)	1.2
Rural population growth rate 2000-2005 (% per annum)	1.1
Foreign born (%)	0.3[g]
Government education expenditure (% of GNP)	5.7
Primary-secondary gross enrolment ratio (w and m per 100)	99/91
Third-level students (women and men, % of total)	62/38
Newspaper circulation (per 1,000 inhabitants)	18
Television receivers (per 1,000 inhabitants)	81
Parliamentary seats (women and men, % of total)	7/93

Environment	2000-2006
Threatened species	40
Forested area (% of land area)	7
CO2 emissions (000s Mt of carbon dioxide/per capita)	7987/3.1
Energy consumption per capita (kilograms oil equiv.)	853
Precipitation (mm)	271[b]
Average minimum and maximum temperatures (centigrade)[b]	—14.5/15.4

a September 2006. b Ulan Bator. c June 2006. d 31st December of each year. e 2004.
f Excluding diplomats and foreign residents in Mongolia. g Estimated data.

Montenegro

Region	Southern Europe
Largest urban agglom. (pop., 000s)	Podgorica (173)[ab]
Currency	euro
Population in 2006 (proj., 000s)	620[bc]
Surface area (square kms)	13812[b]
Population density (per square km)	45[b]
United Nations membership date	28 June 2006

Economic indicators	2000	2005
Exchange rate (national currency per US$)	70.00[d]	0.79[e]
Consumer price index (2000=100)	100	267[f]
Industrial production index (1995=100)[g]	102	116[h]
Unemployment (percentage of labour force)[gi]	12.6	15.2[c]
Tourist arrivals (000s)[g]	239	580[f]
GDP (million current US$)	942	2042
GDP (per capita current US$)	1538	3310
Gross fixed capital formation (% of GDP)	18.0	15.0
Agricultural production index (1999-2001=100)[g]	100	114[f]
Food production index (1999-2001=100)[g]	100	114[f]
Primary energy production (000s Mt oil equiv.)	10596	10730[f]
Motor vehicles (per 1,000 inhabitants)[g]	162.7	162.9[f]
Telephone lines (per 100 inhabitants)[g]	22.6	32.9
Internet users, estimated (000s)[g]	847.0[c]	1517.0[f]

Total trade	Major trading partners			2005	
	(million US$)[fg]	(% of exports)[fg]		(% of imports)[fg]	
Exports	3801.3	Bosnia/Herzeg	17	Germany	13
Imports	11366.4	Italy	15	Russian Fed.	12
		Germany	9	Italy	10

Social indicators	2000-2006
Population growth rate 2000-2005 (% per annum)	−0.1[g]
Population aged 0-14 years (%)[g]	18.3
Population aged 60+ years (women and men, % of total)[g]	21.0/16.3
Sex ratio (women per 100 men)	101[g]
Life expectancy at birth 2000-2005 (women and men, years)	76/13[g]
Total fertility rate 2000-2005 (births per woman)	1.7[g]
Contraceptive use (% of currently married women)	58[j]
Urban population growth rate 2000-2005 (% per annum)[g]	0.2
Rural population growth rate 2000-2005 (% per annum)[g]	−0.3
Foreign born (%)	9.2
Refugees and others of concern to UNHCR[k]	487998[g]
Television receivers (per 1,000 inhabitants)	282[g]
Parliamentary seats (women and men, % of total)	9/81

Environment	2000-2006
CO2 emissions (000s Mt of carbon dioxide/per capita)[g]	50023/4.8
Energy consumption per capita (kilograms oil equiv.)[g]	1451
Precipitation (mm)	1661[l]
Average minimum and maximum temperatures (centigrade)[l]	10.7/20.5

a Estimate. b Statistical Office of Republic of Montenegro. c 2003. d Currency Dinar. e October 2006. f 2004. g Including Serbia. h March 2006. i Oct. of each year, persons aged 15 years and over. j The data refer to the former Yugoslavia, excluding the Province of Kosovo and Metohija. k Provisional. l Podgorcia.

Morocco

Region	Northern Africa
Largest urban agglom. (pop., 000s)	Casablanca (3138)
Currency	dirham
Population in 2006 (proj., 000s)	31943
Surface area (square kms)	446550
Population density (per square km)	72
United Nations membership date	12 November 1956

Economic indicators	2000	2005
Exchange rate (national currency per US$)[a]	10.62	8.71[b]
Consumer price index (2000=100)	100	110[c]
Industrial production index (1995=100)[d]	115	148[e]
Unemployment (percentage of labour force)[f]	13.6	10.8
Balance of payments, current account (million US$)	−501	922[g]
Tourist arrivals (000s)[h]	4278	5477[g]
GDP (million current US$)	33335	51461
GDP (per capita current US$)	1129	1617
Gross fixed capital formation (% of GDP)	24.0	25.0
Labour force participation, adult female pop. (%)	30.3[i]	27.3[j]
Labour force participation, adult male pop. (%)	79.3[i]	77.4[j]
Employment in industrial sector (%)	20.1[k]	20.2[j]
Employment in agricultural sector (%)	44.4[k]	43.9[j]
Agricultural production index (1999-2001=100)	94	132[g]
Food production index (1999-2001=100)	94	132[g]
Primary energy production (000s Mt oil equiv.)	144	212[g]
Motor vehicles (per 1,000 inhabitants)[l]	55.7	58.6[j]
Telephone lines (per 100 inhabitants)[m]	5.0	4.3
Internet users, estimated (000s)[m]	200.0	4600.0

Total trade		Major trading partners			2005
	(million US$)	(% of exports)		(% of imports)	
Exports	10631.9	France	30	France	18
Imports	20342.2	Spain	18	Spain	11
		UK	6	Saudi Arabia	7

Social indicators	2000-2006
Population growth rate 2000-2005 (% per annum)	1.5
Population aged 0-14 years (%)	31.0
Population aged 60+ years (women and men, % of total)	8.0/6.0
Sex ratio (women per 100 men)	101
Life expectancy at birth 2000-2005 (women and men, years)	72/67
Infant mortality rate 2000-2005 (per 1,000 births)	38
Total fertility rate 2000-2005 (births per woman)	2.8
Contraceptive use (% of currently married women)	63[n]
Urban population (%)	59
Urban population growth rate 2000-2005 (% per annum)	2.7
Rural population growth rate 2000-2005 (% per annum)	−0.2
Refugees and others of concern to UNHCR[o]	583
Government education expenditure (% of GNP)	6.4
Primary-secondary gross enrolment ratio (w and m per 100)	71/81
Third-level students (women and men, % of total)	46/54
Newspaper circulation (per 1,000 inhabitants)	29
Television receivers (per 1,000 inhabitants)	168
Parliamentary seats (women and men, % of total)	6/94

Environment	2000-2006
Threatened species	79
Forested area (% of land area)	7
CO_2 emissions (000s Mt of carbon dioxide/per capita)	37968/1.2
Energy consumption per capita (kilograms oil equiv.)	360
Precipitation (mm)	300[p]
Average minimum and maximum temperatures (centigrade)[p]	14.6/22.0

a Official rate. b October 2006. c June 2006. d Calculated by the Statistics Division of the United Nations from component national indices. e 2nd quarter 2006. f Persons aged 15 years and over. g 2004. h Including nationals residing abroad. i 1999. j 2003. k 2002. l Including special-purpose vehicles. m Including data for Western Sahara. n 2003/04. o Provisional. p Casablanca.

Mozambique

Region	Eastern Africa
Largest urban agglom. (pop., 000s)	Maputo (1320)
Currency	metical
Population in 2006 (proj., 000s)	20158
Surface area (square kms)	801590
Population density (per square km)	25
United Nations membership date	16 September 1975

Economic indicators	2000	2005
Exchange rate (national currency per US$)[ab]	17140.50	26.29[c]
Consumer price index (2000=100)[d]	100	195[e]
Balance of payments, current account (million US$)	−764	−761
Tourist arrivals (000s)	323[f]	470[g]
GDP (million current US$)	3832	6682
GDP (per capita current US$)	214	338
Gross fixed capital formation (% of GDP)	31.0	23.0
Agricultural production index (1999-2001=100)	95	105[g]
Food production index (1999-2001=100)	95	104[g]
Primary energy production (000s Mt oil equiv.)	777	2280[g]
Motor vehicles (per 1,000 inhabitants)	8.8	7.8[g]
Telephone lines (per 100 inhabitants)	0.5	0.4
Internet users, estimated (000s)	20.0	138.0[g]

Total trade		Major trading partners	2005
	(million US$)	(% of exports)	(% of imports)
Exports	1783.0	...	...
Imports	2408.2	...	...
		...	

Social indicators	2000-2006
Population growth rate 2000-2005 (% per annum)	2.0
Population aged 0-14 years (%)	44.0
Population aged 60+ years (women and men, % of total)	6.0/5.0
Sex ratio (women per 100 men)	106
Life expectancy at birth 2000-2005 (women and men, years)	43/41
Infant mortality rate 2000-2005 (per 1,000 births)	101
Total fertility rate 2000-2005 (births per woman)	5.5
Contraceptive use (% of currently married women)	26
Urban population (%)	35
Urban population growth rate 2000-2005 (% per annum)	4.3
Rural population growth rate 2000-2005 (% per annum)	0.9
Foreign born (%)	2.1[h]
Refugees and others of concern to UNHCR[i]	2066
Government education expenditure (% of GNP)	2.5[j]
Primary-secondary gross enrolment ratio (w and m per 100)	57/69
Third-level students (women and men, % of total)	32/68
Television receivers (per 1,000 inhabitants)	21
Parliamentary seats (women and men, % of total)	35/65

Environment	2000-2006
Threatened species	143
Forested area (% of land area)	39
CO2 emissions (000s Mt of carbon dioxide/per capita)	1571/0.1
Energy consumption per capita (kilograms oil equiv.)	80
Precipitation (mm)	814[d]
Average minimum and maximum temperatures (centigrade)[d]	18.6/27.2

a As of 1 July 2006, Mozambique redenominates its currency, taking three zero's off.
1 MZN = 1,000 MZM. b Principal rate. c October 2006. d Maputo. e May 2006. f 2001.
g 2004. h Estimated data. i Provisional. j 1999.

Myanmar

Region	South-eastern Asia
Largest urban agglom. (pop., 000s)	Yangon (4107)
Currency	kyat
Population in 2006 (proj., 000s)	51009
Surface area (square kms)	676578
Population density (per square km)	75
United Nations membership date	19 April 1948

Economic indicators	2000	2005
Exchange rate (national currency per US$)[a]	6.53	5.77[b]
Consumer price index (2000=100)	100	315[c]
Balance of payments, current account (million US$)	−212	112[d]
Tourist arrivals (000s)[e]	208	242[d]
GDP (million current US$)	7275	10938
GDP (per capita current US$)	152	217
Gross fixed capital formation (% of GDP)	12.0	11.0
Employment in industrial sector (%)	12.2[f]	...
Employment in agricultural sector (%)	62.7[f]	...
Agricultural production index (1999-2001=100)	100	115[d]
Food production index (1999-2001=100)	100	115[d]
Primary energy production (000s Mt oil equiv.)	6848	9443[d]
Motor vehicles (per 1,000 inhabitants)	5.5	6.4[d]
Telephone lines (per 100 inhabitants)	0.5	0.8[d]
Internet users, estimated (000s)	...	31.5

Social indicators	2000-2006
Population growth rate 2000-2005 (% per annum)	1.1
Population aged 0-14 years (%)	29.0
Population aged 60+ years (women and men, % of total)	8.0/7.0
Sex ratio (women per 100 men)	102
Life expectancy at birth 2000-2005 (women and men, years)	63/57
Infant mortality rate 2000-2005 (per 1,000 births)	75
Total fertility rate 2000-2005 (births per woman)	2.5
Contraceptive use (% of currently married women)	37
Urban population (%)	31
Urban population growth rate 2000-2005 (% per annum)	2.9
Rural population growth rate 2000-2005 (% per annum)	0.4
Foreign born (%)	0.2[g]
Refugees and others of concern to UNHCR[h]	5969
Government education expenditure (% of GNP)	1.3
Primary-secondary gross enrolment ratio (w and m per 100)	66/66
Third-level students (women and men, % of total)	63/37
Television receivers (per 1,000 inhabitants)	7

Environment	2000-2006
Threatened species	177
Forested area (% of land area)	52
CO2 emissions (000s Mt of carbon dioxide/per capita)	9467/0.2
Energy consumption per capita (kilograms oil equiv.)	74
Precipitation (mm)	2681[i]
Average minimum and maximum temperatures (centigrade)[i]	22.6/32.3

a Official rate. b October 2006. c February 2006. d 2004. e Including tourist arrivals through border entry points to Yangon. f 1998. g Estimated data. h Provisional. i Yangon.

Namibia

Region	Southern Africa
Largest urban agglom. (pop., 000s)	Windhoek (289)
Currency	dollar
Population in 2006 (proj., 000s)	2052
Surface area (square kms)	824292
Population density (per square km)	2
United Nations membership date	23 April 1990

Economic indicators	2000	2005
Exchange rate (national currency per US$)[a]	7.57	7.45[b]
Consumer price index (2000=100)[c]	...	118[d]
Unemployment (percentage of labour force)[e]	33.8	...
Balance of payments, current account (million US$)	293	573[f]
Tourist arrivals (000s)	656	695[g]
GDP (million current US$)	3414	6130
GDP (per capita current US$)	1802	3018
Gross fixed capital formation (% of GDP)	19.0	26.0
Labour force participation, adult female pop. (%)	47.5[h]	49.3[i]
Labour force participation, adult male pop. (%)	62.1	60.1[i]
Employment in industrial sector (%)	12.2	...
Employment in agricultural sector (%)	31.1	...
Agricultural production index (1999-2001=100)	106	114[f]
Food production index (1999-2001=100)	106	114[f]
Telephone lines (per 100 inhabitants)	6.2	6.4
Internet users, estimated (000s)[j]	30.0	75.0[f]

Total trade		Major trading partners			2005
	(million US$)[g]		(% of exports)[g]		(% of imports)[g]
Exports	1303.7	South Africa	31	South Africa	80
Imports	1427.9	Angola	25	Germany	2
		Spain	13	Spain	1

Social indicators	2000-2006
Population growth rate 2000-2005 (% per annum)	1.4
Population aged 0-14 years (%)	42.0
Population aged 60+ years (women and men, % of total)	6.0/5.0
Sex ratio (women per 100 men)	102
Life expectancy at birth 2000-2005 (women and men, years)	49/48
Infant mortality rate 2000-2005 (per 1,000 births)	44
Total fertility rate 2000-2005 (births per woman)	4.0
Contraceptive use (% of currently married women)	44
Urban population (%)	35
Urban population growth rate 2000-2005 (% per annum)	3.0
Rural population growth rate 2000-2005 (% per annum)	0.6
Refugees and others of concern to UNHCR[k]	236587
Government education expenditure (% of GNP)	7.1
Primary-secondary gross enrolment ratio (w and m per 100)	87/83
Third-level students (women and men, % of total)	53/47
Television receivers (per 1,000 inhabitants)[l]	81
Parliamentary seats (women and men, % of total)	27/73

Environment	2000-2006
Threatened species	81
Forested area (% of land area)	10
CO2 emissions (000s Mt of carbon dioxide/per capita)	2331/1.2

a Official rate. b October 2006. c Base: 2002=100. d May 2006. e Persons aged 15 to 69 years. f 2004. g 2003. h 1999. i 2001. j From 1993, data refer to year ending 30 September. k Provisional. l Year ending 30 September.

Nauru

Region	Oceania-Micronesia
Largest urban agglom. (pop., 000s)	Nauru (...)
Currency	Australian dollar
Population in 2006 (proj., 000s)	10
Surface area (square kms)	21
Population density (per square km)	479
United Nations membership date	14 September 1999

Economic indicators	2000	2005
Exchange rate (national currency per US$)[a]	1.91	1.33[b]
GDP (million current US$)	33	55
GDP (per capita current US$)	2702	4068
Gross fixed capital formation (% of GDP)	43.0	43.0
Agricultural production index (1999-2001=100)	100	100[c]
Food production index (1999-2001=100)	100	100[c]
Telephone lines (per 100 inhabitants)	...	16.0[d]
Internet users, estimated (000s)	...	0.3[d]

Social indicators	2000-2006
Population growth rate 2000-2005 (% per annum)	2.2
Population aged 0-14 years (%)	42.0
Life expectancy at birth 2000-2005 (women and men, years)	58/57[ef]
Total fertility rate 2000-2005 (births per woman)	4.0[eg]
Urban population (%)	100
Urban population growth rate 2000-2005 (% per annum)	2.2
Rural population growth rate 2000-2005 (% per annum)	−
Foreign born (%)	37.4[h]
Primary-secondary gross enrolment ratio (w and m per 100)	68/67
Television receivers (per 1,000 inhabitants)	44[i]
Parliamentary seats (women and men, % of total)	1/99

Environment	2000-2006
Threatened species	7
CO2 emissions (000s Mt of carbon dioxide/per capita)	141/10.8
Energy consumption per capita (kilograms oil equiv.)	3590[h]
Precipitation (mm)	2236

a Data refer to non-commercial rates derived from the Operational Rates of Exchange for United Nations Programmes. b October 2006. c 2003. d 2001. e Published by the Secretariat of the Pacific Community. f 2002. g 1997-2002. h Estimated data. i 1998.

Nepal

Region	South-central Asia
Largest urban agglom. (pop., 000s)	Kathmandu (815)
Currency	rupee
Population in 2006 (proj., 000s)	27678
Surface area (square kms)	147181
Population density (per square km)	188
United Nations membership date	14 December 1955

Economic indicators	2000	2005
Exchange rate (national currency per US$)[a]	74.30	72.60[b]
Consumer price index (2000=100)	100	115[c]
Balance of payments, current account (million US$)	−299	1
Tourist arrivals (000s)[d]	464	385[c]
GDP (million current US$)	5338	7412
GDP (per capita current US$)	218	273
Gross fixed capital formation (% of GDP)	19.0	19.0
Labour force participation, adult female pop. (%)	81.9	81.9[e]
Labour force participation, adult male pop. (%)	90.2[f]	78.7[g]
Employment in industrial sector (%)	5.5[h]	...
Employment in agricultural sector (%)	78.5[h]	...
Agricultural production index (1999-2001=100)	100	111[c]
Food production index (1999-2001=100)	100	111[c]
Primary energy production (000s Mt oil equiv.)	152	209[c]
Motor vehicles (per 1,000 inhabitants)	4.1	5.3[i]
Telephone lines (per 100 inhabitants)	1.2	1.7
Internet users, estimated (000s)	50.0	225.0

Total trade		Major trading partners			2005
	(million US$)[j]	(% of exports)[j]		(% of imports)[j]	
Exports	652.7	India	52	India	53
Imports	1801.6	USA	29	China	8
		China	3	Singapore	6

Social indicators	2000-2006
Population growth rate 2000-2005 (% per annum)	2.1
Population aged 0-14 years (%)	39.0
Population aged 60+ years (women and men, % of total)	6.0/5.0
Sex ratio (women per 100 men)	102
Life expectancy at birth 2000-2005 (women and men, years)	62/61
Infant mortality rate 2000-2005 (per 1,000 births)	64
Total fertility rate 2000-2005 (births per woman)	3.7
Contraceptive use (% of currently married women)	39
Urban population (%)	16
Urban population growth rate 2000-2005 (% per annum)	5.3
Rural population growth rate 2000-2005 (% per annum)	1.6
Foreign born (%)	2.7
Refugees and others of concern to UNHCR[k]	6433
Government education expenditure (% of GNP)	3.4
Primary-secondary gross enrolment ratio (w and m per 100)	71/80
Third-level students (women and men, % of total)	28/72
Television receivers (per 1,000 inhabitants)	11

Environment	2000-2006
Threatened species	85
Forested area (% of land area)	27
CO2 emissions (000s Mt of carbon dioxide/per capita)	2955/0.1
Energy consumption per capita (kilograms oil equiv.)	42
Precipitation (mm)	1425[l]
Average minimum and maximum temperatures (centigrade)[l]	11.7/24.8

a Official rate. b October 2006. c 2004. d Including arrivals from India. e 1998. f 1999. g 2001. h 1995. i 2002. j 2003. k Provisional. l Kathmandu.

Netherlands

Region	Western Europe
Largest urban agglom. (pop., 000s)	Amsterdam (1147)
Currency	euro[a]
Population in 2006 (proj., 000s)	16367
Surface area (square kms)	41528
Population density (per square km)	394
United Nations membership date	10 December 1945

Economic indicators	2000	2005
Exchange rate (national currency per US$)	1.07	0.79[b]
Consumer price index (2000=100)	100	115[c]
Industrial production index (1995=100)[d]	110	96[c]
Unemployment (percentage of labour force)[e]	2.9	5.2
Balance of payments, current account (million US$)	7264	41388
Tourist arrivals (000s)	10003	9646[f]
GDP (million current US$)	386510	624187
GDP (per capita current US$)	24313	38296
Gross fixed capital formation (% of GDP)	22.0	19.0
Labour force participation, adult female pop. (%)	53.9	55.9[g]
Labour force participation, adult male pop. (%)	73.0	72.8[g]
Employment in industrial sector (%)	20.8	20.0
Employment in agricultural sector (%)	3.1	3.0
Agricultural production index (1999-2001=100)	101	95[f]
Food production index (1999-2001=100)	101	95[f]
Primary energy production (000s Mt oil equiv.)	60516	71881[f]
Motor vehicles (per 1,000 inhabitants)	454.6	491.7[f]
Telephone lines (per 100 inhabitants)	61.9	46.6
Internet users, estimated (000s)	7000.0	12060.0

Total trade		Major trading partners			2005
	(million US$)	(% of exports)		(% of imports)	
Exports	320065.0	Germany	23	Germany	18
Imports	283172.0	Belgium	11	Belgium	10
		France	9	USA	8

Social indicators	2000-2006
Population growth rate 2000-2005 (% per annum)	0.5
Population aged 0-14 years (%)	18.0
Population aged 60+ years (women and men, % of total)	21.0/17.0
Sex ratio (women per 100 men)	101
Life expectancy at birth 2000-2005 (women and men, years)	81/76
Infant mortality rate 2000-2005 (per 1,000 births)	5
Total fertility rate 2000-2005 (births per woman)	1.7
Contraceptive use (% of currently married women)	79[hi]
Urban population (%)	80
Urban population growth rate 2000-2005 (% per annum)	1.4
Rural population growth rate 2000-2005 (% per annum)	−2.7
Foreign born (%)	10.4
Refugees and others of concern to UNHCR[j]	538636
Government education expenditure (% of GNP)	5.5
Primary-secondary gross enrolment ratio (w and m per 100)	112/114
Third-level students (women and men, % of total)	51/49
Newspaper circulation (per 1,000 inhabitants)	279
Television receivers (per 1,000 inhabitants)	761
Intentional homicides (per 100,000 inhabitants)	1
Parliamentary seats (women and men, % of total)	34/66

Environment	2000-2006
Threatened species	39
Forested area (% of land area)	11
CO2 emissions (000s Mt of carbon dioxide/per capita)[k]	176860/11.0
Energy consumption per capita (kilograms oil equiv.)	5586
Precipitation (mm)	780[l]
Average minimum and maximum temperatures (centigrade)[l]	6.1/13.4

a Prior to 1 January 1999, guilder. b September 2006. c August 2006. d Monthly indices are adjusted for differences in the number of working days. e Persons aged 15 to 64 years. f 2004. g 2003. h For women aged 18-42 in union or married. i 1993. j Provisional. k Source: UNFCCC. l Amsterdam (Schiphol).

Netherlands Antilles

Region	Caribbean	
Largest urban agglom. (pop., 000s)	Willemstad (125)	
Currency	guilder	
Population in 2006 (proj., 000s)	184	
Surface area (square kms)	800	
Population density (per square km)	230	

Economic indicators	2000	2005
Exchange rate (national currency per US$)[a]	1.79	1.79[b]
Consumer price index (2000=100)[c]	100	113[d]
Unemployment (percentage of labour force)[e]	14.2[f]	15.1[g]
Balance of payments, current account (million US$)	−51	−92[h]
Tourist arrivals (000s)	672[i]	...
GDP (million current US$)	2798	3204
GDP (per capita current US$)	15931	17540
Gross fixed capital formation (% of GDP)	28.0	23.0
Labour force participation, adult female pop. (%)	53.1	...
Labour force participation, adult male pop. (%)	66.6	...
Employment in industrial sector (%)	18.0	...
Employment in agricultural sector (%)	1.1	...
Agricultural production index (1999-2001=100)	98	88[j]
Food production index (1999-2001=100)	98	88[j]
Telephone lines (per 100 inhabitants)	37.2	37.2[jk]
Internet users, estimated (000s)	2.0[l]	...

Total trade		Major trading partners			2005
	(million US$)[i]	(% of exports)[i]		(% of imports)[i]	
Exports	1699.2	USA	22	Venezuela	60
Imports	2268.5	Venezuela	10	USA	13
		Bahamas	8	Netherlands	8

Social indicators	2000-2006
Population growth rate 2000-2005 (% per annum)	0.8
Population aged 0-14 years (%)	23.0
Population aged 60+ years (women and men, % of total)	15.0/13.0
Sex ratio (women per 100 men)	112
Life expectancy at birth 2000-2005 (women and men, years)	79/73
Infant mortality rate 2000-2005 (per 1,000 births)	13
Total fertility rate 2000-2005 (births per woman)	2.1
Urban population (%)	70
Urban population growth rate 2000-2005 (% per annum)	1.1
Rural population growth rate 2000-2005 (% per annum)	0.1
Foreign born (%)	26.4
Primary-secondary gross enrolment ratio (w and m per 100)	108/105
Third-level students (women and men, % of total)	60/40
Television receivers (per 1,000 inhabitants)	334

Environment	2000-2006
Threatened species	27
Forested area (% of land area)	–
CO2 emissions (000s Mt of carbon dioxide/per capita)	4059/22.7
Energy consumption per capita (kilograms oil equiv.)	10320
Precipitation (mm)	552[c]

a Official rate. b October 2006. c Curaçao. d July 2006. e Persons aged 15 years and over, Curaçao. f October. g 2003, Oct. h 2004. i 2002. j 2003. k Estimate. l 1999.

New Caledonia

Region	Oceania-Melanesia
Largest urban agglom. (pop., 000s)	Noumea (149)
Currency	CFP franc
Population in 2006 (proj., 000s)	241
Surface area (square kms)	18575
Population density (per square km)	13

Economic indicators	2000	2005
Exchange rate (national currency per US$)[c]	121.12[ab]	93.79[d]
Consumer price index (2000=100)[e]	100	111[f]
Unemployment (percentage of labour force)	18.6[g]	...
Tourist arrivals (000s)[h]	110	100[i]
GDP (million current US$)	3003	4341
GDP (per capita current US$)	13949	18328
Gross fixed capital formation (% of GDP)	24.0	24.0
Labour force participation, adult female pop. (%)	42.1[g]	...
Labour force participation, adult male pop. (%)	61.3[g]	...
Employment in industrial sector (%)	23.4[g]	...
Employment in agricultural sector (%)	7.2[g]	...
Agricultural production index (1999-2001=100)	99	103[i]
Food production index (1999-2001=100)	99	103[i]
Primary energy production (000s Mt oil equiv.)	39	29[i]
Motor vehicles (per 1,000 inhabitants)	404.1	451.8
Telephone lines (per 100 inhabitants)	23.8	23.3
Internet users, estimated (000s)	30.0	76.0

Total trade		Major trading partners			2005
	(million US$)	(% of exports)		(% of imports)	
Exports	1113.9	Japan	18	France	32
Imports	1773.6	France	16	Singapore	15
		Korea Rep.	13	Australia	9

Social indicators	2000-2006
Population growth rate 2000-2005 (% per annum)	1.9
Population aged 0-14 years (%)	28.0
Population aged 60+ years (women and men, % of total)	10.0/8.8
Sex ratio (women per 100 men)	95
Life expectancy at birth 2000-2005 (women and men, years)	78/73
Infant mortality rate 2000-2005 (per 1,000 births)	7
Total fertility rate 2000-2005 (births per woman)	2.4
Urban population (%)	64
Urban population growth rate 2000-2005 (% per annum)	2.5
Rural population growth rate 2000-2005 (% per annum)	1.0
Foreign born (%)	21.9[j]
Government education expenditure (% of GNP)	12.3[k]
Third-level students (women and men, % of total)	44/56[l]
Television receivers (per 1,000 inhabitants)	496

Environment	2000-2006
Threatened species	266
Forested area (% of land area)	20
CO2 emissions (000s Mt of carbon dioxide/per capita)	1872/8.2
Energy consumption per capita (kilograms oil equiv.)	3500
Precipitation (mm)	1072[e]
Average minimum and maximum temperatures (centigrade)[e]	20.2/26.0

a 2002. b September 2002. c Data refer to non-commercial rates derived from the Operational Rates of Exchange for United Nations Programmes. d November 2006. e Nouméa. f August 2006. g 1996. h Including nationals residing abroad. i 2004. j 1989. k 1985. l 1995.

New Zealand

Region	Oceania
Largest urban agglom. (pop., 000s)	Auckland (1148)
Currency	dollar
Population in 2006 (proj., 000s)	4063
Surface area (square kms)	270534
Population density (per square km)	15
United Nations membership date	24 October 1945

Economic indicators	2000	2005
Exchange rate (national currency per US$)	2.27	1.50[a]
Consumer price index (2000=100)	100	117[b]
Industrial production index (1995=100)	105	114[b]
Unemployment (percentage of labour force)[c]	6.0	3.7
Balance of payments, current account (million US$)	−2661	−6456[d]
Tourist arrivals (000s)	1787	2334[d]
GDP (million current US$)	52673	109607
GDP (per capita current US$)	13795	27209
Gross fixed capital formation (% of GDP)	20.0	24.0
Labour force participation, adult female pop. (%)	57.6	59.9[d]
Labour force participation, adult male pop. (%)	73.7	74.5[d]
Employment in industrial sector (%)	23.2	22.0
Employment in agricultural sector (%)	8.7	7.1
Agricultural production index (1999-2001=100)	101	115[d]
Food production index (1999-2001=100)	101	116[d]
Primary energy production (000s Mt oil equiv.)	11711	9978[d]
Motor vehicles (per 1,000 inhabitants)[e]	613.8	619.0
Telephone lines (per 100 inhabitants)[f]	47.5	45.1
Internet users, estimated (000s)[f]	1515.0	2754.0

Total trade	(million US$)	Major trading partners				2005
		(% of exports)		(% of imports)		
Exports	21730.1	Australia	21	Australia		21
Imports	26219.4	USA	14	USA		11
		Japan	11	Japan		11

Social indicators	2000-2006
Population growth rate 2000-2005 (% per annum)	1.1
Population aged 0-14 years (%)	21.0
Population aged 60+ years (women and men, % of total)	18.0/16.0
Sex ratio (women per 100 men)	103
Life expectancy at birth 2000-2005 (women and men, years)	81/77
Infant mortality rate 2000-2005 (per 1,000 births)	5
Total fertility rate 2000-2005 (births per woman)	2.0
Urban population (%)	86
Urban population growth rate 2000-2005 (% per annum)	1.2
Rural population growth rate 2000-2005 (% per annum)	0.3
Foreign born (%)	19.5
Refugees and others of concern to UNHCR[g]	139353
Government education expenditure (% of GNP)	7.3
Primary-secondary gross enrolment ratio (w and m per 100)	113/108
Third-level students (women and men, % of total)	58/42
Newspaper circulation (per 1,000 inhabitants)	200
Television receivers (per 1,000 inhabitants)[h]	586
Intentional homicides (per 100,000 inhabitants)	1
Parliamentary seats (women and men, % of total)	32/68

Environment	2000-2006
Threatened species	155
Forested area (% of land area)	30
CO2 emissions (000s Mt of carbon dioxide/per capita)[i]	34700/8.8
Energy consumption per capita (kilograms oil equiv.)	3448
Precipitation (mm)	1137[j]
Average minimum and maximum temperatures (centigrade)[j]	11.6/18.7

a October 2006. b 2nd quarter 2006. c Persons aged 15 years and over. d Data refer to fiscal years ending 31 March. e Data refer to fiscal years ending 31 March. f Year ending 30 June. Prior to 2000, year beginning 1 April. g Provisional. h Year beginning 1 April. i Source: UNFCCC. j Auckland.

Nicaragua

Region	Central America
Largest urban agglom. (pop., 000s)	Managua (1165)
Currency	gold córdoba
Population in 2006 (proj., 000s)	5600
Surface area (square kms)	130000
Population density (per square km)	43
United Nations membership date	24 October 1945

Economic indicators	2000	2005
Exchange rate (national currency per US$)[a]	13.06	17.71[b]
Consumer price index (2000=100)[c]	...	162[d]
Unemployment (percentage of labour force)[e]	9.8	12.2[f]
Balance of payments, current account (million US$)	−839	−800
Tourist arrivals (000s)	486	615[g]
GDP (million current US$)	3907	4910
GDP (per capita current US$)	788	895
Gross fixed capital formation (% of GDP)	26.0	24.0
Labour force participation, adult female pop. (%)	44.5	47.2[h]
Labour force participation, adult male pop. (%)	80.2	70.7[h]
Employment in industrial sector (%)	14.7	18.0[i]
Employment in agricultural sector (%)	43.5	30.5[i]
Agricultural production index (1999-2001=100)	104	120[g]
Food production index (1999-2001=100)	104	123[g]
Primary energy production (000s Mt oil equiv.)	30	50[g]
Motor vehicles (per 1,000 inhabitants)	34.5	37.5[h]
Telephone lines (per 100 inhabitants)	3.2	3.8[g]
Internet users, estimated (000s)	50.0	140.0

Total trade		Major trading partners			2005
	(million US$)	(% of exports)		(% of imports)	
Exports	865.9	USA	35	USA	21
Imports	2520.1	El Salvador	14	Costa Rica	9
		Honduras	8	Mexico	9

Social indicators	2000-2006
Population growth rate 2000-2005 (% per annum)	2.0
Population aged 0-14 years (%)	39.0
Population aged 60+ years (women and men, % of total)	5.0/4.0
Sex ratio (women per 100 men)	100
Life expectancy at birth 2000-2005 (women and men, years)	72/67
Infant mortality rate 2000-2005 (per 1,000 births)	30
Total fertility rate 2000-2005 (births per woman)	3.3
Contraceptive use (% of currently married women)	69
Urban population (%)	59
Urban population growth rate 2000-2005 (% per annum)	2.7
Rural population growth rate 2000-2005 (% per annum)	1.1
Foreign born (%)	0.5[j]
Refugees and others of concern to UNHCR[k]	5703
Government education expenditure (% of GNP)	3.2
Primary-secondary gross enrolment ratio (w and m per 100)	92/90
Third-level students (women and men, % of total)	52/48
Television receivers (per 1,000 inhabitants)	121
Intentional homicides (per 100,000 inhabitants)	7
Parliamentary seats (women and men, % of total)	21/79

Environment	2000-2006
Threatened species	95
Forested area (% of land area)	27
CO2 emissions (000s Mt of carbon dioxide/per capita)	3917/0.7
Energy consumption per capita (kilograms oil equiv.)	253
Precipitation (mm)	1989[l]
Average minimum and maximum temperatures (centigrade)[m]	22.3/33.8

a Principal rate. b August 2006. c Base: 1999=100. d July 2006. e Persons aged 10 years and over. f 2002. g 2004. h 2001. i 2003. j Estimated data. k Provisional. l Chinandega. m Juigalpa.

Niger

Region	Western Africa
Largest urban agglom. (pop., 000s)	Niamey (850)
Currency	CFA franc
Population in 2006 (proj., 000s)	14426
Surface area (square kms)	1267000
Population density (per square km)	11
United Nations membership date	20 September 1960

Economic indicators	2000	2005
Exchange rate (national currency per US$)[a]	704.95	516.66[b]
Consumer price index (2000=100)[c]	100	116[de]
Balance of payments, current account (million US$)	−104	−219[f]
Tourist arrivals (000s)[g]	50	55[f]
GDP (million current US$)	1666	3245
GDP (per capita current US$)	141	232
Gross fixed capital formation (% of GDP)	13.0	17.0
Agricultural production index (1999-2001=100)	92	117[h]
Food production index (1999-2001=100)	92	118[h]
Primary energy production (000s Mt oil equiv.)	123	125[h]
Motor vehicles (per 1,000 inhabitants)	<	1.0[h]
Telephone lines (per 100 inhabitants)	0.2	0.2
Internet users, estimated (000s)	4.0	29.0

Total trade		Major trading partners			2005
	(million US$)[f]		(% of exports)[f]		(% of imports)[f]
Exports	209.1	France	36	France	15
Imports	558.4	Nigeria	27	Côte d'Ivoire	13
		Japan	15	China	9

Social indicators	2000-2006
Population growth rate 2000-2005 (% per annum)	3.4
Population aged 0-14 years (%)	49.0
Population aged 60+ years (women and men, % of total)	4.0/3.0
Sex ratio (women per 100 men)	96
Life expectancy at birth 2000-2005 (women and men, years)	44/44
Infant mortality rate 2000-2005 (per 1,000 births)	153
Total fertility rate 2000-2005 (births per woman)	7.9
Contraceptive use (% of currently married women)	14
Urban population (%)	17
Urban population growth rate 2000-2005 (% per annum)	4.1
Rural population growth rate 2000-2005 (% per annum)	3.2
Foreign born (%)	1.1[i]
Refugees and others of concern to UNHCR[j]	272
Government education expenditure (% of GNP)	2.3
Primary-secondary gross enrolment ratio (w and m per 100)	22/31
Third-level students (women and men, % of total)	27/73
Television receivers (per 1,000 inhabitants)	12
Parliamentary seats (women and men, % of total)	12/88

Environment	2000-2006
Threatened species	21
Forested area (% of land area)	1
CO2 emissions (000s Mt of carbon dioxide/per capita)	1209/0.1
Energy consumption per capita (kilograms oil equiv.)	32[i]
Precipitation (mm)	541[e]
Average minimum and maximum temperatures (centigrade)[e]	22.4/36.2

a Official rate. b October 2006. c Excluding rent. d June 2006. e Niamey. f 2003.
g Arrivals in hotels only. h 2004. i Estimated data. j Provisional.

Nigeria

Region	Western Africa
Largest urban agglom. (pop., 000s)	Lagos (10886)
Currency	naira
Population in 2006 (proj., 000s)	134375
Surface area (square kms)	923768
Population density (per square km)	145
United Nations membership date	7 October 1960

Economic indicators	2000	2005
Exchange rate (national currency per US$)[a]	109.55	127.04[b]
Consumer price index (2000=100)[c]	100	153[d]
Balance of payments, current account (million US$)	7429	12264[e]
Tourist arrivals (000s)	813	962[e]
GDP (million current US$)	67359	113461
GDP (per capita current US$)	573	863
Gross fixed capital formation (% of GDP)	5.0	6.0
Employment in industrial sector (%)	22.0[f]	...
Employment in agricultural sector (%)	2.9[f]	...
Agricultural production index (1999-2001=100)	101	106[e]
Food production index (1999-2001=100)	101	106[e]
Primary energy production (000s Mt oil equiv.)	124335	150196[e]
Motor vehicles (per 1,000 inhabitants)	0.6[g]	...
Telephone lines (per 100 inhabitants)	0.5	0.9
Internet users, estimated (000s)	80.0	5000.0

Total trade		Major trading partners			2005
	(million US$)[d]	(% of exports)[d]		(% of imports)[d]	
Exports	24078.3	USA	38	USA	16
Imports	14892.5	India	10	UK	10
		Brazil	7	Germany	7

Social indicators	2000-2006
Population growth rate 2000-2005 (% per annum)	2.2
Population aged 0-14 years (%)	44.0
Population aged 60+ years (women and men, % of total)	5.0/4.0
Sex ratio (women per 100 men)	97
Life expectancy at birth 2000-2005 (women and men, years)	44/43
Infant mortality rate 2000-2005 (per 1,000 births)	114
Total fertility rate 2000-2005 (births per woman)	5.9
Contraceptive use (% of currently married women)	13
Urban population (%)	48
Urban population growth rate 2000-2005 (% per annum)	4.1
Rural population growth rate 2000-2005 (% per annum)	0.7
Foreign born (%)	0.7[h]
Refugees and others of concern to UNHCR[i]	349
Primary-secondary gross enrolment ratio (w and m per 100)	63/75
Third-level students (women and men, % of total)	35/65
Television receivers (per 1,000 inhabitants)	68
Parliamentary seats (women and men, % of total)	6/94

Environment	2000-2006
Threatened species	253
Forested area (% of land area)	15
CO2 emissions (000s Mt of carbon dioxide/per capita)	52276/0.4
Energy consumption per capita (kilograms oil equiv.)	155
Precipitation (mm)	1538[j]
Average minimum and maximum temperatures (centigrade)[j]	22.7/30.7

a Principal rate. b October 2006. c Rural and urban areas. d 2003. e 2004. f 1995.
g 1997. h Estimated data. i Provisional. j Lagos.

Niue

		2000	2005
Region	Oceania-Polynesia		
Largest urban agglom. (pop., 000s)	Alofi (1)		
Currency	New Zealand dollar		
Population in 2006 (proj., 000s)	1[a]		
Surface area (square kms)	260		
Population density (per square km)	7		

Economic indicators	2000	2005
Exchange rate (national currency per US$)	2.27	1.50[b]
Consumer price index (2000=100)	100	124[c]
Tourist arrivals (000s)[d]	2	3[e]
Agricultural production index (1999-2001=100)	100	100[f]
Food production index (1999-2001=100)	100	100[f]
Internet users, estimated (000s)	0.5	0.9

Social indicators	2000-2006
Population growth rate 2000-2005 (% per annum)	−2.2
Sex ratio (women per 100 men)	99[gh]
Life expectancy at birth 2000-2005 (women and men, years)	71/29[hi]
Total fertility rate 2000-2005 (births per woman)	3.0[ij]
Urban population (%)	37
Urban population growth rate 2000-2005 (% per annum)	−0.4
Rural population growth rate 2000-2005 (% per annum)	−3.1
Foreign born (%)	6.9[k]
Primary-secondary gross enrolment ratio (w and m per 100)	95/89
Television receivers (per 1,000 inhabitants)	277

Environment	2000-2006
Threatened species	13
CO2 emissions (000s Mt of carbon dioxide/per capita)	3/2.0
Energy consumption per capita (kilograms oil equiv.)	507[k]

a 2005. b October 2006. c 1st quarter 2006. d Arrivals by air; including Niueans residing usually in New Zealand. e 2004. f 2003. g Refers to de facto population count. h 2001. i Published by the Secretariat of the Pacific Community. j 1997-2001. k Estimated data.

Northern Mariana Islands

Region	Oceania-Micronesia
Largest urban agglom. (pop., 000s)	Saipan (75)
Currency	US dollar
Population in 2006 (proj., 000s)	81[a]
Surface area (square kms)	464
Population density (per square km)	149

Economic indicators	2000	2005[c]
Consumer price index (2000=100)[b]	100	99[c]
Tourist arrivals (000s)[d]	517	525[c]
Telephone lines (per 100 inhabitants)	30.9	30.9[ef]

Social indicators	2000-2006
Population growth rate 2000-2005 (% per annum)	2.9
Sex ratio (women per 100 men)	116[gh]
Life expectancy at birth 2000-2005 (women and men, years)	78/5[hi]
Urban population (%)	94
Urban population growth rate 2000-2005 (% per annum)	3.2
Rural population growth rate 2000-2005 (% per annum)	−1.0

Environment	2000-2006
Threatened species	32

a 2005. b Saipan. c 2004. d Arrivals by air. e Estimate. f 2002. g Refers to de facto population count. h 2000. i Published by the Secretariat of the Pacific Community.

146

Norway

Region	Northern Europe
Largest urban agglom. (pop., 000s)	Oslo (802)
Currency	krone
Population in 2006 (proj., 000s)	4643[a]
Surface area (square kms)	385155
Population density (per square km)	12
United Nations membership date	27 November 1945

Economic indicators	2000	2005
Exchange rate (national currency per US$)[b]	8.85	6.56[c]
Consumer price index (2000=100)	100	111[d]
Industrial production index (1995=100)[e]	111	103[f]
Unemployment (percentage of labour force)[g]	3.4	4.6
Balance of payments, current account (million US$)	25851	49488
Tourist arrivals (000s)	3104	3600[h]
GDP (million current US$)	166905	295513
GDP (per capita current US$)	37072	63960
Gross fixed capital formation (% of GDP)	19.0	19.0
Labour force participation, adult female pop. (%)	68.9	68.8[h]
Labour force participation, adult male pop. (%)	78.0	76.4[h]
Employment in industrial sector (%)	21.7	20.8
Employment in agricultural sector (%)	4.1	3.3
Agricultural production index (1999-2001=100)	99	100[h]
Food production index (1999-2001=100)	99	100[h]
Primary energy production (000s Mt oil equiv.)[a]	220819	234210[h]
Motor vehicles (per 1,000 inhabitants)	511.5	534.6[h]
Telephone lines (per 100 inhabitants)	53.3	46.1
Internet users, estimated (000s)	1200.0	2702.0

Total trade		Major trading partners			2005
	(million US$)[a]	(% of exports)[a]		(% of imports)[a]	
Exports	103759.2	UK	24	Sweden	14
Imports	55488.2	Germany	12	Germany	13
		Netherlands	10	Denmark	7

Social indicators	2000-2006
Population growth rate 2000-2005 (% per annum)	0.5[a]
Population aged 0-14 years (%)	20.0
Population aged 60+ years (women and men, % of total)	22.0/18.0
Sex ratio (women per 100 men)	101[a]
Life expectancy at birth 2000-2005 (women and men, years)	82/77[a]
Infant mortality rate 2000-2005 (per 1,000 births)	4
Total fertility rate 2000-2005 (births per woman)	1.8[a]
Contraceptive use (% of currently married women)	74[ijk]
Urban population (%)[a]	77
Urban population growth rate 2000-2005 (% per annum)	0.9
Rural population growth rate 2000-2005 (% per annum)	−0.6
Foreign born (%)	6.9
Refugees and others of concern to UNHCR[l]	19840
Government education expenditure (% of GNP)	7.6
Primary-secondary gross enrolment ratio (w and m per 100)	107/106
Third-level students (women and men, % of total)	60/40
Newspaper circulation (per 1,000 inhabitants)	565
Television receivers (per 1,000 inhabitants)	1554
Intentional homicides (per 100,000 inhabitants)	1
Parliamentary seats (women and men, % of total)	38/62

Environment	2000-2006
Threatened species	37
Forested area (% of land area)	29
CO2 emissions (000s Mt of carbon dioxide/per capita)[m]	43220/9.4
Energy consumption per capita (kilograms oil equiv.)[a]	7406
Precipitation (mm)	763[n]
Average minimum and maximum temperatures (centigrade)[n]	2.4/9.6

a Including Svalbard and Jan Mayen Islands. b Official rate. c October 2006. d July 2006. e Monthly indices are adjusted for differences in the number of working days. f August 2006. g Persons aged 16 to 74 years. h 2004. i Cohorts. j From a sample of women born in 1945, 1950, 1955, 1960, 1965 or 1968. k 1988/89. l Provisional. m Source: UNFCCC. n Oslo.

Oman

Region	Western Asia
Largest urban agglom. (pop., 000s)	Muscat (565)
Currency	rial Omani
Population in 2006 (proj., 000s)	2612
Surface area (square kms)	309500
Population density (per square km)	8
United Nations membership date	7 October 1971

Economic indicators	2000	2005
Exchange rate (national currency per US$)[a]	0.38	0.38[b]
Consumer price index (2000=100)[c]	100	107[d]
Balance of payments, current account (million US$)	3129	4717
Tourist arrivals (000s)[e]	571	630[f]
GDP (million current US$)	19868	30269
GDP (per capita current US$)	8136	11792
Gross fixed capital formation (% of GDP)	12.0	18.0
Labour force participation, adult female pop. (%)	13.2	25.1[f]
Labour force participation, adult male pop. (%)	62.1	78.3[f]
Employment in industrial sector (%)	11.2	...
Employment in agricultural sector (%)	6.4	...
Agricultural production index (1999-2001=100)	98	92[g]
Food production index (1999-2001=100)	97	92[g]
Primary energy production (000s Mt oil equiv.)	56633	56115[g]
Motor vehicles (per 1,000 inhabitants)[hi]	191.9	172.5[f]
Telephone lines (per 100 inhabitants)	9.1	10.3
Internet users, estimated (000s)	90.0	285.0

Total trade		Major trading partners		2005
	(million US$)	(% of exports)		(% of imports)
Exports	20326.6	...	Untd Arab Em	26
Imports	8827.4	...	Japan	16
		...	Germany	7

Social indicators	2000-2006
Population growth rate 2000-2005 (% per annum)	1.0
Population aged 0-14 years (%)	34.0
Population aged 60+ years (women and men, % of total)	5.0/4.0
Sex ratio (women per 100 men)	79
Life expectancy at birth 2000-2005 (women and men, years)	76/73
Infant mortality rate 2000-2005 (per 1,000 births)	16
Total fertility rate 2000-2005 (births per woman)	3.8
Urban population (%)	71
Urban population growth rate 2000-2005 (% per annum)	1.0
Rural population growth rate 2000-2005 (% per annum)	1.1
Foreign born (%)	23.9
Refugees and others of concern to UNHCR[j]	11
Government education expenditure (% of GNP)	4.8
Primary-secondary gross enrolment ratio (w and m per 100)	86/88
Third-level students (women and men, % of total)	56/44
Television receivers (per 1,000 inhabitants)	620
Parliamentary seats (women and men, % of total)	8/92

Environment	2000-2006
Threatened species	62
CO2 emissions (000s Mt of carbon dioxide/per capita)	32309/12.9
Energy consumption per capita (kilograms oil equiv.)	4784
Precipitation (mm)	98[c]
Average minimum and maximum temperatures (centigrade)[c]	24.0/33.2

a Official rate. b October 2006. c Muscat. d June 2006. e Arrivals of non-resident tourists in hotels and similar establishments. f 2003. g 2004. h Excluding taxis. i Trucks only. j Provisional.

Pakistan

Region	South-central Asia
Largest urban agglom. (pop., 000s)	Karachi (11608)
Currency	rupee
Population in 2006 (proj., 000s)	161209
Surface area (square kms)	796095
Population density (per square km)	202
United Nations membership date	30 September 1947

Economic indicators	2000	2005
Exchange rate (national currency per US$)	58.03	60.60[a]
Consumer price index (2000=100)	100	138[b]
Industrial production index (1995=100)[c]	114	225[d]
Unemployment (percentage of labour force)[e]	7.8	7.7
Balance of payments, current account (million US$)	−85	−3463
Tourist arrivals (000s)	557	648[f]
GDP (million current US$)	70709	110017
GDP (per capita current US$)	496	697
Gross fixed capital formation (% of GDP)	16.0	15.0
Labour force participation, adult female pop. (%)	16.3	15.9[f]
Labour force participation, adult male pop. (%)	83.2	70.6[f]
Employment in industrial sector (%)	18.0	20.8[g]
Employment in agricultural sector (%)	48.4	42.1[g]
Agricultural production index (1999-2001=100)	102	113[f]
Food production index (1999-2001=100)	102	111[f]
Primary energy production (000s Mt oil equiv.)	24218	34762[f]
Motor vehicles (per 1,000 inhabitants)	10.5	13.1[f]
Telephone lines (per 100 inhabitants)	2.2	3.4
Internet users, estimated (000s)	300.0	10500.0

Total trade	(million US$)	Major trading partners			2005
		(% of exports)		(% of imports)	
Exports	16050.2	USA	25	Saudi Arabia	11
Imports	25096.6	Untd Arab Em	8	Untd Arab Em	10
		Afghanistan	7	China	9

Social indicators	2000-2006
Population growth rate 2000-2005 (% per annum)	2.0
Population aged 0-14 years (%)	38.3
Population aged 60+ years (women and men, % of total)	6.0/6.0
Sex ratio (women per 100 men)	94
Life expectancy at birth 2000-2005 (women and men, years)	63/63
Infant mortality rate 2000-2005 (per 1,000 births)	79
Total fertility rate 2000-2005 (births per woman)	4.3
Contraceptive use (% of currently married women)	28[h]
Urban population (%)	35
Urban population growth rate 2000-2005 (% per annum)	3.0
Rural population growth rate 2000-2005 (% per annum)	1.5
Foreign born (%)	3.0[i]
Refugees and others of concern to UNHCR[j]	1088121
Government education expenditure (% of GNP)	2.0
Primary-secondary gross enrolment ratio (w and m per 100)	42/58
Third-level students (women and men, % of total)	43/57
Newspaper circulation (per 1,000 inhabitants)	39
Television receivers (per 1,000 inhabitants)	82
Parliamentary seats (women and men, % of total)	21/79

Environment	2000-2006
Threatened species	86
Forested area (% of land area)	3
CO$_2$ emissions (000s Mt of carbon dioxide/per capita)	114356/0.8
Energy consumption per capita (kilograms oil equiv.)	319
Precipitation (mm)	1142[k]
Average minimum and maximum temperatures (centigrade)[k]	14.1/28.6

a October 2006. b May 2006. c Calculated by the Statistics Division of the United Nations from component national indices. d March 2006. e Persons aged 10 years and over. f 2004. g 2002. h 2000/01. i Estimated data. j Provisional. k Islamabad.

Palau

Region	Oceania-Micronesia
Largest urban agglom. (pop., 000s)	Koror (14)
Currency	US dollar
Population in 2006 (proj., 000s)	20[a]
Surface area (square kms)	459
Population density (per square km)	42
United Nations membership date	15 December 1994

Economic indicators	2000	2005
Tourist arrivals (000s)[b]	58	95[c]
GDP (million current US$)	117	123
GDP (per capita current US$)	6076	6150
Gross fixed capital formation (% of GDP)	31.0	42.0
Primary energy production (000s Mt oil equiv.)	2	2[c]
Internet users, estimated (000s)	4.0[d]	5.4[c]

Total trade		Major trading partners		2005
	(million US$)	(% of exports)		(% of imports)[e]
Exports	...	...	USA	33
Imports	123.2[e]	...	Japan	16
		...	Singapore	7

Social indicators	2000-2006
Population growth rate 2000-2005 (% per annum)	0.7
Population aged 0-14 years (%)	24.0
Sex ratio (women per 100 men)	83[e]
Life expectancy at birth 2000-2005 (women and men, years)	75/65[fg]
Infant mortality rate 2000-2005 (per 1,000 births)	18
Total fertility rate 2000-2005 (births per woman)	3.0[h]
Urban population (%)	70
Urban population growth rate 2000-2005 (% per annum)	0.7
Rural population growth rate 2000-2005 (% per annum)	0.6
Foreign born (%)	13.6[i]
Government education expenditure (% of GNP)	9.7
Primary-secondary gross enrolment ratio (w and m per 100)	105/105
Third-level students (women and men, % of total)	63/37
Parliamentary seats (women and men, % of total)	—/100

Environment	2000-2006
Threatened species	23
CO_2 emissions (000s Mt of carbon dioxide/per capita)	243/12.3
Energy consumption per capita (kilograms oil equiv.)	4106[i]
Precipitation (mm)	3746

a 2005. b Air arrivals (Palau International Airport). c 2004. d 2002. e 2000. f Published by the Secretariat of the Pacific Community. g 2001. h Data refer to a year between 1985 and 1990. i Estimated data.

Panama

Region	Central America
Largest urban agglom. (pop., 000s)	Panama City (1216)
Currency	balboa
Population in 2006 (proj., 000s)	3288
Surface area (square kms)	75517
Population density (per square km)	44
United Nations membership date	13 November 1945

Economic indicators	2000	2005
Exchange rate (national currency per US$)	1.00	1.00
Consumer price index (2000=100)[a]	...	106[bc]
Industrial production index (1995=100)[d]	112	122[e]
Unemployment (percentage of labour force)[f]	13.5	10.3
Balance of payments, current account (million US$)	−673	−782
Tourist arrivals (000s)	484	621[g]
GDP (million current US$)	11621	15241
GDP (per capita current US$)	3939	4716
Gross fixed capital formation (% of GDP)	21.0	17.0
Labour force participation, adult female pop. (%)	41.6	46.5[g]
Labour force participation, adult male pop. (%)	78.8	80.2[g]
Employment in industrial sector (%)	17.4	17.2
Employment in agricultural sector (%)	17.0	15.7
Agricultural production index (1999-2001=100)	100	103[g]
Food production index (1999-2001=100)	101	104[g]
Primary energy production (000s Mt oil equiv.)	294	325[g]
Motor vehicles (per 1,000 inhabitants)	100.8	102.4[g]
Telephone lines (per 100 inhabitants)	15.1	13.6
Internet users, estimated (000s)	107.5	206.2

Total trade		Major trading partners			2005
	(million US$)	(% of exports)		(% of imports)	
Exports	963.8	USA	45	USA	27
Imports	4154.9	Spain	9	Neth.Antilles	11
		Sweden	6	Costa Rica	5

Social indicators	2000-2006
Population growth rate 2000-2005 (% per annum)	1.8
Population aged 0-14 years (%)	30.0
Population aged 60+ years (women and men, % of total)	9.0/8.0
Sex ratio (women per 100 men)	98
Life expectancy at birth 2000-2005 (women and men, years)	77/72
Infant mortality rate 2000-2005 (per 1,000 births)	21
Total fertility rate 2000-2005 (births per woman)	2.7
Urban population (%)	71
Urban population growth rate 2000-2005 (% per annum)	3.3
Rural population growth rate 2000-2005 (% per annum)	−1.3
Foreign born (%)	2.8[h]
Refugees and others of concern to UNHCR[i]	12434
Government education expenditure (% of GNP)	4.2
Primary-secondary gross enrolment ratio (w and m per 100)	92/91
Third-level students (women and men, % of total)	62/38
Television receivers (per 1,000 inhabitants)	195
Intentional homicides (per 100,000 inhabitants)	10
Parliamentary seats (women and men, % of total)	17/83

Environment	2000-2006
Threatened species	322
Forested area (% of land area)	39
CO2 emissions (000s Mt of carbon dioxide/per capita)	6035/1.9
Energy consumption per capita (kilograms oil equiv.)	650
Precipitation (mm)	1907[j]
Average minimum and maximum temperatures (centigrade)[j]	20.0/33.8

a Base: 2003=100. b April 2006. c Urban areas. d Calculated by the Statistics Division of the United Nations from component national indices. e 2nd quarter 2006. f Persons aged 15 years and over, Aug. of each year. g 2004. h Estimated data. i Provisional. j Panama City.

Papua New Guinea

Region	Oceania-Melanesia
Largest urban agglom. (pop., 000s)	Port Moresby (289)
Currency	kina
Population in 2006 (proj., 000s)	6001
Surface area (square kms)	462840
Population density (per square km)	13
United Nations membership date	10 October 1975

Economic indicators	2000	2005
Exchange rate (national currency per US$)[a]	3.07	3.01[b]
Consumer price index (2000=100)	100	140[c]
Unemployment (percentage of labour force)	2.8	...
Balance of payments, current account (million US$)	345	282[d]
Tourist arrivals (000s)	58	59[e]
GDP (million current US$)	3864	5330
GDP (per capita current US$)	729	905
Gross fixed capital formation (% of GDP)	20.0	20.0
Labour force participation, adult female pop. (%)	71.3	...
Labour force participation, adult male pop. (%)	73.5	...
Employment in industrial sector (%)	3.6	...
Employment in agricultural sector (%)	72.3	...
Agricultural production index (1999-2001=100)	102	106[e]
Food production index (1999-2001=100)	101	108[e]
Primary energy production (000s Mt oil equiv.)[f]	3652	2433[e]
Motor vehicles (per 1,000 inhabitants)	21.3	20.3[g]
Telephone lines (per 100 inhabitants)	1.3	1.1
Internet users, estimated (000s)	45.0	135.0

Total trade	Major trading partners		2005
	(million US$)[c]	(% of exports)	(% of imports)[c]
Exports	2260.2	... Australia	56
Imports	1302.4	... USA	9
		... Singapore	7

Social indicators	2000-2006
Population growth rate 2000-2005 (% per annum)	2.1
Population aged 0-14 years (%)	40.0
Population aged 60+ years (women and men, % of total)	4.0/4.0
Sex ratio (women per 100 men)	94
Life expectancy at birth 2000-2005 (women and men, years)	56/55
Infant mortality rate 2000-2005 (per 1,000 births)	71
Total fertility rate 2000-2005 (births per woman)	4.1
Urban population (%)	13
Urban population growth rate 2000-2005 (% per annum)	2.4
Rural population growth rate 2000-2005 (% per annum)	2.1
Foreign born (%)	0.4[h]
Refugees and others of concern to UNHCR[i]	10003
Primary-secondary gross enrolment ratio (w and m per 100)	49/57
Third-level students (women and men, % of total)	35/65[j]
Television receivers (per 1,000 inhabitants)	24
Parliamentary seats (women and men, % of total)	1/99

Environment	2000-2006
Threatened species	301
Forested area (% of land area)	68
CO2 emissions (000s Mt of carbon dioxide/per capita)	2515/0.4
Energy consumption per capita (kilograms oil equiv.)	163[h]
Precipitation (mm)	1150
Average minimum and maximum temperatures (centigrade)	25.4/27.7

a Official rate. b September 2006. c 2003. d 2001. e 2004. f Source: World Automotive Market Report, Auto and Truck International (Illinois). g 2002. h Estimated data. i Provisional. j 1999.

Paraguay

Region	South America
Largest urban agglom. (pop., 000s)	Asunción (1858)
Currency	guaraní
Population in 2006 (proj., 000s)	6301
Surface area (square kms)	406752
Population density (per square km)	15
United Nations membership date	24 October 1945

Economic indicators	2000	2005
Exchange rate (national currency per US$)	3526.90	5400.00[a]
Consumer price index (2000=100)[b]	100	163[c]
Unemployment (percentage of labour force)[d]	7.6	8.1[e]
Balance of payments, current account (million US$)	−163	20[f]
Tourist arrivals (000s)	289[g]	309[fh]
GDP (million current US$)	7095	7684
GDP (per capita current US$)	1297	1248
Gross fixed capital formation (% of GDP)	17.0	21.0
Labour force participation, adult female pop. (%)	48.4[i]	53.5[j]
Labour force participation, adult male pop. (%)	85.5[i]	85.3[j]
Employment in industrial sector (%)	18.1[i]	15.8[e]
Employment in agricultural sector (%)	30.4[i]	31.5[e]
Agricultural production index (1999-2001=100)	96	117[f]
Food production index (1999-2001=100)	96	115[f]
Primary energy production (000s Mt oil equiv.)	4600	4464[f]
Motor vehicles (per 1,000 inhabitants)	86.3	68.8
Telephone lines (per 100 inhabitants)	5.2	5.2
Internet users, estimated (000s)	40.0	200.0

Total trade		Major trading partners			2005
	(million US$)[f]	(% of exports)[f]		(% of imports)[f]	
Exports	1625.7	Uruguay	28	Brazil	28
Imports	3097.4	Brazil	19	Argentina	21
		Cayman Islands	12	China	16

Social indicators	2000-2006
Population growth rate 2000-2005 (% per annum)	2.4
Population aged 0-14 years (%)	38.0
Population aged 60+ years (women and men, % of total)	6.0/5.0
Sex ratio (women per 100 men)	99
Life expectancy at birth 2000-2005 (women and men, years)	73/69
Infant mortality rate 2000-2005 (per 1,000 births)	37
Total fertility rate 2000-2005 (births per woman)	3.9
Contraceptive use (% of currently married women)	73
Urban population (%)	58
Urban population growth rate 2000-2005 (% per annum)	3.5
Rural population growth rate 2000-2005 (% per annum)	0.9
Foreign born (%)	3.4
Refugees and others of concern to UNHCR[k]	58
Government education expenditure (% of GNP)	4.3
Primary-secondary gross enrolment ratio (w and m per 100)	85/86
Third-level students (women and men, % of total)	57/43
Television receivers (per 1,000 inhabitants)	216
Parliamentary seats (women and men, % of total)	10/90

Environment	2000-2006
Threatened species	54
Forested area (% of land area)	59
CO2 emissions (000s Mt of carbon dioxide/per capita)	4143/0.7
Energy consumption per capita (kilograms oil equiv.)	310
Precipitation (mm)	1401[b]
Average minimum and maximum temperatures (centigrade)[b]	18.2/28.4

a October 2006. b Asunción. c August 2006. d Persons aged 10 years and over. e 2003. f 2004. g Arrivals of non-resident tourists at national borders. Excluding nationals residing abroad and crew members. Inbound and outbound tourism survey - Central Bank of Paraguay. h Arrivals of non-resident tourists at national borders. Excluding nationals residing abroad and crew members. i 1999. j 2001. k Provisional.

Peru

Region	South America
Largest urban agglom. (pop., 000s)	Lima (7186)
Currency	new sol
Population in 2006 (proj., 000s)	28380
Surface area (square kms)	1285216
Population density (per square km)	22
United Nations membership date	31 October 1945

Economic indicators	2000	2005
Exchange rate (national currency per US$)	3.53	3.22[a]
Consumer price index (2000=100)[b]	100	112[c]
Industrial production index (1995=100)	113	148[d]
Unemployment (percentage of labour force)[e]	7.4[f]	11.4[g]
Balance of payments, current account (million US$)	−1526	1030
Tourist arrivals (000s)	800	1208[h]
GDP (million current US$)	53131	76607
GDP (per capita current US$)	2047	2739
Gross fixed capital formation (% of GDP)	20.0	19.0
Labour force participation, adult female pop. (%)	54.0	49.9[h]
Labour force participation, adult male pop. (%)	78.8	75.6[h]
Employment in industrial sector (%)	18.8	23.8
Employment in agricultural sector (%)	6.8	0.7
Agricultural production index (1999-2001=100)	102	110[h]
Food production index (1999-2001=100)	101	110[h]
Primary energy production (000s Mt oil equiv.)	7611	8097[h]
Motor vehicles (per 1,000 inhabitants)	44.8	53.8[i]
Telephone lines (per 100 inhabitants)	6.7	8.1
Internet users, estimated (000s)	800.0	4600.0

Total trade		Major trading partners		2005	
	(million US$)	(% of exports)		(% of imports)	
Exports	17114.3	USA	31	USA	18
Imports	12501.8	China	11	China	8
		Chile	7	Brazil	8

Social indicators	2000-2006
Population growth rate 2000-2005 (% per annum)	1.5
Population aged 0-14 years (%)	32.0
Population aged 60+ years (women and men, % of total)	8.0/7.0
Sex ratio (women per 100 men)	99
Life expectancy at birth 2000-2005 (women and men, years)	72/67
Infant mortality rate 2000-2005 (per 1,000 births)	33
Total fertility rate 2000-2005 (births per woman)	2.9
Contraceptive use (% of currently married women)	69
Urban population (%)	73
Urban population growth rate 2000-2005 (% per annum)	1.8
Rural population growth rate 2000-2005 (% per annum)	0.8
Foreign born (%)	0.2[i]
Refugees and others of concern to UNHCR[k]	1187
Government education expenditure (% of GNP)	3.1
Primary-secondary gross enrolment ratio (w and m per 100)	104/104
Third-level students (women and men, % of total)	50/50
Television receivers (per 1,000 inhabitants)	199
Intentional homicides (per 100,000 inhabitants)	2
Parliamentary seats (women and men, % of total)	29/71

Environment	2000-2006
Threatened species	524
Forested area (% of land area)	51
CO2 emissions (000s Mt of carbon dioxide/per capita)	26198/1.0
Energy consumption per capita (kilograms oil equiv.)	410
Precipitation (mm)	13[b]
Average minimum and maximum temperatures (centigrade)[b]	16.7/22.1

a October 2006. b Lima. c July 2006. d August 2006. e Persons aged 14 years and over, urban areas. f Oct., Metropolitan Lima. g Sept., Metropolitan Lima. h 2004. i 2003. j Estimated data. k Provisional.

Philippines

Region	South-eastern Asia	
Largest urban agglom. (pop., 000s)	Metro Manila (10686)	
Currency	peso	
Population in 2006 (proj., 000s)	84477	
Surface area (square kms)	300000	
Population density (per square km)	282	
United Nations membership date	24 October 1945	

Economic indicators	2000	2005
Exchange rate (national currency per US$)	50.00	49.81[a]
Consumer price index (2000=100)	100	139[b]
Unemployment (percentage of labour force)[c]	10.1	7.4
Balance of payments, current account (million US$)	−2225	2338
Tourist arrivals (000s)[d]	1992	2291[e]
GDP (million current US$)	75031	97653
GDP (per capita current US$)	990	1176
Gross fixed capital formation (% of GDP)	20.0	16.0
Labour force participation, adult female pop. (%)	48.4	50.2[e]
Labour force participation, adult male pop. (%)	80.3	82.9[e]
Employment in industrial sector (%)	16.0	14.9
Employment in agricultural sector (%)	37.4	37.0
Agricultural production index (1999-2001=100)	100	115[e]
Food production index (1999-2001=100)	100	116[e]
Primary energy production (000s Mt oil equiv.)	2377	5127[e]
Motor vehicles (per 1,000 inhabitants)	32.2	18.8[e]
Telephone lines (per 100 inhabitants)	4.0	4.2
Internet users, estimated (000s)	1540.0	4400.0[e]

Total trade		Major trading partners			2005
	(million US$)	(% of exports)		(% of imports)	
Exports	41221.3	USA	18	USA	17
Imports	46953.8	Japan	17	Japan	17
		China	10	Singapore	8

Social indicators	2000-2006
Population growth rate 2000-2005 (% per annum)	1.8
Population aged 0-14 years (%)	35.0
Population aged 60+ years (women and men, % of total)	7.0/6.0
Sex ratio (women per 100 men)	99
Life expectancy at birth 2000-2005 (women and men, years)	72/68
Infant mortality rate 2000-2005 (per 1,000 births)	28
Total fertility rate 2000-2005 (births per woman)	3.2
Contraceptive use (% of currently married women)	49
Urban population (%)	63
Urban population growth rate 2000-2005 (% per annum)	3.2
Rural population growth rate 2000-2005 (% per annum)	−0.3
Foreign born (%)	0.2[f]
Refugees and others of concern to UNHCR[g]	904
Government education expenditure (% of GNP)	3.0
Primary-secondary gross enrolment ratio (w and m per 100)	103/101
Third-level students (women and men, % of total)	55/45
Television receivers (per 1,000 inhabitants)	192
Parliamentary seats (women and men, % of total)	15/85

Environment	2000-2006
Threatened species	475
Forested area (% of land area)	19
CO2 emissions (000s Mt of carbon dioxide/per capita)	77095/1.0
Energy consumption per capita (kilograms oil equiv.)	303
Precipitation (mm)	2201[h]
Average minimum and maximum temperatures (centigrade)[h]	25.2/31.2

a October 2006. b September 2006. c Persons aged 15 years and over, Oct. of each year. d Including nationals residing abroad. e 2004. f Estimated data. g Provisional. h Metro Manila.

Poland

Region	Eastern Europe
Largest urban agglom. (pop., 000s)	Warsaw (1680)
Currency	zloty
Population in 2006 (proj., 000s)	38499
Surface area (square kms)	323250
Population density (per square km)	119
United Nations membership date	24 October 1945

Economic indicators	2000	2005
Exchange rate (national currency per US$)[a]	4.14	3.06[b]
Consumer price index (2000=100)	100	116[c]
Industrial production index (1995=100)	144	206[c]
Unemployment (percentage of labour force)[d]	16.1	17.7
Balance of payments, current account (million US$)	−9981	−4314
Tourist arrivals (000s)	17400	14290[e]
GDP (million current US$)	166561	290006
GDP (per capita current US$)	4310	7527
Gross fixed capital formation (% of GDP)	24.0	18.0
Labour force participation, adult female pop. (%)	49.7	47.8[e]
Labour force participation, adult male pop. (%)	64.1	62.3[e]
Employment in industrial sector (%)	30.8	29.2
Employment in agricultural sector (%)	18.8	17.4
Agricultural production index (1999-2001=100)	98	107[e]
Food production index (1999-2001=100)	98	107[e]
Primary energy production (000s Mt oil equiv.)	75786	75844[e]
Motor vehicles (per 1,000 inhabitants)	309.3	382.1
Telephone lines (per 100 inhabitants)	28.3	30.6
Internet users, estimated (000s)	2800.0	10000.0

Total trade		Major trading partners			2005
	(million US$)	(% of exports)		(% of imports)	
Exports	89378.1	Germany	28	Germany	25
Imports	101538.8	France	6	Russian Fed.	9
		Italy	6	Italy	7

Social indicators	2000-2006
Population growth rate 2000-2005 (% per annum)	−0.1
Population aged 0-14 years (%)	16.0
Population aged 60+ years (women and men, % of total)	20.0/14.0
Sex ratio (women per 100 men)	106
Life expectancy at birth 2000-2005 (women and men, years)	78/70
Infant mortality rate 2000-2005 (per 1,000 births)	9
Total fertility rate 2000-2005 (births per woman)	1.3
Contraceptive use (% of currently married women)	49[f,g]
Urban population (%)	62
Urban population growth rate 2000-2005 (% per annum)	<
Rural population growth rate 2000-2005 (% per annum)	−0.2
Foreign born (%)	2.1
Refugees and others of concern to UNHCR[h]	6305
Government education expenditure (% of GNP)	5.9
Primary-secondary gross enrolment ratio (w and m per 100)	98/97
Third-level students (women and men, % of total)	58/42
Newspaper circulation (per 1,000 inhabitants)	102
Television receivers (per 1,000 inhabitants)	229
Intentional homicides (per 100,000 inhabitants)	2
Parliamentary seats (women and men, % of total)	19/81

Environment	2000-2006
Threatened species	51
Forested area (% of land area)	30
CO2 emissions (000s Mt of carbon dioxide/per capita)[i]	308280/8.0
Energy consumption per capita (kilograms oil equiv.)	2329
Precipitation (mm)	520[j]
Average minimum and maximum temperatures (centigrade)[j]	4.0/12.3

a Official rate. b October 2006. c August 2006. d Persons aged 15 to 74 years. e 2004.
f For women aged 20-49 in union or married. g 1991. h Provisional. i Source: UNFCCC.
j Warsaw.

Portugal

Region	Southern Europe
Largest urban agglom. (pop., 000s)	Lisbon (2761)
Currency	euro[a]
Population in 2006 (proj., 000s)	10545
Surface area (square kms)	91982
Population density (per square km)	115
United Nations membership date	14 December 1955

Economic indicators	2000	2005
Exchange rate (national currency per US$)	1.07	0.79[b]
Consumer price index (2000=100)	100	120[c]
Industrial production index (1995=100)	118	100[d]
Unemployment (percentage of labour force)[e]	3.9	7.6
Balance of payments, current account (million US$)	−11748	−17007
Tourist arrivals (000s)[f]	12097	11617[g]
GDP (million current US$)	112650	183300
GDP (per capita current US$)	11017	17466
Gross fixed capital formation (% of GDP)	27.0	22.0
Labour force participation, adult female pop. (%)	52.9	54.8[g]
Labour force participation, adult male pop. (%)	69.7	69.7[g]
Employment in industrial sector (%)	34.4	32.2[h]
Employment in agricultural sector (%)	12.6	12.5[h]
Agricultural production index (1999-2001=100)	101	99[g]
Food production index (1999-2001=100)	101	99[g]
Primary energy production (000s Mt oil equiv.)	1029	950[g]
Motor vehicles (per 1,000 inhabitants)[i]	683.4	767.2[h]
Telephone lines (per 100 inhabitants)	43.1	40.4
Internet users, estimated (000s)	1680.2	2939.0

Total trade		Major trading partners			2005
	(million US$)	(% of exports)		(% of imports)	
Exports	38085.7	Spain	26	Spain	29
Imports	61167.1	France	13	Germany	13
		Germany	12	France	8

Social indicators	2000-2006
Population growth rate 2000-2005 (% per annum)	0.5
Population aged 0-14 years (%)	16.0
Population aged 60+ years (women and men, % of total)	25.0/20.0
Sex ratio (women per 100 men)	107
Life expectancy at birth 2000-2005 (women and men, years)	81/74
Infant mortality rate 2000-2005 (per 1,000 births)	6
Total fertility rate 2000-2005 (births per woman)	1.5
Urban population (%)	58
Urban population growth rate 2000-2005 (% per annum)	1.7
Rural population growth rate 2000-2005 (% per annum)	−0.9
Refugees and others of concern to UNHCR[j]	363
Government education expenditure (% of GNP)	6.0
Primary-secondary gross enrolment ratio (w and m per 100)	107/105
Third-level students (women and men, % of total)	56/44
Newspaper circulation (per 1,000 inhabitants)	100
Television receivers (per 1,000 inhabitants)	421
Intentional homicides (per 100,000 inhabitants)	2
Parliamentary seats (women and men, % of total)	21/79

Environment	2000-2006
Threatened species	167
Forested area (% of land area)	40
CO2 emissions (000s Mt of carbon dioxide/per capita)[k]	64290/6.2
Energy consumption per capita (kilograms oil equiv.)	2018
Precipitation (mm)	751[l]
Average minimum and maximum temperatures (centigrade)[l]	12.8/20.8

a Prior to 1 January 1999, escudo. b September 2006. c July 2006. d August 2006. e Persons aged 15 years and over. f Excluding nationals residing abroad. Including arrivals from abroad in Madeira and Azores. g 2004. h 2003. i Excluding Madeira and Azores. j Provisional. k Source: UNFCCC. l Lisbon.

Puerto Rico

Region	Caribbean	
Largest urban agglom. (pop., 000s)	San Juan (2605)	
Currency	US dollar	
Population in 2006 (proj., 000s)	3977	
Surface area (square kms)	8875	
Population density (per square km)	448	

Economic indicators	2000	2005
Consumer price index (2000=100)	100	179[a]
Unemployment (percentage of labour force)[b]	10.1	11.3
Tourist arrivals (000s)[c]	3341	3541[d]
GDP (million current US$)	69208	86882
GDP (per capita current US$)	18047	21970
Gross fixed capital formation (% of GDP)	17.0	16.0
Labour force participation, adult female pop. (%)	34.9	36.6[d]
Labour force participation, adult male pop. (%)	59.1	58.9[d]
Employment in industrial sector (%)	22.3	19.0
Employment in agricultural sector (%)	1.9	2.1
Agricultural production index (1999-2001=100)	106	99[d]
Food production index (1999-2001=100)	106	98[d]
Primary energy production (000s Mt oil equiv.)	14	12[d]
Motor vehicles (per 1,000 inhabitants)[e]	556.1	578.6
Telephone lines (per 100 inhabitants)[f]	34.1	28.5
Internet users, estimated (000s)	400.0	862.0[d]

Social indicators	2000-2006
Population growth rate 2000-2005 (% per annum)	0.6
Population aged 0-14 years (%)	22.0
Population aged 60+ years (women and men, % of total)	18.0/15.0
Sex ratio (women per 100 men)	108
Life expectancy at birth 2000-2005 (women and men, years)	80/72
Infant mortality rate 2000-2005 (per 1,000 births)	10
Total fertility rate 2000-2005 (births per woman)	1.9
Urban population (%)	98
Urban population growth rate 2000-2005 (% per annum)	1.2
Rural population growth rate 2000-2005 (% per annum)	−15.8
Third-level students (women and men, % of total)	53/47[g]
Television receivers (per 1,000 inhabitants)[h]	339
Intentional homicides (per 100,000 inhabitants)	18

Environment	2000-2006
Threatened species	102
Forested area (% of land area)	26
CO2 emissions (000s Mt of carbon dioxide/per capita)	2105/0.5
Energy consumption per capita (kilograms oil equiv.)	166
Precipitation (mm)	1631
Average minimum and maximum temperatures (centigrade)	23.6/27.2

a August 2006. b Persons aged 16 years and over, excl. persons temporarily laid off. c Arrivals of non-resident tourists by air. Source: 'Junta de Planificación de Puerto Rico'. d 2004. e Data refer to fiscal years beginning 1 July. f Switched access lines. g 1996. h Data refer to Puerto Rico Telephone Authority.

Qatar

Region	Western Asia
Largest urban agglom. (pop., 000s)	Doha (357)
Currency	riyal
Population in 2006 (proj., 000s)	839
Surface area (square kms)	11000
Population density (per square km)	76
United Nations membership date	21 September 1971

Economic indicators	2000	2005
Exchange rate (national currency per US$)[a]	3.64	3.64[b]
Consumer price index (2000=100)[c]	...	131[d]
Unemployment (percentage of labour force)[e]	2.3[f]	3.9[g]
Tourist arrivals (000s)[h]	378	732[i]
GDP (million current US$)	17760	42113
GDP (per capita current US$)	29290	51809
Gross fixed capital formation (% of GDP)	19.0	34.0
Labour force participation, adult female pop. (%)	33.4[f]	...
Labour force participation, adult male pop. (%)	89.2[f]	...
Employment in industrial sector (%)	33.1[f]	41.0[i]
Employment in agricultural sector (%)	3.7[f]	2.7[i]
Agricultural production index (1999-2001=100)	112	144[i]
Food production index (1999-2001=100)	112	144[i]
Primary energy production (000s Mt oil equiv.)	60825	81622[i]
Motor vehicles (per 1,000 inhabitants)	482.4	495.1[j]
Telephone lines (per 100 inhabitants)	26.4	26.4
Internet users, estimated (000s)	30.0	219.0

Total trade		Major trading partners			2005
	(million US$)	(% of exports)		(% of imports)	
Exports	25762.5	Japan	40	Japan	12
Imports	10060.9	Korea Rep.	16	USA	12
		Singapore	8	Germany	9

Social indicators	2000-2006
Population growth rate 2000-2005 (% per annum)	5.9
Population aged 0-14 years (%)	22.0
Population aged 60+ years (women and men, % of total)	2.0/3.0
Sex ratio (women per 100 men)	48
Life expectancy at birth 2000-2005 (women and men, years)	76/71
Infant mortality rate 2000-2005 (per 1,000 births)	12
Total fertility rate 2000-2005 (births per woman)	3.0
Urban population (%)	95
Urban population growth rate 2000-2005 (% per annum)	6.0
Rural population growth rate 2000-2005 (% per annum)	3.9
Foreign born (%)	70.4[k]
Refugees and others of concern to UNHCR[l]	74
Government education expenditure (% of GNP)	3.1[m]
Primary-secondary gross enrolment ratio (w and m per 100)	98/100
Third-level students (women and men, % of total)	71/29
Television receivers (per 1,000 inhabitants)	423

Environment	2000-2006
Threatened species	16
Forested area (% of land area)	<
CO2 emissions (000s Mt of carbon dioxide/per capita)	46262/63.1
Energy consumption per capita (kilograms oil equiv.)	33324
Precipitation (mm)	75[n]
Average minimum and maximum temperatures (centigrade)[n]	21.6/32.7

a Official rate. b October 2006. c Base: 2002=100. d June 2006. e Persons aged 15 years and over. f 1997. g 2001, April. h Arrivals in hotels only. i 2004. j 2002. k Estimated data. l Provisional. m 1994. n Doha.

Réunion

	Region	Eastern Africa
	Largest urban agglom. (pop., 000s)	Saint-Denis (137)
	Currency	euro[ab]
	Population in 2006 (proj., 000s)	796
	Surface area (square kms)	2510
	Population density (per square km)	317

Economic indicators	2000	2005
Exchange rate (national currency per US$)	1.07	0.79[cd]
Consumer price index (2000=100)	100	114[e]
Unemployment (percentage of labour force)[f]	36.5	31.9
Tourist arrivals (000s)	430	430[g]
Labour force participation, adult female pop. (%)	41.9[h]	...
Labour force participation, adult male pop. (%)	62.6[h]	...
Agricultural production index (1999-2001=100)	100	103[g]
Food production index (1999-2001=100)	100	103[g]
Primary energy production (000s Mt oil equiv.)	48	50[g]
Motor vehicles (per 1,000 inhabitants)	342.1	399.5[g]
Telephone lines (per 100 inhabitants)	40.1	41.0[ij]
Internet users, estimated (000s)	100.0	220.0

Social indicators	2000-2006
Population growth rate 2000-2005 (% per annum)	1.6
Population aged 0-14 years (%)	27.0
Population aged 60+ years (women and men, % of total)	11.0/8.0
Sex ratio (women per 100 men)	105
Life expectancy at birth 2000-2005 (women and men, years)	80/71
Infant mortality rate 2000-2005 (per 1,000 births)	8
Total fertility rate 2000-2005 (births per woman)	2.5
Contraceptive use (% of currently married women)	67[kl]
Urban population (%)	92
Urban population growth rate 2000-2005 (% per annum)	2.2
Rural population growth rate 2000-2005 (% per annum)	−4.1
Television receivers (per 1,000 inhabitants)	192

Environment	2000-2006
Threatened species	67
Forested area (% of land area)	28
CO2 emissions (000s Mt of carbon dioxide/per capita)	2479/3.3
Energy consumption per capita (kilograms oil equiv.)	980[m]
Precipitation (mm)	1526[n]
Average minimum and maximum temperatures (centigrade)[n]	20.7/21.6

a 2005. b Prior to 1 January 1999, French franc. c November 2006. d Data refer to non-commercial rates derived from the Operational Rates of Exchange for United Nations Programmes. e August 2006. f Persons aged 15 years and over, second quarter of each year. g 2004. h 1993. i Estimate. j 2003. k From a sample that included women in visiting unions. l 1990. m Estimated data. n Saint-Denis/Gillot.

Romania

Region	Eastern Europe
Largest urban agglom. (pop., 000s)	Bucharest (1934)
Currency	leu
Population in 2006 (proj., 000s)	21629
Surface area (square kms)	238391
Population density (per square km)	91
United Nations membership date	14 December 1955

Economic indicators	2000	2005
Exchange rate (national currency per US$) [ab]	2.59	2.79[c]
Consumer price index (2000=100)	100	247[d]
Industrial production index (1995=100) [e]	83	104[f]
Unemployment (percentage of labour force) [g]	7.1	7.2
Balance of payments, current account (million US$)	−1355	−8312
Tourist arrivals (000s)	5264	6600[h]
GDP (million current US$)	37025	98566
GDP (per capita current US$)	1674	4540
Gross fixed capital formation (% of GDP)	19.0	23.0
Labour force participation, adult female pop. (%)	56.4	47.8[h]
Labour force participation, adult male pop. (%)	70.6	62.4[h]
Employment in industrial sector (%)	26.2	30.3
Employment in agricultural sector (%)	42.8	32.1
Agricultural production index (1999-2001=100)	88	123[h]
Food production index (1999-2001=100)	88	123[h]
Primary energy production (000s Mt oil equiv.)	25363	24654[h]
Motor vehicles (per 1,000 inhabitants)	163.9	172.1[h]
Telephone lines (per 100 inhabitants)	17.4	20.2
Internet users, estimated (000s)	800.0	4500.0[h]

Total trade		Major trading partners			2005
	(million US$)	(% of exports)		(% of imports)	
Exports	27729.6	Italy	19	Italy	15
Imports	40462.9	Germany	14	Germany	14
		Turkey	8	Russian Fed.	8

Social indicators	2000-2006
Population growth rate 2000-2005 (% per annum)	−0.4
Population aged 0-14 years (%)	15.0
Population aged 60+ years (women and men, % of total)	22.0/17.0
Sex ratio (women per 100 men)	105
Life expectancy at birth 2000-2005 (women and men, years)	75/68
Infant mortality rate 2000-2005 (per 1,000 births)	18
Total fertility rate 2000-2005 (births per woman)	1.3
Urban population (%)	54
Urban population growth rate 2000-2005 (% per annum)	−0.7
Rural population growth rate 2000-2005 (% per annum)	<
Foreign born (%)	0.6
Refugees and others of concern to UNHCR [i]	2720
Government education expenditure (% of GNP)	3.7
Primary-secondary gross enrolment ratio (w and m per 100)	91/91
Third-level students (women and men, % of total)	55/45
Television receivers (per 1,000 inhabitants)	893
Intentional homicides (per 100,000 inhabitants)	4
Parliamentary seats (women and men, % of total)	11/89

Environment	2000-2006
Threatened species	161
Forested area (% of land area)	28
CO2 emissions (000s Mt of carbon dioxide/per capita) [j]	111390/5.1
Energy consumption per capita (kilograms oil equiv.)	1553
Precipitation (mm)	595[k]
Average minimum and maximum temperatures (centigrade) [k]	5.7/16.5

a As of 1 July 2005, Romania redenominated the Leu. The (new) Leu (ISO Code: RON) is equivalent to 10,000 (old) Lei (ROL). b Principal rate. c September 2006. d July 2006. e Monthly indices are adjusted for differences in the number of working days. f August 2006. g Persons aged 15 years and over. h 2004. i Provisional. j Source: UNFCCC. k Bucharest.

Russian Federation

Region	Eastern Europe
Largest urban agglom. (pop., 000s)	Moscow (10654)
Currency	ruble
Population in 2006 (proj., 000s)	142537
Surface area (square kms)	17075400
Population density (per square km)	8
United Nations membership date	24 October 1945

Economic indicators	2000	2005
Exchange rate (national currency per US$)[a]	28.16	26.75[b]
Consumer price index (2000=100)	100	219[c]
Industrial production index (1995=100)	105	146[d]
Unemployment (percentage of labour force)[e]	9.8	7.8[f]
Balance of payments, current account (million US$)	46839	83559
Tourist arrivals (000s)	19457[g]	19892[f]
GDP (million current US$)	259718	765968
GDP (per capita current US$)	1772	5349
Gross fixed capital formation (% of GDP)	17.0	18.0
Labour force participation, adult female pop. (%)	53.3	60.4[f]
Labour force participation, adult male pop. (%)	68.1	70.0[f]
Employment in industrial sector (%)	28.4	29.8
Employment in agricultural sector (%)	14.5	10.2
Agricultural production index (1999-2001=100)	100	111[f]
Food production index (1999-2001=100)	100	111[f]
Primary energy production (000s Mt oil equiv.)	970849	1160610[f]
Motor vehicles (per 1,000 inhabitants)	166.3	198.5[f]
Telephone lines (per 100 inhabitants)	21.9	27.9
Internet users, estimated (000s)	2900.0	21800.0

Total trade		Major trading partners			2005
	(million US$)	(% of exports)		(% of imports)	
Exports	241243.8	Netherlands	10	Germany	13
Imports	98576.5	Italy	6	Ukraine	8
		Germany	5	China	7

Social indicators	2000-2006
Population growth rate 2000-2005 (% per annum)	−0.5
Population aged 0-14 years (%)	15.0
Population aged 60+ years (women and men, % of total)	21.0/12.0
Sex ratio (women per 100 men)	116
Life expectancy at birth 2000-2005 (women and men, years)	72/59
Infant mortality rate 2000-2005 (per 1,000 births)	17
Total fertility rate 2000-2005 (births per woman)	1.3
Urban population (%)	73
Urban population growth rate 2000-2005 (% per annum)	−0.6
Rural population growth rate 2000-2005 (% per annum)	−0.2
Refugees and others of concern to UNHCR[h]	483029
Government education expenditure (% of GNP)	3.8
Primary-secondary gross enrolment ratio (w and m per 100)	99/100
Third-level students (women and men, % of total)	57/43
Television receivers (per 1,000 inhabitants)	350
Intentional homicides (per 100,000 inhabitants)	30
Parliamentary seats (women and men, % of total)	8/92

Environment	2000-2006
Threatened species	54
Forested area (% of land area)	50
CO2 emissions (000s Mt of carbon dioxide/per capita)	1495867/10.3
Energy consumption per capita (kilograms oil equiv.)	4232
Precipitation (mm)	637[i]
Average minimum and maximum temperatures (centigrade)[i]	1.8/8.3

a Official rate. b October 2006. c June 2006. d August 2006. e Persons aged 15 to 72 years. f 2004. g 2001. h Provisional. i St. Petersburg.

Rwanda

Region	Eastern Africa	
Largest urban agglom. (pop., 000s)	Kigali (779)	
Currency	franc	
Population in 2006 (proj., 000s)	9230	
Surface area (square kms)	26338	
Population density (per square km)	350	
United Nations membership date	18 September 1962	

Economic indicators	2000	2005
Exchange rate (national currency per US$)[a]	430.32	549.85[b]
Consumer price index (2000=100)[c]	100	150[d]
Unemployment (percentage of labour force)	0.6[e]	...
Balance of payments, current account (million US$)	−94	−90
Tourist arrivals (000s)	104	113[f]
GDP (million current US$)	1732	2118
GDP (per capita current US$)	216	234
Gross fixed capital formation (% of GDP)	18.0	22.0
Labour force participation, adult female pop. (%)	85.1[g]	87.9[f]
Labour force participation, adult male pop. (%)	87.1[g]	84.6[f]
Employment in industrial sector (%)	2.9[h]	...
Employment in agricultural sector (%)	90.1[h]	...
Agricultural production index (1999-2001=100)	104	113[i]
Food production index (1999-2001=100)	104	113[i]
Primary energy production (000s Mt oil equiv.)	14	15[i]
Motor vehicles (per 1,000 inhabitants)	3.4	3.5[j]
Telephone lines (per 100 inhabitants)	0.2	0.3
Internet users, estimated (000s)	5.0	50.0

Total trade		Major trading partners			2005
	(million US$)[k]	(% of exports)[k]		(% of imports)[k]	
Exports	50.4	Kenya	41	Kenya	28
Imports	261.2	Uganda	27	Belgium	12
		Tanzania	8	Uganda	8

Social indicators	2000-2006
Population growth rate 2000-2005 (% per annum)	2.4
Population aged 0-14 years (%)	43.0
Population aged 60+ years (women and men, % of total)	4.0/4.0
Sex ratio (women per 100 men)	106
Life expectancy at birth 2000-2005 (women and men, years)	45/42
Infant mortality rate 2000-2005 (per 1,000 births)	116
Total fertility rate 2000-2005 (births per woman)	5.7
Contraceptive use (% of currently married women)	13
Urban population (%)	19
Urban population growth rate 2000-2005 (% per annum)	9.2
Rural population growth rate 2000-2005 (% per annum)	1.0
Foreign born (%)	1.2[l]
Refugees and others of concern to UNHCR[m]	59361
Government education expenditure (% of GNP)	2.8[n]
Primary-secondary gross enrolment ratio (w and m per 100)	68/68
Third-level students (women and men, % of total)	39/61
Television receivers (per 1,000 inhabitants)	8
Parliamentary seats (women and men, % of total)	45/55

Environment	2000-2006
Threatened species	53
Forested area (% of land area)	12
CO2 emissions (000s Mt of carbon dioxide/per capita)	602/0.1
Energy consumption per capita (kilograms oil equiv.)	22[l]
Precipitation (mm)	1028[c]

a Official rate. b September 2006. c Kigali. d May 2006. e 1996, persons aged 10 to 65 years. f 2001. g 1996. h 1989. i 2004. j 2002. k 2003. l Estimated data. m Provisional. n 1999.

Saint Kitts and Nevis

Region	Caribbean
Largest urban agglom. (pop., 000s)	Basse-Terre (13)
Currency	EC dollar
Population in 2006 (proj., 000s)	43[a]
Surface area (square kms)	261
Population density (per square km)	176
United Nations membership date	23 September 1983

Economic indicators	2000	2005
Exchange rate (national currency per US$)[b]	2.70	2.70[c]
Balance of payments, current account (million US$)	−66	−124[d]
Tourist arrivals (000s)[e]	73	118[f]
GDP (million current US$)	329	453
GDP (per capita current US$)	8138	10612
Gross fixed capital formation (% of GDP)	50.0	45.0
Agricultural production index (1999-2001=100)	100	100[f]
Food production index (1999-2001=100)	100	100[f]
Motor vehicles (per 1,000 inhabitants)	296.6	284.4[f]
Telephone lines (per 100 inhabitants)	54.2	59.3
Internet users, estimated (000s)[g]	2.7	10.0[d]

Total trade		Major trading partners			2005
	(million US$)	(% of exports)		(% of imports)	
Exports	34.3	USA	92	USA	58
Imports	210.2	UK	2	Trinidad Tbg	14
		Trinidad Tbg	2	UK	6

Social indicators	2000-2006
Population growth rate 2000-2005 (% per annum)	1.1
Population aged 0-14 years (%)	31.0
Population aged 60+ years (women and men, % of total)	12.0/10.0[h]
Sex ratio (women per 100 men)	101[ij]
Life expectancy at birth 2000-2005 (women and men, years)	71/68[k]
Total fertility rate 2000-2005 (births per woman)	2.6
Urban population (%)	32
Urban population growth rate 2000-2005 (% per annum)	0.7
Rural population growth rate 2000-2005 (% per annum)	1.3
Foreign born (%)	10.2[l]
Government education expenditure (% of GNP)	3.7
Primary-secondary gross enrolment ratio (w and m per 100)	101/96
Third-level students (women and men, % of total)	27/73[m]
Television receivers (per 1,000 inhabitants)	312
Parliamentary seats (women and men, % of total)	13/87

Environment	2000-2006
Threatened species	22
Forested area (% of land area)	11
CO2 emissions (000s Mt of carbon dioxide/per capita)	126/3.0
Energy consumption per capita (kilograms oil equiv.)	1080[l]

a 2005. b Official rate. c October 2006. d 2002. e Arrivals of non-resident tourists by air. f 2004. g From 2002: Ministry of Communications, Works, Utilities, Posts and Telegraph. h De facto estimate. i Refers to de facto population count. j 2001. k 1998. l Estimated data. m 1997.

Saint Lucia

Region	Caribbean
Largest urban agglom. (pop., 000s)	Castries (13)
Currency	EC dollar
Population in 2006 (proj., 000s)	162
Surface area (square kms)	539
Population density (per square km)	301
United Nations membership date	18 September 1979

Economic indicators	2000	2005
Exchange rate (national currency per US$)[a]	2.70	2.70[b]
Consumer price index (2000=100)	100	115[c]
Unemployment (percentage of labour force)	16.4[d]	...
Balance of payments, current account (million US$)	−79	−104[e]
Tourist arrivals (000s)[f]	270	298[g]
GDP (million current US$)	707	851
GDP (per capita current US$)	4575	5292
Gross fixed capital formation (% of GDP)	26.0	21.0
Labour force participation, adult female pop. (%)	62.0	...
Labour force participation, adult male pop. (%)	75.0	...
Employment in industrial sector (%)	20.3	17.7[h]
Employment in agricultural sector (%)	20.8	11.4[h]
Agricultural production index (1999-2001=100)	100	92[g]
Food production index (1999-2001=100)	100	92[g]
Motor vehicles (per 1,000 inhabitants)	207.4	213.9[g]
Telephone lines (per 100 inhabitants)	31.7	32.6[gi]
Internet users, estimated (000s)	8.0	55.0[g]

Total trade		Major trading partners			2005
	(million US$)	(% of exports)		(% of imports)	
Exports	64.2	UK	26	USA	44
Imports	475.4	Trinidad Tbg	22	Trinidad Tbg	14
		USA	14	UK	7

Social indicators	2000-2006
Population growth rate 2000-2005 (% per annum)	0.8
Population aged 0-14 years (%)	29.0
Population aged 60+ years (women and men, % of total)	11.0/9.0
Sex ratio (women per 100 men)	103
Life expectancy at birth 2000-2005 (women and men, years)	74/71
Infant mortality rate 2000-2005 (per 1,000 births)	15
Total fertility rate 2000-2005 (births per woman)	2.2
Contraceptive use (% of currently married women)	47[jk]
Urban population (%)	28
Urban population growth rate 2000-2005 (% per annum)	0.5
Rural population growth rate 2000-2005 (% per annum)	0.9
Foreign born (%)	5.0
Government education expenditure (% of GNP)	8.2
Primary-secondary gross enrolment ratio (w and m per 100)	96/94
Third-level students (women and men, % of total)	74/26
Television receivers (per 1,000 inhabitants)	341
Intentional homicides (per 100,000 inhabitants)	16
Parliamentary seats (women and men, % of total)	21/79

Environment	2000-2006
Threatened species	30
Forested area (% of land area)	15
CO2 emissions (000s Mt of carbon dioxide/per capita)	326/2.1
Energy consumption per capita (kilograms oil equiv.)	749
Precipitation (mm)	1399[l]
Average minimum and maximum temperatures (centigrade)[l]	24.6/30.2

a Official rate. b October 2006. c June 2006. d Persons aged 15 years and over.
e 2002. f Excluding nationals residing abroad. g 2004. h 2003. i Estimate. j For women
aged 15-44 in union or married. k 1988. l Vieux-Fort.

Saint Vincent and the Grenadines

Region	Caribbean
Largest urban agglom. (pop., 000s)	Kingstown (41)
Currency	EC dollar
Population in 2006 (proj., 000s)	120
Surface area (square kms)	388
Population density (per square km)	308
United Nations membership date	16 September 1980

Economic indicators	2000	2005
Exchange rate (national currency per US$)[a]	2.70	2.70[b]
Consumer price index (2000=100)	100	113[c]
Balance of payments, current account (million US$)	−29	−42[d]
Tourist arrivals (000s)[e]	73	87[f]
GDP (million current US$)	335	428
GDP (per capita current US$)	2891	3596
Gross fixed capital formation (% of GDP)	27.0	40.0
Employment in industrial sector (%)	21.3[g]	19.7[h]
Employment in agricultural sector (%)	24.8[g]	15.4[h]
Agricultural production index (1999-2001=100)	100	104[f]
Food production index (1999-2001=100)	100	104[f]
Primary energy production (000s Mt oil equiv.)	2	3[f]
Motor vehicles (per 1,000 inhabitants)	113.0	127.3[i]
Telephone lines (per 100 inhabitants)	22.0	18.9
Internet users, estimated (000s)	3.5	10.0

Total trade		Major trading partners			2005
	(million US$)	(% of exports)		(% of imports)	
Exports	39.9	UK	27	USA	33
Imports	240.4	Barbados	13	Trinidad Tbg	24
		Trinidad Tbg	12	UK	9

Social indicators	2000-2006
Population growth rate 2000-2005 (% per annum)	0.5
Population aged 0-14 years (%)	29.0
Population aged 60+ years (women and men, % of total)	10.0/8.0
Sex ratio (women per 100 men)	101
Life expectancy at birth 2000-2005 (women and men, years)	74/68
Infant mortality rate 2000-2005 (per 1,000 births)	26
Total fertility rate 2000-2005 (births per woman)	2.3
Contraceptive use (% of currently married women)	58[jk]
Urban population (%)	46
Urban population growth rate 2000-2005 (% per annum)	1.2
Rural population growth rate 2000-2005 (% per annum)	−<
Government education expenditure (% of GNP)	10.5
Primary-secondary gross enrolment ratio (w and m per 100)	93/94
Third-level students (women and men, % of total)	68/32[l]
Television receivers (per 1,000 inhabitants)	234
Parliamentary seats (women and men, % of total)	18/82

Environment	2000-2006
Threatened species	27
Forested area (% of land area)	15
CO2 emissions (000s Mt of carbon dioxide/per capita)	194/1.6
Energy consumption per capita (kilograms oil equiv.)	595[m]

a Official rate. b October 2006. c July 2006. d 2002. e Arrivals of non-resident tourists by air. f 2004. g 1991. h 2001. i 2003. j For women aged 15-44 in union or married. k 1988. l 1995. m Estimated data.

Samoa

Region	Oceania-Polynesia
Largest urban agglom. (pop., 000s)	Apia (41)
Currency	tala
Population in 2006 (proj., 000s)	186
Surface area (square kms)	2831
Population density (per square km)	66
United Nations membership date	15 December 1976

Economic indicators	2000	2005
Exchange rate (national currency per US$)[a]	3.34	2.74[b]
Balance of payments, current account (million US$)	...	−29
Tourist arrivals (000s)	88	98[c]
GDP (million current US$)	231	406
GDP (per capita current US$)	1301	2196
Gross fixed capital formation (% of GDP)	13.0	12.0
Agricultural production index (1999-2001=100)	101	103[c]
Food production index (1999-2001=100)	101	103[c]
Primary energy production (000s Mt oil equiv.)	3	3[c]
Motor vehicles (per 1,000 inhabitants)	36.9[d]	...
Telephone lines (per 100 inhabitants)	4.8	7.3
Internet users, estimated (000s)	1.0	6.0

Total trade		Major trading partners			2005
	(million US$)	(% of exports)		(% of imports)	
Exports	87.2	Australia	74	New Zealand	27
Imports	238.6	Amer Samoa	23	Australia	21
			...	USA	12

Social indicators	2000-2006
Population growth rate 2000-2005 (% per annum)	0.8
Population aged 0-14 years (%)	41.0
Population aged 60+ years (women and men, % of total)	7.0/6.0
Sex ratio (women per 100 men)	92
Life expectancy at birth 2000-2005 (women and men, years)	74/67
Infant mortality rate 2000-2005 (per 1,000 births)	26
Total fertility rate 2000-2005 (births per woman)	4.4
Urban population (%)	22
Urban population growth rate 2000-2005 (% per annum)	1.3
Rural population growth rate 2000-2005 (% per annum)	0.7
Foreign born (%)	4.6[e]
Government education expenditure (% of GNP)	4.3
Primary-secondary gross enrolment ratio (w and m per 100)	93/88
Third-level students (women and men, % of total)	44/56
Television receivers (per 1,000 inhabitants)	148
Parliamentary seats (women and men, % of total)	8/92

Environment	2000-2006
Threatened species	21
Forested area (% of land area)	37
CO2 emissions (000s Mt of carbon dioxide/per capita)	151/0.8
Energy consumption per capita (kilograms oil equiv.)	304[e]
Precipitation (mm)	2965[f]
Average minimum and maximum temperatures (centigrade)[f]	23.5/30.2

a Official rate. b October 2006. c 2004. d 1989. e Estimated data. f Apia.

San Marino

Region	Southern Europe
Largest urban agglom. (pop., 000s)	San Marino (27)
Currency	euro[a]
Population in 2006 (proj., 000s)	29
Surface area (square kms)	61
Population density (per square km)	475
United Nations membership date	2 March 1992

Economic indicators	2000	2005
Exchange rate (national currency per US$)	1.07	0.79[b]
Consumer price index (2000=100)[c]	...	101[d]
Unemployment (percentage of labour force)[e]	2.8	2.8[d]
Tourist arrivals (000s)[f]	43	42[d]
GDP (million current US$)	774	1315
GDP (per capita current US$)	28708	46781
Gross fixed capital formation (% of GDP)	43.0	53.0
Labour force participation, adult female pop. (%)	57.3[g]	...
Labour force participation, adult male pop. (%)	76.0[g]	...
Employment in industrial sector (%)	40.2[g]	40.9[d]
Employment in agricultural sector (%)	1.3[g]	0.4[d]
Telephone lines (per 100 inhabitants)	...	76.6[d]
Internet users, estimated (000s)	13.2	15.2

Social indicators	2000-2006
Population growth rate 2000-2005 (% per annum)	0.9
Population aged 0-14 years (%)	15.0
Population aged 60+ years (women and men, % of total)	23.0/19.0
Life expectancy at birth 2000-2005 (women and men, years)	84/0[h]
Total fertility rate 2000-2005 (births per woman)	1.3
Urban population (%)	97
Urban population growth rate 2000-2005 (% per annum)	1.6
Rural population growth rate 2000-2005 (% per annum)	−16.0
Foreign born (%)	34.6[i]
Third-level students (women and men, % of total)	58/42
Television receivers (per 1,000 inhabitants)	904
Intentional homicides (per 100,000 inhabitants)	—
Parliamentary seats (women and men, % of total)	12/88

Environment	2000-2006
Threatened species	2

a Prior to 1 January 1999, Italian lire. b October 2006. c Base: 2003=100. d 2004. e 31st December of each year. f Including Italian tourists. g 1999. h 2000. i Estimated data.

Sao Tome and Principe

Region	Middle Africa
Largest urban agglom. (pop., 000s)	Sao Tome (57)
Currency	dobra
Population in 2006 (proj., 000s)	160
Surface area (square kms)	964
Population density (per square km)	166
United Nations membership date	16 September 1975

Economic indicators	2000	2005
Exchange rate (national currency per US$)[a]	8610.65	11748.80[b]
Balance of payments, current account (million US$)	−19	−23[c]
Tourist arrivals (000s)	7	8[d]
GDP (million current US$)	46	73
GDP (per capita current US$)	332	464
Gross fixed capital formation (% of GDP)	36.0	34.0
Labour force participation, adult female pop. (%)	33.5	...
Labour force participation, adult male pop. (%)	62.2	...
Agricultural production index (1999-2001=100)	101	109[e]
Food production index (1999-2001=100)	101	109[e]
Primary energy production (000s Mt oil equiv.)	1	1[e]
Motor vehicles (per 1,000 inhabitants)	26.9[f]	...
Telephone lines (per 100 inhabitants)	3.3	4.6
Internet users, estimated (000s)	6.5	20.0[e]

Total trade		Major trading partners			2005
	(million US$)[g]	(% of exports)[g]		(% of imports)[g]	
Exports	6.6	Netherlands	39	Portugal	63
Imports	42.2	Portugal	36	Belgium	11
		Belgium	15	Angola	10

Social indicators	2000-2006
Population growth rate 2000-2005 (% per annum)	2.3
Population aged 0-14 years (%)	39.0
Population aged 60+ years (women and men, % of total)	6.0/5.0
Sex ratio (women per 100 men)	101
Life expectancy at birth 2000-2005 (women and men, years)	64/62
Infant mortality rate 2000-2005 (per 1,000 births)	82
Total fertility rate 2000-2005 (births per woman)	4.1
Contraceptive use (% of currently married women)	29
Urban population (%)	58
Urban population growth rate 2000-2005 (% per annum)	3.9
Rural population growth rate 2000-2005 (% per annum)	0.2
Foreign born (%)	4.9[h]
Primary-secondary gross enrolment ratio (w and m per 100)	91/91
Television receivers (per 1,000 inhabitants)	127
Parliamentary seats (women and men, % of total)	7/93

Environment	2000-2006
Threatened species	61[e]
Forested area (% of land area)	28
CO2 emissions (000s Mt of carbon dioxide/per capita)	92/0.6
Energy consumption per capita (kilograms oil equiv.)	174[h]

a Official rate. b 1st quarter 2006. c 2002. d 2001. e 2004. f 1987. g 2003. h Estimated data.

Saudi Arabia

Region	Western Asia
Largest urban agglom. (pop., 000s)	Riyadh (4193)
Currency	riyal
Population in 2006 (proj., 000s)	25193
Surface area (square kms)	2149690
Population density (per square km)	12
United Nations membership date	24 October 1945

Economic indicators	2000	2005
Exchange rate (national currency per US$)[a]	3.75	3.75[b]
Consumer price index (2000=100)[c]	100	100[d]
Unemployment (percentage of labour force)[e]	4.6	5.2[f]
Balance of payments, current account (million US$)	14317	87132
Tourist arrivals (000s)	6585	8599[g]
GDP (million current US$)	188442	314021
GDP (per capita current US$)	8771	12779
Gross fixed capital formation (% of GDP)	17.0	16.0
Employment in industrial sector (%)	19.9	21.0[f]
Employment in agricultural sector (%)	6.1	4.7[f]
Agricultural production index (1999-2001=100)	93	119[g]
Food production index (1999-2001=100)	93	119[g]
Primary energy production (000s Mt oil equiv.)[h]	486267	554465[g]
Motor vehicles (per 1,000 inhabitants)	374.6	415.3[g]
Telephone lines (per 100 inhabitants)	13.8	15.5
Internet users, estimated (000s)[i]	460.0	1586.0[g]

Total trade		Major trading partners		2005
	(million US$)	(% of exports)		(% of imports)
Exports	180737.2	...	USA	15
Imports	59510.8	...	Japan	9
		...	Germany	8

Social indicators	2000-2006
Population growth rate 2000-2005 (% per annum)	2.7
Population aged 0-14 years (%)	37.0
Population aged 60+ years (women and men, % of total)	5.0/4.0
Sex ratio (women per 100 men)	86
Life expectancy at birth 2000-2005 (women and men, years)	74/70
Infant mortality rate 2000-2005 (per 1,000 births)	23
Total fertility rate 2000-2005 (births per woman)	4.1
Urban population (%)	81
Urban population growth rate 2000-2005 (% per annum)	3.0
Rural population growth rate 2000-2005 (% per annum)	1.5
Foreign born (%)	23.7[j]
Refugees and others of concern to UNHCR[k]	310913
Primary-secondary gross enrolment ratio (w and m per 100)	65/70
Third-level students (women and men, % of total)	59/41
Television receivers (per 1,000 inhabitants)	275
Parliamentary seats (women and men, % of total)	—/100

Environment	2000-2006
Threatened species	50
Forested area (% of land area)	1
CO2 emissions (000s Mt of carbon dioxide/per capita)[l]	302884/13.0
Energy consumption per capita (kilograms oil equiv.)[h]	6020
Precipitation (mm)	99[m]
Average minimum and maximum temperatures (centigrade)[m]	19.7/32.9

a Official rate. b October 2006. c All cities. d March 2006. e Persons aged 15 years and over. f 2002. g 2004. h Part Neutral Zone. i Data refer to Heigirian year, correspondence to Gregorian year is assumed. j Estimated data. k Provisional. l Including part of the Neutral Zone. m Riyadh.

Senegal

Region	Western Africa
Largest urban agglom. (pop., 000s)	Dakar (2159)
Currency	CFA franc
Population in 2006 (proj., 000s)	11936
Surface area (square kms)	196722
Population density (per square km)	61
United Nations membership date	28 September 1960

Economic indicators	2000	2005
Exchange rate (national currency per US$)[a]	704.95	516.66[b]
Consumer price index (2000=100)[c]	100	109[d]
Industrial production index (1995=100)	100	154[e]
Balance of payments, current account (million US$)	−332	−437[f]
Tourist arrivals (000s)	389	363[g]
GDP (million current US$)	4374	8274
GDP (per capita current US$)	423	710
Gross fixed capital formation (% of GDP)	17.0	23.0
Agricultural production index (1999-2001=100)	101	84[g]
Food production index (1999-2001=100)	101	82[g]
Primary energy production (000s Mt oil equiv.)	2	37[g]
Motor vehicles (per 1,000 inhabitants)	23.2	17.0[g]
Telephone lines (per 100 inhabitants)	2.2	2.3
Internet users, estimated (000s)	40.0	540.0

Total trade		Major trading partners			2005
	(million US$)	(% of exports)		(% of imports)	
Exports	1470.8	Mali	19	France	21
Imports	3497.7	India	13	Nigeria	10
		France	9	UK	5

Social indicators	2000-2006
Population growth rate 2000-2005 (% per annum)	2.4
Population aged 0-14 years (%)	43.0
Population aged 60+ years (women and men, % of total)	5.0/4.0
Sex ratio (women per 100 men)	103
Life expectancy at birth 2000-2005 (women and men, years)	57/54
Infant mortality rate 2000-2005 (per 1,000 births)	83
Total fertility rate 2000-2005 (births per woman)	5.1
Urban population (%)	42
Urban population growth rate 2000-2005 (% per annum)	2.8
Rural population growth rate 2000-2005 (% per annum)	2.1
Foreign born (%)	3.0[h]
Refugees and others of concern to UNHCR[i]	23341
Government education expenditure (% of GNP)	4.1
Primary-secondary gross enrolment ratio (w and m per 100)	45/50
Third-level students (women and men, % of total)	54/46[j]
Television receivers (per 1,000 inhabitants)	45
Parliamentary seats (women and men, % of total)	19/81

Environment	2000-2006
Threatened species	58
Forested area (% of land area)	32
CO2 emissions (000s Mt of carbon dioxide/per capita)	4846/0.4
Energy consumption per capita (kilograms oil equiv.)	120
Precipitation (mm)	514[c]
Average minimum and maximum temperatures (centigrade)[c]	21.7/27.6

a Official rate. b October 2006. c Dakar. d July 2006. e 1st quarter 2006. f 2003.
g 2004. h 1988. i Provisional. j 1998.

Serbia

Region	Southern Europe
Largest urban agglom. (pop., 000s)	Belgrade (1576)[a]
Currency	dinar
Population in 2006 (proj., 000s)	7441[abc]
Surface area (square kms)	88361[a]
Population density (per square km)	84[a]
United Nations membership date	1 November 2000

Economic indicators	2000	2005
Exchange rate (national currency per US$)[d]	70.00	60.00[e]
Consumer price index (2000=100)	100	267[f]
Industrial production index (1995=100)[g]	102	116[h]
Unemployment (percentage of labour force)[gi]	12.6	15.2[j]
Tourist arrivals (000s)[g]	239	580[f]
GDP (million current US$)	8068	24207
GDP (per capita current US$)	1053	3244
Gross fixed capital formation (% of GDP)	14.0	13.0
Agricultural production index (1999-2001=100)[g]	100	114[f]
Food production index (1999-2001=100)[g]	100	114[f]
Primary energy production (000s Mt oil equiv.)	10596	10730[f]
Motor vehicles (per 1,000 inhabitants)[g]	162.7	162.9[f]
Telephone lines (per 100 inhabitants)[g]	22.6	32.9
Internet users, estimated (000s)	847.0[gj]	776.8

Total trade		Major trading partners			2005
	(million US$)[fg]	(% of exports)[fg]		(% of imports)[fg]	
Exports	3801.3	Bosnia/Herzeg	17	Germany	13
Imports	11366.4	Italy	15	Russian Fed.	12
		Germany	9	Italy	10

Social indicators	2000-2006
Population growth rate 2000-2005 (% per annum)	-0.1[g]
Population aged 0-14 years (%)[g]	18.0
Population aged 60+ years (women and men, % of total)[g]	21.0/101.0
Life expectancy at birth 2000-2005 (women and men, years)	76/13[g]
Total fertility rate 2000-2005 (births per woman)	1.7[g]
Urban population growth rate 2000-2005 (% per annum)[g]	0.2
Rural population growth rate 2000-2005 (% per annum)[g]	-0.3
Foreign born (%)	9.2
Refugees and others of concern to UNHCR[k]	487998[g]
Television receivers (per 1,000 inhabitants)	282[g]
Parliamentary seats (women and men, % of total)	12/88

Environment	2000-2006
CO_2 emissions (000s Mt of carbon dioxide/per capita)[g]	50023/4.8
Energy consumption per capita (kilograms oil equiv.)[g]	1451
Precipitation (mm)	683[l]
Average minimum and maximum temperatures (centigrade)[l]	7.8/16.7

a Statistical Office of Republic of Serbia. b Data on Kosovo and Metohija not included. c 2005. d Data refer to non-commercial rates derived from the Operational Rates of Exchange for United Nations Programmes. e November 2006. f 2004. g Including Montenegro. h March 2006. i Oct. of each year, persons aged 15 years and over. j 2003. k Provisional. l Belgrade.

Seychelles

Region	Eastern Africa
Largest urban agglom. (pop., 000s)	Victoria (25)[a]
Currency	rupee
Population in 2006 (proj., 000s)	83
Surface area (square kms)	455
Population density (per square km)	182
United Nations membership date	21 September 1976

Economic indicators	2000	2005
Exchange rate (national currency per US$)[b]	6.27	5.52[c]
Balance of payments, current account (million US$)	−54	−179
Tourist arrivals (000s)	130	121[d]
GDP (million current US$)	618	699
GDP (per capita current US$)	8008	8668
Gross fixed capital formation (% of GDP)	35.0	12.0
Agricultural production index (1999-2001=100)	101	92[d]
Food production index (1999-2001=100)	100	92[d]
Motor vehicles (per 1,000 inhabitants)	112.1[e]	108.6[a]
Telephone lines (per 100 inhabitants)	26.7	26.5
Internet users, estimated (000s)	6.0	21.0

Total trade		Major trading partners			2005
	(million US$)	(% of exports)		(% of imports)	
Exports	339.7	Saudi Arabia	36	Saudi Arabia	23
Imports	674.9	UK	29	Germany	14
		France	15	Spain	8

Social indicators	2000-2006
Population growth rate 2000-2005 (% per annum)	0.9
Population aged 0-14 years (%)	28.0
Total fertility rate 2000-2005 (births per woman)	2.0
Urban population (%)	53
Urban population growth rate 2000-2005 (% per annum)	1.6
Rural population growth rate 2000-2005 (% per annum)	0.1
Government education expenditure (% of GNP)	5.7
Primary-secondary gross enrolment ratio (w and m per 100)	108/104
Third-level students (women and men, % of total)	88/12[f]
Television receivers (per 1,000 inhabitants)	278
Parliamentary seats (women and men, % of total)	29/71

Environment	2000-2006
Threatened species	95
Forested area (% of land area)	67
CO2 emissions (000s Mt of carbon dioxide/per capita)	547/6.9
Energy consumption per capita (kilograms oil equiv.)	2230[g]
Precipitation (mm)	2172[h]
Average minimum and maximum temperatures (centigrade)[h]	24.4/29.8

a 2003. b Official rate. c October 2006. d 2004. e 1999. f 1995. g Estimated data.
h Victoria.

Sierra Leone

Region	Western Africa
Largest urban agglom. (pop., 000s)	Freetown (799)
Currency	leone
Population in 2006 (proj., 000s)	5679
Surface area (square kms)	71740
Population density (per square km)	79
United Nations membership date	27 September 1961

Economic indicators	2000	2005
Exchange rate (national currency per US$)	1666.67	2986.91[a]
Consumer price index (2000=100)[b]	100	106[c]
Balance of payments, current account (million US$)	−112	−129
Tourist arrivals (000s)[d]	16	44[e]
GDP (million current US$)	636	1162
GDP (per capita current US$)	141	210
Gross fixed capital formation (% of GDP)	6.0	21.0
Agricultural production index (1999-2001=100)	94	114[e]
Food production index (1999-2001=100)	94	114[e]
Motor vehicles (per 1,000 inhabitants)[f]	8.0	3.9[g]
Telephone lines (per 100 inhabitants)	0.4	0.5[eh]
Internet users, estimated (000s)	5.0	10.0[e]

Total trade		Major trading partners			2005
	(million US$)[g]	(% of exports)		(% of imports)[g]	
Exports	41.4	...	Côte d'Ivoire	37	
Imports	352.0	...	Canada	7	
		...	Netherlands	6	

Social indicators	2000-2006
Population growth rate 2000-2005 (% per annum)	4.1
Population aged 0-14 years (%)	43.0
Population aged 60+ years (women and men, % of total)	6.0/5.0
Sex ratio (women per 100 men)	103
Life expectancy at birth 2000-2005 (women and men, years)	42/39
Infant mortality rate 2000-2005 (per 1,000 births)	165
Total fertility rate 2000-2005 (births per woman)	6.5
Contraceptive use (% of currently married women)	4
Urban population (%)	41
Urban population growth rate 2000-2005 (% per annum)	5.9
Rural population growth rate 2000-2005 (% per annum)	2.9
Foreign born (%)	1.1[i]
Refugees and others of concern to UNHCR[j]	60352
Government education expenditure (% of GNP)	3.8
Primary-secondary gross enrolment ratio (w and m per 100)	72/97
Third-level students (women and men, % of total)	29/71
Television receivers (per 1,000 inhabitants)	13
Parliamentary seats (women and men, % of total)	15/85

Environment	2000-2006
Threatened species	91
Forested area (% of land area)	15
CO2 emissions (000s Mt of carbon dioxide/per capita)	653/0.1
Energy consumption per capita (kilograms oil equiv.)	44
Precipitation (mm)	2946[b]
Average minimum and maximum temperatures (centigrade)[b]	23.8/29.9

a October 2006. b Freetown. c 2003. d Arrivals by air. e 2004. f Source: World Automotive Market Report, Auto and Truck International (Illinois). g 2002. h Estimate. i Estimated data. j Provisional.

Singapore

Region	South-eastern Asia
Largest urban agglom. (pop., 000s)	Singapore (4326)
Currency	dollar
Population in 2006 (proj., 000s)	4380
Surface area (square kms)	683
Population density (per square km)	6412
United Nations membership date	21 September 1965

Economic indicators	2000	2005
Exchange rate (national currency per US$)	1.73	1.56[a]
Consumer price index (2000=100)	100	104[b]
Industrial production index (1995=100)[c]	141	186[d]
Unemployment (percentage of labour force)[e]	4.6[f]	5.3[g]
Balance of payments, current account (million US$)	11936	27897[g]
Tourist arrivals (000s)	6917	5705[h]
GDP (million current US$)	92717	116775
GDP (per capita current US$)	23079	26997
Gross fixed capital formation (% of GDP)	31.0	22.0
Labour force participation, adult female pop. (%)	55.5	53.9[hi]
Labour force participation, adult male pop. (%)	81.1	75.8[hi]
Employment in industrial sector (%)	34.2	29.5
Employment in agricultural sector (%)	0.2	0.3[g]
Agricultural production index (1999-2001=100)	66	70[g]
Food production index (1999-2001=100)	66	70[g]
Motor vehicles (per 1,000 inhabitants)	137.1	169.9[gi]
Telephone lines (per 100 inhabitants)[j]	48.4	43.5
Internet users, estimated (000s)[k]	1300.0	1749.9

Total trade		Major trading partners			2005
	(million US$)	(% of exports)		(% of imports)	
Exports	229652.3	Malaysia	13	Malaysia	14
Imports	200050.3	USA	10	USA	12
		Indonesia	10	China	10

Social indicators	2000-2006
Population growth rate 2000-2005 (% per annum)	1.5
Population aged 0-14 years (%)	20.0
Population aged 60+ years (women and men, % of total)	13.0/11.0
Sex ratio (women per 100 men)	99
Life expectancy at birth 2000-2005 (women and men, years)	81/77
Infant mortality rate 2000-2005 (per 1,000 births)	3
Total fertility rate 2000-2005 (births per woman)	1.4
Contraceptive use (% of currently married women)	75[i]
Urban population (%)	100
Urban population growth rate 2000-2005 (% per annum)	1.5
Rural population growth rate 2000-2005 (% per annum)	–
Foreign born (%)	17.6
Refugees and others of concern to UNHCR[l]	4
Government education expenditure (% of GNP)	3.7
Third-level students (women and men, % of total)	45/55
Newspaper circulation (per 1,000 inhabitants)	408[i]
Television receivers (per 1,000 inhabitants)[j]	307
Intentional homicides (per 100,000 inhabitants)	1[i]
Parliamentary seats (women and men, % of total)	21/79

Environment	2000-2006
Threatened species	100
Forested area (% of land area)	3
CO2 emissions (000s Mt of carbon dioxide/per capita)	47885/11.3
Energy consumption per capita (kilograms oil equiv.)	3482
Precipitation (mm)	2150[m]
Average minimum and maximum temperatures (centigrade)[m]	23.9/30.9

a October 2006. b June 2006. c Calculated by the Statistics Division of the United Nations from component national indices. d August 2006. e Persons aged 15 years and over, June of each year. f 1999. g 2004. h 2003. i Singapore Department of Statistics. j Year beginning 1 April. k From 1983, year beginning 1 April. l Provisional. m Singapore.

Slovakia

Region	Eastern Europe
Largest urban agglom. (pop., 000s)	Bratislava (424)
Currency	koruna
Population in 2006 (proj., 000s)	5401
Surface area (square kms)	49033
Population density (per square km)	110
United Nations membership date	19 January 1993

Economic indicators	2000	2005
Exchange rate (national currency per US$)[a]	47.39	28.63[b]
Consumer price index (2000=100)	100	139[c]
Industrial production index (1995=100)[d]	115	160[e]
Unemployment (percentage of labour force)[f]	18.6	16.2
Balance of payments, current account (million US$)	−694	−282[g]
Tourist arrivals (000s)	1053	1401[h]
GDP (million current US$)	20291	46417
GDP (per capita current US$)	3757	8594
Gross fixed capital formation (% of GDP)	26.0	27.0
Labour force participation, adult female pop. (%)	52.6	52.5[h]
Labour force participation, adult male pop. (%)	68.2	68.3[h]
Employment in industrial sector (%)	37.3	38.8
Employment in agricultural sector (%)	6.7	4.7
Agricultural production index (1999-2001=100)	92	107[h]
Food production index (1999-2001=100)	92	107[h]
Primary energy production (000s Mt oil equiv.)	3133	2948[h]
Motor vehicles (per 1,000 inhabitants)	264.3	254.0[h]
Telephone lines (per 100 inhabitants)	31.4	22.2
Internet users, estimated (000s)	507.0	1905.2

Total trade		Major trading partners		2005
	(million US$)	(% of exports)		(% of imports)
Exports	31997.2	Germany	26	Germany	21
Imports	34445.9	Czech Republic	14	Czech Republic	13
		Austria	7	Russian Fed.	11

Social indicators	2000-2006
Population growth rate 2000-2005 (% per annum)	<
Population aged 0-14 years (%)	17.0
Population aged 60+ years (women and men, % of total)	19.0/13.0
Sex ratio (women per 100 men)	106
Life expectancy at birth 2000-2005 (women and men, years)	78/70
Infant mortality rate 2000-2005 (per 1,000 births)	8
Total fertility rate 2000-2005 (births per woman)	1.2
Urban population (%)	56
Urban population growth rate 2000-2005 (% per annum)	−<
Rural population growth rate 2000-2005 (% per annum)	<
Foreign born (%)	2.2
Refugees and others of concern to UNHCR[i]	3075
Government education expenditure (% of GNP)	4.4
Primary-secondary gross enrolment ratio (w and m per 100)	96/95
Third-level students (women and men, % of total)	54/46
Newspaper circulation (per 1,000 inhabitants)	131
Television receivers (per 1,000 inhabitants)	424
Intentional homicides (per 100,000 inhabitants)	2
Parliamentary seats (women and men, % of total)	20/80

Environment	2000-2006
Threatened species	52
Forested area (% of land area)	45
CO2 emissions (000s Mt of carbon dioxide/per capita)[j]	43050/8.0
Energy consumption per capita (kilograms oil equiv.)	2784
Precipitation (mm)	557[k]
Average minimum and maximum temperatures (centigrade)[k]	5.7/15.2

a Principal rate. b October 2006. c May 2006. d Monthly indices are adjusted for differences in the number of working days. e August 2006. f Persons aged 15 years and over, excl. persons on child-care leave. g 2003. h 2004. i Provisional. j Source: UNFCCC. k Bratislava.

Slovenia

Region	Southern Europe
Largest urban agglom. (pop., 000s)	Ljubljana (263)
Currency	tolar
Population in 2006 (proj., 000s)	1966
Surface area (square kms)	20256
Population density (per square km)	97
United Nations membership date	22 May 1992

Economic indicators	2000	2005
Exchange rate (national currency per US$)[a]	227.38	188.32[b]
Consumer price index (2000=100)[c]	100	135[d]
Industrial production index (1995=100)	112	145[e]
Unemployment (percentage of labour force)[f]	7.2	5.8
Balance of payments, current account (million US$)	−548	−362
Tourist arrivals (000s)	1090	1499[g]
GDP (million current US$)	19098	34030
GDP (per capita current US$)	9710	17302
Gross fixed capital formation (% of GDP)	25.0	25.0
Labour force participation, adult female pop. (%)	51.8[h]	52.8[g]
Labour force participation, adult male pop. (%)	64.9[h]	65.3[g]
Employment in industrial sector (%)	37.4	37.2
Employment in agricultural sector (%)	9.5	8.8
Agricultural production index (1999-2001=100)	102	109[g]
Food production index (1999-2001=100)	102	109[g]
Primary energy production (000s Mt oil equiv.)	1720	1872[g]
Motor vehicles (per 1,000 inhabitants)	467.3	505.8[g]
Telephone lines (per 100 inhabitants)	39.5	41.5
Internet users, estimated (000s)	300.0	1090.0

Total trade		Major trading partners			2005
	(million US$)	(% of exports)		(% of imports)	
Exports	17896.0	Germany	20	Germany	20
Imports	19626.3	Italy	13	Italy	19
		Croatia	9	Austria	12

Social indicators	2000-2006
Population growth rate 2000-2005 (% per annum)	−
Population aged 0-14 years (%)	14.0
Population aged 60+ years (women and men, % of total)	24.0/17.0
Sex ratio (women per 100 men)	105
Life expectancy at birth 2000-2005 (women and men, years)	80/73
Infant mortality rate 2000-2005 (per 1,000 births)	5
Total fertility rate 2000-2005 (births per woman)	1.2
Urban population (%)	51
Urban population growth rate 2000-2005 (% per annum)	0.1
Rural population growth rate 2000-2005 (% per annum)	−0.1
Foreign born (%)	8.6
Refugees and others of concern to UNHCR[i]	881
Government education expenditure (% of GNP)	6.1
Primary-secondary gross enrolment ratio (w and m per 100)	106/107
Third-level students (women and men, % of total)	57/43
Newspaper circulation (per 1,000 inhabitants)	170
Television receivers (per 1,000 inhabitants)	366
Intentional homicides (per 100,000 inhabitants)	1
Parliamentary seats (women and men, % of total)	11/89

Environment	2000-2006
Threatened species	81
Forested area (% of land area)	55
CO2 emissions (000s Mt of carbon dioxide/per capita)[j]	16100/8.2
Energy consumption per capita (kilograms oil equiv.)	2745
Precipitation (mm)	1368[k]
Average minimum and maximum temperatures (centigrade)[k]	5.9/15.0

a Official rate. b October 2006. c Urban areas. d August 2006. e June 2006. f Persons aged 15 years and over, second quarter of each year. g 2004. h 1999. i Provisional. j Source: UNFCCC. k Ljubljana.

Solomon Islands

Region	Oceania-Melanesia
Largest urban agglom. (pop., 000s)	Honiara (61)
Currency	dollar
Population in 2006 (proj., 000s)	490
Surface area (square kms)	28896
Population density (per square km)	17
United Nations membership date	19 September 1978

Economic indicators	2000	2005
Exchange rate (national currency per US$)[a]	5.10	7.62[b]
Consumer price index (2000=100)	100	162[c]
Tourist arrivals (000s)	21[d]	...
GDP (million current US$)	338	299
GDP (per capita current US$)	808	626
Gross fixed capital formation (% of GDP)	18.0	19.0
Agricultural production index (1999-2001=100)	98	143[e]
Food production index (1999-2001=100)	98	143[e]
Telephone lines (per 100 inhabitants)[f]	1.8	1.6
Internet users, estimated (000s)	2.0	4.0

Social indicators	2000-2006
Population growth rate 2000-2005 (% per annum)	2.6
Population aged 0-14 years (%)	41.0
Population aged 60+ years (women and men, % of total)	4.0/4.0
Sex ratio (women per 100 men)	94
Life expectancy at birth 2000-2005 (women and men, years)	63/62
Infant mortality rate 2000-2005 (per 1,000 births)	34
Total fertility rate 2000-2005 (births per woman)	4.3
Urban population (%)	17
Urban population growth rate 2000-2005 (% per annum)	4.2
Rural population growth rate 2000-2005 (% per annum)	2.3
Foreign born (%)	1.5[g]
Government education expenditure (% of GNP)	3.3[d]
Primary-secondary gross enrolment ratio (w and m per 100)	59/65
Television receivers (per 1,000 inhabitants)	11
Parliamentary seats (women and men, % of total)	—/100

Environment	2000-2006
Threatened species	75
Forested area (% of land area)	89
CO2 emissions (000s Mt of carbon dioxide/per capita)	178/0.4
Energy consumption per capita (kilograms oil equiv.)	111[h]
Precipitation (mm)	3290[i]
Average minimum and maximum temperatures (centigrade)[i]	23.2/30.1

a Official rate. b September 2006. c 2nd quarter 2006. d 1999. e 2004. f Billable lines. g 1986. h Estimated data. i Auki.

Somalia

Region	Eastern Africa
Largest urban agglom. (pop., 000s)	Mogadishu (1320)
Currency	shilling
Population in 2006 (proj., 000s)	8496
Surface area (square kms)	637657
Population density (per square km)	13
United Nations membership date	20 September 1960

Economic indicators	2000	2005
Exchange rate (national currency per US$)[a]	10280.00	14406.00[b]
Tourist arrivals (000s)	10[c]	...
GDP (million current US$)	2070	2182
GDP (per capita current US$)	295	265
Gross fixed capital formation (% of GDP)	20.0	20.0
Motor vehicles (per 1,000 inhabitants)	3.8[de]	...
Telephone lines (per 100 inhabitants)	0.4	1.2
Internet users, estimated (000s)	15.0	90.0

Social indicators	2000-2006
Population growth rate 2000-2005 (% per annum)	3.2
Population aged 0-14 years (%)	44.0
Population aged 60+ years (women and men, % of total)	4.0/4.0
Sex ratio (women per 100 men)	102
Life expectancy at birth 2000-2005 (women and men, years)	47/45
Infant mortality rate 2000-2005 (per 1,000 births)	126
Total fertility rate 2000-2005 (births per woman)	6.4
Urban population (%)	35
Urban population growth rate 2000-2005 (% per annum)	4.3
Rural population growth rate 2000-2005 (% per annum)	2.6
Refugees and others of concern to UNHCR[f]	412543
Government education expenditure (% of GNP)	0.3[g]
Television receivers (per 1,000 inhabitants)	26
Parliamentary seats (women and men, % of total)	8/92

Environment	2000-2006
Threatened species	71
Forested area (% of land area)	12
CO2 emissions (000s Mt of carbon dioxide/per capita)	—/0.0
Precipitation (mm)	422
Average minimum and maximum temperatures (centigrade)	26.0/29.3

a Data refer to non-commercial rates derived from the Operational Rates of Exchange for United Nations Programmes. b October 2006. c 1998. d Source: AAMA Motor Vehicle Facts and Figures, American Automobile Manufacturers Association (Michigan). e 1995. f Provisional. g 1986.

South Africa

Region	Southern Africa
Largest urban agglom. (pop., 000s)	Johannesburg (3254)
Currency	rand
Population in 2006 (proj., 000s)	47594
Surface area (square kms)	1221037
Population density (per square km)	39
United Nations membership date	7 November 1945

Economic indicators	2000	2005
Exchange rate (national currency per US$)[a]	7.57	7.45[b]
Consumer price index (2000=100)	100	136[c]
Industrial production index (1995=100)[d]	105	128[e]
Unemployment (percentage of labour force)[f]	26.3[g]	26.6[h]
Balance of payments, current account (million US$)	−191	−10079
Tourist arrivals (000s)[i]	5872	6678[j]
GDP (million current US$)	132878	238825
GDP (per capita current US$)	2913	5035
Gross fixed capital formation (% of GDP)	15.0	17.0
Labour force participation, adult female pop. (%)	43.9[k]	46.2[j]
Labour force participation, adult male pop. (%)	57.7[k]	62.0[j]
Employment in industrial sector (%)	24.1	24.5[l]
Employment in agricultural sector (%)	14.5	10.3[l]
Agricultural production index (1999-2001=100)	105	105[j]
Food production index (1999-2001=100)	105	106[j]
Primary energy production (000s Mt oil equiv.)[m]	134667	137277[j]
Motor vehicles (per 1,000 inhabitants)	138.0[k]	...
Telephone lines (per 100 inhabitants)[n]	10.9	10.0
Internet users, estimated (000s)[n]	2400.0	5100.0

Total trade		Major trading partners			2005
	(million US$)	(% of exports)		(% of imports)	
Exports	46994.8	Japan	11	Germany	14
Imports	55032.6	UK	11	China	9
		USA	10	USA	8

Social indicators	2000-2006
Population growth rate 2000-2005 (% per annum)	0.8
Population aged 0-14 years (%)	33.0
Population aged 60+ years (women and men, % of total)	8.0/6.0
Sex ratio (women per 100 men)	104
Life expectancy at birth 2000-2005 (women and men, years)	51/47
Infant mortality rate 2000-2005 (per 1,000 births)	43
Total fertility rate 2000-2005 (births per woman)	2.8
Urban population (%)	59
Urban population growth rate 2000-2005 (% per annum)	1.6
Rural population growth rate 2000-2005 (% per annum)	−0.4
Refugees and others of concern to UNHCR[o]	169809
Government education expenditure (% of GNP)	5.5
Primary-secondary gross enrolment ratio (w and m per 100)	99/99
Third-level students (women and men, % of total)	54/46
Newspaper circulation (per 1,000 inhabitants)	25
Television receivers (per 1,000 inhabitants)[n]	195
Parliamentary seats (women and men, % of total)	33/67

Environment	2000-2006
Threatened species	379
Forested area (% of land area)	7
CO2 emissions (000s Mt of carbon dioxide/per capita)	364853/7.8
Energy consumption per capita (kilograms oil equiv.)[p]	2328
Precipitation (mm)	515[q]
Average minimum and maximum temperatures (centigrade)[q]	11.4/22.0

a Principal rate. b October 2006. c September 2006. d Calculated by the Statistics Division of the United Nations from component national indices. e August 2006. f Persons aged 15 to 65 years. g Feb. and Sept. h March and September. i Excluding arrivals by work and contract workers. j 2004. k 1999. l 2003. m Refers to Southern Africa Customs Union. n Year beginning 1 April. o Provisional. p Southern Africa Customs Union. q Cape Town.

Spain

Region	Southern Europe
Largest urban agglom. (pop., 000s)	Madrid (5608)
Currency	euro[a]
Population in 2006 (proj., 000s)	43379
Surface area (square kms)	505992
Population density (per square km)	86
. United Nations membership date	14 December 1955

Economic indicators	2000	2005
Exchange rate (national currency per US$)	1.07	0.79[b]
Consumer price index (2000=100)[c]	...	118[d]
Industrial production index (1995=100)	119	95[e]
Unemployment (percentage of labour force)[f]	13.9	9.2
Balance of payments, current account (million US$)	−23076	−83136
Tourist arrivals (000s)	47898	52430[g]
GDP (million current US$)	580673	1124612
GDP (per capita current US$)	14261	26115
Gross fixed capital formation (% of GDP)	26.0	29.0
Labour force participation, adult female pop. (%)	41.3	45.2[g]
Labour force participation, adult male pop. (%)	67.1	68.1[g]
Employment in industrial sector (%)	31.0	29.7
Employment in agricultural sector (%)	6.6	5.3
Agricultural production index (1999-2001=100)	103	106[g]
Food production index (1999-2001=100)	103	106[g]
Primary energy production (000s Mt oil equiv.)	16851	17126[g]
Motor vehicles (per 1,000 inhabitants)	526.2	567.6[g]
Telephone lines (per 100 inhabitants)	42.2	42.9
Internet users, estimated (000s)	5486.0	15119.0

Total trade		Major trading partners			2005
	(million US$)		(% of exports)		(% of imports)
Exports	192798.4	France	19	Germany	15
Imports	289610.8	Germany	11	France	14
		Portugal	10	Italy	9

Social indicators	2000-2006
Population growth rate 2000-2005 (% per annum)	1.1
Population aged 0-14 years (%)	14.0
Population aged 60+ years (women and men, % of total)	24.0/19.0
Sex ratio (women per 100 men)	104
Life expectancy at birth 2000-2005 (women and men, years)	83/76
Infant mortality rate 2000-2005 (per 1,000 births)	5
Total fertility rate 2000-2005 (births per woman)	1.3
Urban population (%)	77
Urban population growth rate 2000-2005 (% per annum)	1.2
Rural population growth rate 2000-2005 (% per annum)	0.7
Refugees and others of concern to UNHCR[h]	5392
Government education expenditure (% of GNP)	4.6
Primary-secondary gross enrolment ratio (w and m per 100)	115/112
Third-level students (women and men, % of total)	54/46
Newspaper circulation (per 1,000 inhabitants)	98
Television receivers (per 1,000 inhabitants)	549
Intentional homicides (per 100,000 inhabitants)	1
Parliamentary seats (women and men, % of total)	33/67

Environment	2000-2006
Threatened species	219
Forested area (% of land area)	29
CO2 emissions (000s Mt of carbon dioxide/per capita)[i]	331760/7.9
Energy consumption per capita (kilograms oil equiv.)	2730
Precipitation (mm)	456[j]
Average minimum and maximum temperatures (centigrade)[j]	9.7/19.5

a Prior to 1 January 1999, peseta. b September 2006. c Base: 2001=100, beginning 2002. d May 2006. e August 2006. f Persons aged 16 years and over. g 2004. h Provisional. i Source: UNFCCC. j Madrid.

Sri Lanka

Region	South-central Asia
Largest urban agglom. (pop., 000s)	Colombo (652)
Currency	rupee
Population in 2006 (proj., 000s)	20912
Surface area (square kms)	65610
Population density (per square km)	319
United Nations membership date	14 December 1955

Economic indicators	2000	2005
Exchange rate (national currency per US$)	82.58	107.28[a]
Consumer price index (2000=100)[b]	100	183[c]
Unemployment (percentage of labour force)[d]	8.0[e]	7.7[f]
Balance of payments, current account (million US$)	−1044	−677[g]
Tourist arrivals (000s)[h]	400	566[g]
GDP (million current US$)	16717	23927
GDP (per capita current US$)	842	1154
Gross fixed capital formation (% of GDP)	25.0	25.0
Labour force participation, adult female pop. (%)	36.5	34.7[g]
Labour force participation, adult male pop. (%)	75.7	74.8[g]
Employment in industrial sector (%)	22.5[i]	23.4[j]
Employment in agricultural sector (%)	41.6[i]	34.3[j]
Agricultural production index (1999-2001=100)	102	97[g]
Food production index (1999-2001=100)	102	96[g]
Primary energy production (000s Mt oil equiv.)	275	255[g]
Motor vehicles (per 1,000 inhabitants)	28.9[k]	38.3[g]
Telephone lines (per 100 inhabitants)	4.2	6.0
Internet users, estimated (000s)	121.5	280.0[g]

Total trade		Major trading partners		2005
	(million US$)	(% of exports)		(% of imports)
Exports	6159.9	USA	32	India	17
Imports	8307.1	UK	13	Singapore	9
		India	9	China, HK SAR	8

Social indicators	2000-2006
Population growth rate 2000-2005 (% per annum)	0.9
Population aged 0-14 years (%)	24.0
Population aged 60+ years (women and men, % of total)	12.0/10.0
Sex ratio (women per 100 men)	97
Life expectancy at birth 2000-2005 (women and men, years)	77/71
Infant mortality rate 2000-2005 (per 1,000 births)	17
Total fertility rate 2000-2005 (births per woman)	2.0
Urban population (%)	15
Urban population growth rate 2000-2005 (% per annum)	0.1
Rural population growth rate 2000-2005 (% per annum)	1.0
Foreign born (%)	2.1[l]
Refugees and others of concern to UNHCR[m]	354811
Government education expenditure (% of GNP)	3.1
Primary-secondary gross enrolment ratio (w and m per 100)	90/88
Third-level students (women and men, % of total)	44/56[n]
Newspaper circulation (per 1,000 inhabitants)	27
Television receivers (per 1,000 inhabitants)	125
Parliamentary seats (women and men, % of total)	5/95

Environment	2000-2006
Threatened species	459
Forested area (% of land area)	30
CO2 emissions (000s Mt of carbon dioxide/per capita)	10321/0.5
Energy consumption per capita (kilograms oil equiv.)	203
Precipitation (mm)	2524[b]
Average minimum and maximum temperatures (centigrade)[b]	24.1/30.6

a October 2006. b Colombo. c August 2006. d Persons aged 10 years and over, excl. Northern province. e First quarter of each year. f August. g 2004. h Excluding nationals residing abroad. i 1998. j 2003. k Including three-wheelers. l Estimated data. m Provisional. n 1995.

Sudan

Region	Northern Africa
Largest urban agglom. (pop., 000s)	Khartoum (4518)
Currency	dinar
Population in 2006 (proj., 000s)	36992
Surface area (square kms)	2505813
Population density (per square km)	15
United Nations membership date	12 November 1956

Economic indicators	2000	2005
Exchange rate (national currency per US$)[a]	257.35	205.96[b]
Balance of payments, current account (million US$)	−557	−3013
Tourist arrivals (000s)	38	61[c]
GDP (million current US$)	11549	24667
GDP (per capita current US$)	351	681
Gross fixed capital formation (% of GDP)	18.0	21.0
Labour force participation, adult female pop. (%)	29.1[d]	...
Labour force participation, adult male pop. (%)	74.7[d]	...
Agricultural production index (1999-2001=100)	97	109[c]
Food production index (1999-2001=100)	97	108[c]
Primary energy production (000s Mt oil equiv.)	8959	15091[c]
Motor vehicles (per 1,000 inhabitants)[e]	3.2	3.2[f]
Telephone lines (per 100 inhabitants)	1.2	1.9
Internet users, estimated (000s)	30.0	2800.0

Total trade		Major trading partners			2005
	(million US$)	(% of exports)		(% of imports)	
Exports	4354.9	China	76	China	19
Imports	6976.0	Japan	9	Japan	8
		Saudi Arabia	3	Saudi Arabia	8

Social indicators	2000-2006
Population growth rate 2000-2005 (% per annum)	1.9
Population aged 0-14 years (%)	39.0
Population aged 60+ years (women and men, % of total)	6.0/5.0
Sex ratio (women per 100 men)	99
Life expectancy at birth 2000-2005 (women and men, years)	58/55
Infant mortality rate 2000-2005 (per 1,000 births)	72
Total fertility rate 2000-2005 (births per woman)	4.5
Urban population (%)	41
Urban population growth rate 2000-2005 (% per annum)	4.4
Rural population growth rate 2000-2005 (% per annum)	0.4
Refugees and others of concern to UNHCR[g]	1048262
Primary-secondary gross enrolment ratio (w and m per 100)	45/51
Third-level students (women and men, % of total)	47/53
Television receivers (per 1,000 inhabitants)	387
Parliamentary seats (women and men, % of total)	14/86

Environment	2000-2006
Threatened species	63
Forested area (% of land area)	26
CO2 emissions (000s Mt of carbon dioxide/per capita)	9007/0.3
Energy consumption per capita (kilograms oil equiv.)	95
Precipitation (mm)	162[h]
Average minimum and maximum temperatures (centigrade)[h]	22.7/37.1

a Principal rate. b October 2006. c 2004. d 1996. e Source: World Automotive Market Report, Auto and Truck International (Illinois). f 2002. g Provisional. h Khartoum.

Suriname

Region	South America
Largest urban agglom. (pop., 000s)	Paramaribo (268)
Currency	dollar
Population in 2006 (proj., 000s)	452
Surface area (square kms)	163820
Population density (per square km)	3
United Nations membership date	4 December 1975

Economic indicators	2000	2005
Exchange rate (national currency per US$)[a]	2.18	2.75[b]
Consumer price index (2000=100)[c]	...	187[d]
Unemployment (percentage of labour force)[e]	14.0[f]	
Balance of payments, current account (million US$)	32	−144
Tourist arrivals (000s)	58[g]	138[h]
GDP (million current US$)	775	1503
GDP (per capita current US$)	1785	3346
Gross fixed capital formation (% of GDP)	12.0	33.0
Labour force participation, adult female pop. (%)	33.1[i]	...
Labour force participation, adult male pop. (%)	61.1[i]	...
Employment in industrial sector (%)	14.5[i]	...
Employment in agricultural sector (%)	6.1[i]	...
Agricultural production index (1999-2001=100)	95	101[h]
Food production index (1999-2001=100)	95	101[h]
Primary energy production (000s Mt oil equiv.)	715	730[h]
Motor vehicles (per 1,000 inhabitants)	195.6	276.4[hj]
Telephone lines (per 100 inhabitants)	17.4	18.0
Internet users, estimated (000s)	11.7	32.0

Total trade		Major trading partners			2005
	(million US$)	(% of exports)		(% of imports)	
Exports	306.2[k]	Norway	29	USA	23
Imports	742.5[h]	USA	21	Netherlands	20
		France	14	Trinidad Tbg	18

Social indicators	2000-2006
Population growth rate 2000-2005 (% per annum)	0.7
Population aged 0-14 years (%)	30.0
Population aged 60+ years (women and men, % of total)	10.0/8.0
Sex ratio (women per 100 men)	100
Life expectancy at birth 2000-2005 (women and men, years)	73/66
Infant mortality rate 2000-2005 (per 1,000 births)	26
Total fertility rate 2000-2005 (births per woman)	2.6
Contraceptive use (% of currently married women)	42[l]
Urban population (%)	74
Urban population growth rate 2000-2005 (% per annum)	1.2
Rural population growth rate 2000-2005 (% per annum)	−0.7
Foreign born (%)	1.5[m]
Government education expenditure (% of GNP)	6.4[j]
Primary-secondary gross enrolment ratio (w and m per 100)	102/90
Third-level students (women and men, % of total)	62/38
Television receivers (per 1,000 inhabitants)	266
Intentional homicides (per 100,000 inhabitants)	2[no]
Parliamentary seats (women and men, % of total)	26/74

Environment	2000-2006
Threatened species	65
Forested area (% of land area)	91
CO2 emissions (000s Mt of carbon dioxide/per capita)	2242/5.1
Energy consumption per capita (kilograms oil equiv.)	1631
Precipitation (mm)	2476[p]
Average minimum and maximum temperatures (centigrade)	26.4/28.5

a Suriname changed its currency to the Surinamese Dollar effective 1 Jan. 2004. 1 Surinamese Dollar equals 1000 Surinamese Guilders. b October 2006. c Base: 2001=100, Paramaribo. d May 2006. e Persons aged 15 years and over. f 1999, first semester. g Arrivals at Zanderij airport. h 2004. i 1999. j Suriname General Bureau of Statistics. k 2001. l Excluding areas containing roughly 15 per cent of the population. m Estimated data. n Based on 30 or fewer deaths. o 1992. p Paramaribo.

Swaziland

Region	Southern Africa
Largest urban agglom. (pop., 000s)	Mbabane (73)
Currency	lilangeni
Population in 2006 (proj., 000s)	1029
Surface area (square kms)	17364
Population density (per square km)	59
United Nations membership date	24 September 1968

Economic indicators	2000	2005
Exchange rate (national currency per US$)[a]	7.57	7.45[b]
Consumer price index (2000=100)	100	134[c]
Balance of payments, current account (million US$)	−75	114[c]
Tourist arrivals (000s)	281	459[c]
GDP (million current US$)	1388	2588
GDP (per capita current US$)	1356	2507
Gross fixed capital formation (% of GDP)	20.0	19.0
Labour force participation, adult female pop. (%)	36.1[d]	...
Labour force participation, adult male pop. (%)	57.7[d]	...
Agricultural production index (1999-2001=100)	98	103[c]
Food production index (1999-2001=100)	98	106[c]
Motor vehicles (per 1,000 inhabitants)[e]	83.6	108.3[c]
Telephone lines (per 100 inhabitants)[f]	3.2	3.4
Internet users, estimated (000s)[f]	10.0	36.0[c]

Total trade		Major trading partners			2005
	(million US$)[g]	(% of exports)[g]		(% of imports)[g]	
Exports	974.3	South Africa	67	South Africa	86
Imports	890.7	USA	8	China, HK SAR	2
		Mozambique	5	China	1

Social indicators	2000-2006
Population growth rate 2000-2005 (% per annum)	0.2
Population aged 0-14 years (%)	41.0
Population aged 60+ years (women and men, % of total)	6.0/5.0
Sex ratio (women per 100 men)	107
Life expectancy at birth 2000-2005 (women and men, years)	33/32
Infant mortality rate 2000-2005 (per 1,000 births)	73
Total fertility rate 2000-2005 (births per woman)	4.0
Contraceptive use (% of currently married women)	28[h]
Urban population (%)	24
Urban population growth rate 2000-2005 (% per annum)	0.9
Rural population growth rate 2000-2005 (% per annum)	−<
Foreign born (%)	4.7[i]
Refugees and others of concern to UNHCR[j]	1016
Government education expenditure (% of GNP)	6.3
Primary-secondary gross enrolment ratio (w and m per 100)	74/78
Third-level students (women and men, % of total)	52/48
Television receivers (per 1,000 inhabitants)[f]	36
Parliamentary seats (women and men, % of total)	17/83

Environment	2000-2006
Threatened species	28
Forested area (% of land area)	30
CO2 emissions (000s Mt of carbon dioxide/per capita)	957/0.9
Precipitation (mm)	1442[k]
Average minimum and maximum temperatures (centigrade)[k]	10.5/22.5

a Official rate. b October 2006. c 2004. d 1997. e Excluding government vehicles.
f Year beginning 1 April. g 2002. h North Sudan only. i 1986. j Provisional. k Mbabane.

Sweden

Region	Northern Europe
Largest urban agglom. (pop., 000s)	Stockholm (1708)
Currency	krone
Population in 2006 (proj., 000s)	9070
Surface area (square kms)	449964
Population density (per square km)	20
United Nations membership date	19 November 1946

Economic indicators	2000	2005
Exchange rate (national currency per US$) [a]	9.54	7.26 [b]
Consumer price index (2000=100)	100	109 [c]
Industrial production index (1995=100)	120	138 [d]
Unemployment (percentage of labour force) [e]	4.7	6.0
Balance of payments, current account (million US$)	6617	23643
Tourist arrivals (000s)	2746 [fg]	7627 [hi]
GDP (million current US$)	242003	357683
GDP (per capita current US$)	27261	39561
Gross fixed capital formation (% of GDP)	18.0	17.0
Labour force participation, adult female pop. (%)	67.5	67.9 [j]
Labour force participation, adult male pop. (%)	74.0	73.3 [j]
Employment in industrial sector (%)	24.5	22.0
Employment in agricultural sector (%)	2.4	2.0
Agricultural production index (1999-2001=100)	101	99 [j]
Food production index (1999-2001=100)	102	99 [j]
Primary energy production (000s Mt oil equiv.)	11951	12319 [j]
Motor vehicles (per 1,000 inhabitants)	494.2	512.0
Telephone lines (per 100 inhabitants)	75.8	71.5
Internet users, estimated (000s)	4048.0	6890.0

Total trade		Major trading partners			2005
	(million US$)		(% of exports)		(% of imports)
Exports	130263.7	USA	11	Germany	18
Imports	111351.3	Germany	10	Denmark	10
		Norway	9	Norway	8

Social indicators	2000-2006
Population growth rate 2000-2005 (% per annum)	0.4
Population aged 0-14 years (%)	17.0
Population aged 60+ years (women and men, % of total)	25.0/21.0
Sex ratio (women per 100 men)	101
Life expectancy at birth 2000-2005 (women and men, years)	82/78
Infant mortality rate 2000-2005 (per 1,000 births)	3
Total fertility rate 2000-2005 (births per woman)	1.6
Urban population (%)	84
Urban population growth rate 2000-2005 (% per annum)	0.4
Rural population growth rate 2000-2005 (% per annum)	0.1
Foreign born (%)	12.0
Refugees and others of concern to UNHCR [k]	95916
Government education expenditure (% of GNP)	7.5
Primary-secondary gross enrolment ratio (w and m per 100)	102/100
Third-level students (women and men, % of total)	60/40
Newspaper circulation (per 1,000 inhabitants)	409
Television receivers (per 1,000 inhabitants)	965
Intentional homicides (per 100,000 inhabitants)	1
Parliamentary seats (women and men, % of total)	47/53

Environment	2000-2006
Threatened species	41
Forested area (% of land area)	66
CO_2 emissions (000s Mt of carbon dioxide/per capita) [l]	56000/6.2
Energy consumption per capita (kilograms oil equiv.)	3093
Precipitation (mm)	539 [m]
Average minimum and maximum temperatures (centigrade) [m]	3.6/10.0

a Official rate. b October 2006. c July 2006. d June 2006. e Persons aged 16 to 64 years. f Arrivals of non-resident tourists in all types of accommodation establishments. g Excluding camping. h Arrivals of non-resident tourists at national borders (excluding same-day visitors). i 2003. j 2004. k Provisional. l Source: UNFCCC. m Stockholm.

Switzerland

Region	Western Europe
Largest urban agglom. (pop., 000s)	Zurich (1144)
Currency	franc
Population in 2006 (proj., 000s)	7264
Surface area (square kms)	41284
Population density (per square km)	176
United Nations membership date	10 September 2002

Economic indicators	2000	2005
Exchange rate (national currency per US$)[a]	1.64	1.25[b]
Consumer price index (2000=100)	100	106[c]
Industrial production index (1995=100)[d]	122	130[e]
Unemployment (percentage of labour force)[f]	2.7	4.4
Balance of payments, current account (million US$)	33562	53859
Tourist arrivals (000s)	7821	6530[g]
GDP (million current US$)	246044	365887
GDP (per capita current US$)	34328	50451
Gross fixed capital formation (% of GDP)	23.0	21.0
Labour force participation, adult female pop. (%)	57.6	59.1[h]
Labour force participation, adult male pop. (%)	77.8	76.0[h]
Employment in industrial sector (%)	24.7	22.6[h]
Employment in agricultural sector (%)	4.7	3.9[h]
Agricultural production index (1999-2001=100)	103	100[h]
Food production index (1999-2001=100)	103	100[h]
Primary energy production (000s Mt oil equiv.)	5589	5398[h]
Motor vehicles (per 1,000 inhabitants)[i]	539.1	559.4[j]
Telephone lines (per 100 inhabitants)[k]	72.6	68.7
Internet users, estimated (000s)	2096.0	3800.0

Total trade		Major trading partners			2005
	(million US$)[l]		(% of exports)[l]		(% of imports)[l]
Exports	125926.8	Germany	20	Germany	32
Imports	121215.7	USA	11	Italy	11
		France	9	France	9

Social indicators	2000-2006
Population growth rate 2000-2005 (% per annum)	0.2
Population aged 0-14 years (%)	16.0
Population aged 60+ years (women and men, % of total)	24.0/19.0
Sex ratio (women per 100 men)	107
Life expectancy at birth 2000-2005 (women and men, years)	83/78
Infant mortality rate 2000-2005 (per 1,000 births)	4
Total fertility rate 2000-2005 (births per woman)	1.4
Urban population (%)	75
Urban population growth rate 2000-2005 (% per annum)	0.8
Rural population growth rate 2000-2005 (% per annum)	−1.3
Foreign born (%)	20.5
Refugees and others of concern to UNHCR[m]	62486
Government education expenditure (% of GNP)	5.8
Primary-secondary gross enrolment ratio (w and m per 100)	95/100
Third-level students (women and men, % of total)	45/55
Newspaper circulation (per 1,000 inhabitants)	372
Television receivers (per 1,000 inhabitants)	580
Intentional homicides (per 100,000 inhabitants)	1
Parliamentary seats (women and men, % of total)	25/75

Environment	2000-2006
Threatened species	54
Forested area (% of land area)	30
CO2 emissions (000s Mt of carbon dioxide/per capita)[n]	44720/6.2
Energy consumption per capita (kilograms oil equiv.)[l]	2589
Precipitation (mm)	1086[o]
Average minimum and maximum temperatures (centigrade)[o]	4.9/12.7

a Official rate. b October 2006. c August 2006. d Monthly indices are adjusted for differences in the number of working days. e Second quarter 2006. f Second quarter of each year, persons aged 15 years and over. g 2003. h 2004. i Data refer to fiscal years ending 30 September. j 2002. k Including ISDN channels. l Including Liechtenstein. m Provisional. n Source: UNFCCC. o Zurich.

Syrian Arab Republic

Region	Western Asia
Largest urban agglom. (pop., 000s)	Aleppo (2520)
Currency	pound
Population in 2006 (proj., 000s)	19512
Surface area (square kms)	185180
Population density (per square km)	105
United Nations membership date	24 October 1945

Economic indicators	2000	2005
Exchange rate (national currency per US$)[a]	11.23	11.23[b]
Consumer price index (2000=100)[c]	100	129[d]
Unemployment (percentage of labour force)	...	11.7[e]
Balance of payments, current account (million US$)	1061	205[f]
Tourist arrivals (000s)	1416[g]	3032[fhi]
GDP (million current US$)	19651	25812
GDP (per capita current US$)	1169	1355
Gross fixed capital formation (% of GDP)	17.0	21.0
Labour force participation, adult female pop. (%)	20.6	18.7[j]
Labour force participation, adult male pop. (%)	79.9	75.3[j]
Employment in industrial sector (%)	...	26.9[k]
Employment in agricultural sector (%)	...	30.3[k]
Agricultural production index (1999-2001=100)	103	119[f]
Food production index (1999-2001=100)	103	122[f]
Primary energy production (000s Mt oil equiv.)	34155	30177[f]
Motor vehicles (per 1,000 inhabitants)	31.4	36.3[f]
Telephone lines (per 100 inhabitants)	10.4	15.2
Internet users, estimated (000s)	30.0	1100.0

Total trade		Major trading partners			2005
	(million US$)[f]	(% of exports)[f]		(% of imports)[f]	
Exports	5382.6	Italy	24	Ukraine	8
Imports	7048.8	France	19	China	7
		Iraq	10	Russian Fed.	5

Social indicators	2000-2006
Population growth rate 2000-2005 (% per annum)	2.5
Population aged 0-14 years (%)	37.0
Population aged 60+ years (women and men, % of total)	5.0/4.0
Sex ratio (women per 100 men)	99
Life expectancy at birth 2000-2005 (women and men, years)	75/71
Infant mortality rate 2000-2005 (per 1,000 births)	18
Total fertility rate 2000-2005 (births per woman)	3.5
Urban population (%)	51
Urban population growth rate 2000-2005 (% per annum)	2.7
Rural population growth rate 2000-2005 (% per annum)	2.3
Foreign born (%)	5.5[l]
Refugees and others of concern to UNHCR[m]	328006
Primary-secondary gross enrolment ratio (w and m per 100)	80/86
Third-level students (women and men, % of total)	41/59[n]
Television receivers (per 1,000 inhabitants)	192
Parliamentary seats (women and men, % of total)	12/88

Environment	2000-2006
Threatened species	58
Forested area (% of land area)	3
CO2 emissions (000s Mt of carbon dioxide/per capita)	49036/2.7
Energy consumption per capita (kilograms oil equiv.)	1001
Precipitation (mm)	134[c]
Average minimum and maximum temperatures (centigrade)[c]	8.0/25.3

a Principal rate. b October 2006. c Damascus. d 2005. e 2002, persons aged 15 years and over. f 2004. g Arrivals of non-resident tourists in all types of accommodation establishments. Excluding private accommodation. h Arrivals of non-resident tourists at national borders (excluding same-day visitors). i Including nationals residing abroad. j 2003. k 2002. l Estimated data. m Provisional. n 1995.

Tajikistan

Region	South-central Asia
Largest urban agglom. (pop., 000s)	Dushanbe (549)
Currency	somoni[a]
Population in 2006 (proj., 000s)	6591
Surface area (square kms)	143100
Population density (per square km)	46
United Nations membership date	2 March 1992

Economic indicators	2000	2005
Exchange rate (national currency per US$)[b]	2.20	3.39[c]
Unemployment (percentage of labour force)	2.7[d]	...
Balance of payments, current account (million US$)	−15[e]	−19
Tourist arrivals (000s)	4	4[f]
GDP (million current US$)	870	2342
GDP (per capita current US$)	141	360
Gross fixed capital formation (% of GDP)	9.0	10.0
Agricultural production index (1999-2001=100)	99	148[g]
Food production index (1999-2001=100)	104	146[g]
Primary energy production (000s Mt oil equiv.)	1262	1528[g]
Motor vehicles (per 1,000 inhabitants)	21.7	181.6[g]
Telephone lines (per 100 inhabitants)	3.6	3.8
Internet users, estimated (000s)	3.0	5.0[g]

Total trade		Major trading partners			2005
	(million US$)[h]	(% of exports)[h]		(% of imports)[h]	
Exports	692.3	Russian Fed.	37	Uzbekistan	29
Imports	644.0	Netherlands	26	Russian Fed.	16
		Uzbekistan	14	Ukraine	13

Social indicators	2000-2006
Population growth rate 2000-2005 (% per annum)	1.1
Population aged 0-14 years (%)	39.0
Population aged 60+ years (women and men, % of total)	5.0/5.0
Sex ratio (women per 100 men)	102
Life expectancy at birth 2000-2005 (women and men, years)	66/61
Infant mortality rate 2000-2005 (per 1,000 births)	89
Total fertility rate 2000-2005 (births per woman)	3.8
Contraceptive use (% of currently married women)	34
Urban population (%)	25
Urban population growth rate 2000-2005 (% per annum)	0.2
Rural population growth rate 2000-2005 (% per annum)	1.4
Refugees and others of concern to UNHCR[i]	1080
Government education expenditure (% of GNP)	2.9
Primary-secondary gross enrolment ratio (w and m per 100)	83/94
Third-level students (women and men, % of total)	25/75
Television receivers (per 1,000 inhabitants)	375
Intentional homicides (per 100,000 inhabitants)	2
Parliamentary seats (women and men, % of total)	20/80

Environment	2000-2006
Threatened species	30
Forested area (% of land area)	3
CO2 emissions (000s Mt of carbon dioxide/per capita)	4662/0.7
Energy consumption per capita (kilograms oil equiv.)	520
Precipitation (mm)	653[j]
Average minimum and maximum temperatures (centigrade)[j]	8.1/22.0

a Prior to November 2000, Ruble per US dollar. b Data refer to non-commercial rates derived from the Operational Rates of Exchange for United Nations Programmes. c October 2006. d 1997. e 2002. f 2001. g 2004. h 2000. i Provisional. j Dushanbe.

Thailand

Region	South-eastern Asia
Largest urban agglom. (pop., 000s)	Bangkok (6593)
Currency	baht
Population in 2006 (proj., 000s)	64762
Surface area (square kms)	513115
Population density (per square km)	126
United Nations membership date	16 December 1946

Economic indicators	2000	2005
Exchange rate (national currency per US$)[a]	43.27	36.74[b]
Consumer price index (2000=100)	100	118[c]
Industrial production index (1995=100)[d]	112	167[e]
Unemployment (percentage of labour force)[f]	2.4[g]	1.4[h]
Balance of payments, current account (million US$)	9313	−3719
Tourist arrivals (000s)[i]	9579	11737[j]
GDP (million current US$)	3587	5651
GDP (per capita current US$)	1998	2749
Gross fixed capital formation (% of GDP)	22.0	29.0
Labour force participation, adult female pop. (%)	64.9	65.1[j]
Labour force participation, adult male pop. (%)	80.6	81.8[j]
Employment in industrial sector (%)	17.9	20.2
Employment in agricultural sector (%)	48.5	42.6
Agricultural production index (1999-2001=100)	101	107[j]
Food production index (1999-2001=100)	100	105[j]
Primary energy production (000s Mt oil equiv.)	30885	36439[j]
Motor vehicles (per 1,000 inhabitants)	165.1	138.1
Telephone lines (per 100 inhabitants)[k]	9.1	11.0
Internet users, estimated (000s)[k]	2300.0	7084.2

Total trade		Major trading partners			2005
	(million US$)	(% of exports)			(% of imports)
Exports	110110.0	USA	15	Japan	22
Imports	118164.3	Japan	14	China	9
		China	8	USA	7

Social indicators	2000-2006
Population growth rate 2000-2005 (% per annum)	0.9
Population aged 0-14 years (%)	24.0
Population aged 60+ years (women and men, % of total)	11.0/10.0
Sex ratio (women per 100 men)	104
Life expectancy at birth 2000-2005 (women and men, years)	74/66
Infant mortality rate 2000-2005 (per 1,000 births)	20
Total fertility rate 2000-2005 (births per woman)	1.9
Urban population (%)	32
Urban population growth rate 2000-2005 (% per annum)	1.6
Rural population growth rate 2000-2005 (% per annum)	0.5
Foreign born (%)	0.6[l]
Refugees and others of concern to UNHCR[m]	149351
Government education expenditure (% of GNP)	4.3
Primary-secondary gross enrolment ratio (w and m per 100)	84/86
Third-level students (women and men, % of total)	54/46
Television receivers (per 1,000 inhabitants)[k]	285
Intentional homicides (per 100,000 inhabitants)	5
Parliamentary seats (women and men, % of total)	11/89

Environment	2000-2006
Threatened species	250
Forested area (% of land area)	29
CO2 emissions (000s Mt of carbon dioxide/per capita)	246372/3.9
Energy consumption per capita (kilograms oil equiv.)	1268
Precipitation (mm)	1498[n]
Average minimum and maximum temperatures (centigrade)[n]	24.1/32.7

a Official rate. b October 2006. c May 2006. d Manufacturing. e August 2006. f Third round (Aug.) of each year. g Persons aged 15 years and over. h Persons aged 13 years and over. i Including arrivals of nationals residing abroad. j 2004. k Year ending 30 September. l Estimated data. m Provisional. n Bangkok.

The FYR of Macedonia

Region	Southern Europe
Largest urban agglom. (pop., 000s)	Skopje (475)
Currency	denar
Population in 2006 (proj., 000s)	2037
Surface area (square kms)	25713
Population density (per square km)	79
United Nations membership date	8 April 1993

Economic indicators	2000	2005
Exchange rate (national currency per US$)	66.33	48.10[a]
Consumer price index (2000=100)	100	112[b]
Industrial production index (1995=100)	110	109[c]
Unemployment (percentage of labour force)[d]	30.5[e]	37.3
Balance of payments, current account (million US$)	−72	−415[f]
Tourist arrivals (000s)	224	165[f]
GDP (million current US$)	122725	176602
GDP (per capita current US$)	1785	2778
Gross fixed capital formation (% of GDP)	16.0	18.0
Labour force participation, adult male pop. (%)	64.4	62.5[f]
Employment in industrial sector (%)	33.3[g]	32.3
Employment in agricultural sector (%)	23.9[g]	19.5
Agricultural production index (1999-2001=100)	103	107[f]
Food production index (1999-2001=100)	104	109[f]
Primary energy production (000s Mt oil equiv.)	2126	2080[f]
Motor vehicles (per 1,000 inhabitants)[h]	216.8	197.3
Telephone lines (per 100 inhabitants)	25.3	26.2
Internet users, estimated (000s)	50.0	159.9

Total trade		Major trading partners			2005
	(million US$)	(% of exports)		(% of imports)	
Exports	2041.3	Serbia, Mtneg	23	Russian Fed.	13
Imports	3228.0	Germany	18	Germany	10
		Greece	15	Greece	9

Social indicators	2000-2006
Population growth rate 2000-2005 (% per annum)	0.2
Population aged 0-14 years (%)	20.0
Population aged 60+ years (women and men, % of total)	17.0/14.0
Sex ratio (women per 100 men)	100
Life expectancy at birth 2000-2005 (women and men, years)	76/71
Infant mortality rate 2000-2005 (per 1,000 births)	16
Total fertility rate 2000-2005 (births per woman)	1.5
Urban population (%)	69
Urban population growth rate 2000-2005 (% per annum)	1.4
Rural population growth rate 2000-2005 (% per annum)	−2.1
Refugees and others of concern to UNHCR[i]	4320
Government education expenditure (% of GNP)	3.4
Primary-secondary gross enrolment ratio (w and m per 100)	88/89
Third-level students (women and men, % of total)	52/48
Newspaper circulation (per 1,000 inhabitants)	54
Television receivers (per 1,000 inhabitants)	250
Intentional homicides (per 100,000 inhabitants)	3
Parliamentary seats (women and men, % of total)	28/72

Environment	2000-2006
Threatened species	35
Forested area (% of land area)	28
CO2 emissions (000s Mt of carbon dioxide/per capita)	10545/5.2
Energy consumption per capita (kilograms oil equiv.)	1558
Precipitation (mm)	504[j]
Average minimum and maximum temperatures (centigrade)[j]	6.0/18.2

a October 2006. b April 2006. c August 2006. d Persons aged 15 years and over.
e 2001. f 2004. g 2002. h Data refer to fiscal years ending 30 September. i Provisional.
j Skopje.

Timor Leste

Region	South-eastern Asia
Largest urban agglom. (pop., 000s)	Dili (50)
Currency	US dollar
Population in 2006 (proj., 000s)	1007
Surface area (square kms)	14874
Population density (per square km)	68
United Nations membership date	27 September 2002

Economic indicators	2000	2005
GDP (million current US$)	316	354
GDP (per capita current US$)[a]	438	374
Gross fixed capital formation (% of GDP)	41.0	28.0
Agricultural production index (1999-2001=100)	99	112[b]
Food production index (1999-2001=100)	99	113[b]
Primary energy production (000s Mt oil equiv.)	...	7315[c]

Social indicators	2000-2006
Population growth rate 2000-2005 (% per annum)	5.4
Population aged 0-14 years (%)	41.0
Population aged 60+ years (women and men, % of total)	5.0/5.0
Sex ratio (women per 100 men)	93
Life expectancy at birth 2000-2005 (women and men, years)	56/54
Infant mortality rate 2000-2005 (per 1,000 births)	94
Total fertility rate 2000-2005 (births per woman)	7.8
Urban population (%)	26
Urban population growth rate 2000-2005 (% per annum)	7.0
Rural population growth rate 2000-2005 (% per annum)	4.9
Foreign born (%)	0.8[d]
Refugees and others of concern to UNHCR[e]	13
Third-level students (women and men, % of total)	53/47
Parliamentary seats (women and men, % of total)	25/75

Environment	2000-2006
Threatened species	11[b]
CO2 emissions (000s Mt of carbon dioxide/per capita)	163/0.2
Energy consumption per capita (kilograms oil equiv.)	57[d]
Precipitation (mm)	975
Average minimum and maximum temperatures (centigrade)	23.6/27.1

a Indonesia includes Timor Leste up to 1998. b 2004. c 2003 estimate. d Estimated data. e Provisional.

Togo

Region	Western Africa
Largest urban agglom. (pop., 000s)	Lomé (1337)
Currency	CFA franc
Population in 2006 (proj., 000s)	6306
Surface area (square kms)	56785
Population density (per square km)	111
United Nations membership date	20 September 1960

Economic indicators	2000	2005
Exchange rate (national currency per US$)[a]	704.95	516.66[b]
Consumer price index (2000=100)[c]	100	115[d]
Balance of payments, current account (million US$)	−140	−162[e]
Tourist arrivals (000s)	60	83[f]
GDP (million current US$)	1329	2187
GDP (per capita current US$)	248	356
Gross fixed capital formation (% of GDP)	18.0	20.0
Agricultural production index (1999-2001=100)	97	110[f]
Food production index (1999-2001=100)	98	104[f]
Primary energy production (000s Mt oil equiv.)	17	14[f]
Motor vehicles (per 1,000 inhabitants)[g]	14.2	13.4[h]
Telephone lines (per 100 inhabitants)	0.9	1.0
Internet users, estimated (000s)	100.0	300.0

Total trade		Major trading partners			2005
	(million US$)	(% of exports)		(% of imports)	
Exports	359.9	Ghana	20	France	18
Imports	592.6	Burkina Faso	18	China	13
		Benin	12	Côte d'Ivoire	6

Social indicators	2000-2006
Population growth rate 2000-2005 (% per annum)	2.7
Population aged 0-14 years (%)	43.0
Population aged 60+ years (women and men, % of total)	5.0/4.0
Sex ratio (women per 100 men)	102
Life expectancy at birth 2000-2005 (women and men, years)	56/52
Infant mortality rate 2000-2005 (per 1,000 births)	93
Total fertility rate 2000-2005 (births per woman)	5.4
Contraceptive use (% of currently married women)	26
Urban population (%)	40
Urban population growth rate 2000-2005 (% per annum)	4.5
Rural population growth rate 2000-2005 (% per annum)	1.6
Foreign born (%)	3.9[i]
Refugees and others of concern to UNHCR[j]	18710
Government education expenditure (% of GNP)	2.7
Primary-secondary gross enrolment ratio (w and m per 100)	59/81
Third-level students (women and men, % of total)	17/83
Newspaper circulation (per 1,000 inhabitants)	2
Television receivers (per 1,000 inhabitants)	130
Parliamentary seats (women and men, % of total)	7/93

Environment	2000-2006
Threatened species	43
Forested area (% of land area)	9
CO2 emissions (000s Mt of carbon dioxide/per capita)	2200/0.4
Energy consumption per capita (kilograms oil equiv.)	113
Precipitation (mm)	877[c]
Average minimum and maximum temperatures (centigrade)[c]	23.1/30.7

a Official rate. b October 2006. c Lomé. d July 2006. e 2003. f 2004. g Source: World Automotive Market Report, Auto and Truck International (Illinois). h 2002. i Estimated data. j Provisional.

Tonga

Region	Oceania-Polynesia
Largest urban agglom. (pop., 000s)	Nuku'alofa (25)
Currency	pa'anga
Population in 2006 (proj., 000s)	103
Surface area (square kms)	650
Population density (per square km)	158
United Nations membership date	14 September 1999

Economic indicators	2000	2005
Exchange rate (national currency per US$)[a]	1.98	2.01[b]
Consumer price index (2000=100)[c]	100	174[d]
Balance of payments, current account (million US$)	−10804[e]	−3319[f]
Tourist arrivals (000s)[g]	35	41[h]
GDP (million current US$)	148	214
GDP (per capita current US$)	1481	2089
Gross fixed capital formation (% of GDP)	18.0	15.0
Labour force participation, adult female pop. (%)	41.7[i]	...
Labour force participation, adult male pop. (%)	75.3[i]	...
Agricultural production index (1999-2001=100)	99	102[h]
Food production index (1999-2001=100)	99	102[h]
Motor vehicles (per 1,000 inhabitants)	91.8	137.8[j]
Telephone lines (per 100 inhabitants)	9.8	11.3[hk]
Internet users, estimated (000s)	2.4	3.0

Total trade		Major trading partners			2005
	(million US$)[l]	(% of exports)[l]		(% of imports)[l]	
Exports	8.9	Japan	44	New Zealand	37
Imports	69.4	USA	26	Australia	25
		New Zealand	13	Fiji	15

Social indicators	2000-2006
Population growth rate 2000-2005 (% per annum)	0.4
Population aged 0-14 years (%)	36.0
Population aged 60+ years (women and men, % of total)	9.0/9.0
Sex ratio (women per 100 men)	96
Life expectancy at birth 2000-2005 (women and men, years)	73/71
Infant mortality rate 2000-2005 (per 1,000 births)	21
Total fertility rate 2000-2005 (births per woman)	3.5
Urban population (%)	24
Urban population growth rate 2000-2005 (% per annum)	1.1
Rural population growth rate 2000-2005 (% per annum)	0.2
Government education expenditure (% of GNP)	4.9
Primary-secondary gross enrolment ratio (w and m per 100)	107/106
Third-level students (women and men, % of total)	60/40
Television receivers (per 1,000 inhabitants)	70
Parliamentary seats (women and men, % of total)	3/97

Environment	2000-2006
Threatened species	19
Forested area (% of land area)	6
CO2 emissions (000s Mt of carbon dioxide/capita)	114/1.1
Energy consumption per capita (kilograms oil equiv.)	355[m]

a Official rate. b September 2006. c Excluding rent. d June 2006. e 2001. f 2002.
g Arrivals by air. h 2004. i 1996. j 2003. k Estimate. l 2000. m Estimated data.

Trinidad and Tobago

Region	Caribbean
Largest urban agglom. (pop., 000s)	Port of Spain (52)
Currency	dollar
Population in 2006 (proj., 000s)	1309
Surface area (square kms)	5130
Population density (per square km)	255
United Nations membership date	18 September 1962

Economic indicators	2000	2005
Exchange rate (national currency per US$)[a]	6.30	6.29[b]
Consumer price index (2000=100)	100	138[c]
Industrial production index (1995=100)	146	259[d]
Unemployment (percentage of labour force)[e]	12.2	10.4
Balance of payments, current account (million US$)	544	985[f]
Tourist arrivals (000s)[g]	399	443[h]
GDP (million current US$)	8154	14763
GDP (per capita current US$)	6347	11311
Gross fixed capital formation (% of GDP)	16.0	18.0
Labour force participation, adult female pop. (%)	47.0	47.5[i]
Labour force participation, adult male pop. (%)	75.3	74.6[i]
Employment in industrial sector (%)	28.0	28.4[i]
Employment in agricultural sector (%)	7.2	6.9[i]
Agricultural production index (1999-2001=100)	105	118[h]
Food production index (1999-2001=100)	105	118[h]
Primary energy production (000s Mt oil equiv.)	19582	30798[h]
Motor vehicles (per 1,000 inhabitants)	238.2	299.6
Telephone lines (per 100 inhabitants)	24.5	24.8
Internet users, estimated (000s)	100.0	160.0[h]

Total trade		Major trading partners			2005
	(million US$)		(% of exports)		(% of imports)
Exports	9611.5	USA	59	USA	29
Imports	5693.9	Jamaica	7	Brazil	14
		France	4	Venezuela	6

Social indicators	2000-2006
Population growth rate 2000-2005 (% per annum)	0.3
Population aged 0-14 years (%)	22.0
Population aged 60+ years (women and men, % of total)	12.0/10.0
Sex ratio (women per 100 men)	103
Life expectancy at birth 2000-2005 (women and men, years)	73/67
Infant mortality rate 2000-2005 (per 1,000 births)	14
Total fertility rate 2000-2005 (births per woman)	1.6
Contraceptive use (% of currently married women)	38
Urban population (%)	12
Urban population growth rate 2000-2005 (% per annum)	2.7
Rural population growth rate 2000-2005 (% per annum)	<
Foreign born (%)	3.2[i]
Government education expenditure (% of GNP)	4.6
Primary-secondary gross enrolment ratio (w and m per 100)	94/93
Third-level students (women and men, % of total)	55/45
Television receivers (per 1,000 inhabitants)	354
Parliamentary seats (women and men, % of total)	25/75

Environment	2000-2006
Threatened species	38
Forested area (% of land area)	51
CO2 emissions (000s Mt of carbon dioxide/per capita)	28699/22.1
Energy consumption per capita (kilograms oil equiv.)	9929
Precipitation (mm)	1408[k]
Average minimum and maximum temperatures (centigrade)[k]	22.0/31.9

a Official rate. b October 2006. c July 2006. d 1st quarter 2006. e Persons aged 15 years and over. f 2003. g Arrivals by air. h 2004. i 2002. j Estimated data. k Port of Spain (Trinidad).

Tunisia

Region	Northern Africa
Largest urban agglom. (pop., 000s)	Tunis (734)
Currency	dinar
Population in 2006 (proj., 000s)	10210
Surface area (square kms)	163610
Population density (per square km)	62
United Nations membership date	12 November 1956

Economic indicators	2000	2005
Exchange rate (national currency per US$)	1.39	1.33[a]
Consumer price index (2000=100)	100	120[b]
Industrial production index (1995=100)	128	150[c]
Unemployment (percentage of labour force)[d]	15.7[e]	14.2
Balance of payments, current account (million US$)	−821	−303
Tourist arrivals (000s)[f]	5058	5998[g]
GDP (million current US$)	19444	29049
GDP (per capita current US$)	2033	2875
Gross fixed capital formation (% of GDP)	26.0	23.0
Labour force participation, adult female pop. (%)	23.7[h]	...
Labour force participation, adult male pop. (%)	73.4[h]	...
Agricultural production index (1999-2001=100)	98	102[g]
Food production index (1999-2001=100)	98	102[g]
Primary energy production (000s Mt oil equiv.)	5488	5702[g]
Motor vehicles (per 1,000 inhabitants)	81.8	86.3[i]
Telephone lines (per 100 inhabitants)	10.0	12.5
Internet users, estimated (000s)	260.0	953.8

Total trade		Major trading partners			2005
	(million US$)	(% of exports)		(% of imports)	
Exports	10493.6	France	33	France	23
Imports	13174.3	Italy	24	Italy	21
		Germany	8	Germany	8

Social indicators	2000-2006
Population growth rate 2000-2005 (% per annum)	1.1
Population aged 0-14 years (%)	26.0
Population aged 60+ years (women and men, % of total)	9.0/8.0
Sex ratio (women per 100 men)	99
Life expectancy at birth 2000-2005 (women and men, years)	75/71
Infant mortality rate 2000-2005 (per 1,000 births)	22
Total fertility rate 2000-2005 (births per woman)	2.0
Urban population (%)	65
Urban population growth rate 2000-2005 (% per annum)	1.7
Rural population growth rate 2000-2005 (% per annum)	<
Foreign born (%)	0.4[j]
Refugees and others of concern to UNHCR[k]	113
Government education expenditure (% of GNP)	8.5
Primary-secondary gross enrolment ratio (w and m per 100)	92/91
Third-level students (women and men, % of total)	57/43
Newspaper circulation (per 1,000 inhabitants)	19
Television receivers (per 1,000 inhabitants)	218
Parliamentary seats (women and men, % of total)	19/81

Environment	2000-2006
Threatened species	51
Forested area (% of land area)	3
CO2 emissions (000s Mt of carbon dioxide/per capita)	20909/2.1
Energy consumption per capita (kilograms oil equiv.)	770
Precipitation (mm)	466[l]
Average minimum and maximum temperatures (centigrade)[l]	13.3/23.5

a October 2006. b July 2006. c March 2006. d Persons aged 15 years and over.
e Figures revised on the basis of the 2004 census results. f Excluding nationals residing abroad. g 2004. h 1997. i 2001. j Estimated data. k Provisional. l Tunis-Carthage.

Turkey

Region	Western Asia/Southern Europe
Largest urban agglom. (pop., 000s)	Istanbul (9712)
Currency	new lira
Population in 2006 (proj., 000s)	74175
Surface area (square kms)	783562
Population density (per square km)	95
United Nations membership date	24 October 1945

Economic indicators	2000	2005
Exchange rate (national currency per US$)[a]	0.67	1.46[b]
Consumer price index (2000=100)	100	415[c]
Industrial production index (1995=100)	121	163[d]
Unemployment (percentage of labour force)[e]	6.5[f]	10.3
Balance of payments, current account (million US$)	−9821	−23082
Tourist arrivals (000s)	9586	16826[g]
GDP (million current US$)	199264	362614
GDP (per capita current US$)	2920	4954
Gross fixed capital formation (% of GDP)	22.0	20.0
Labour force participation, adult female pop. (%)	26.6	25.4[g]
Labour force participation, adult male pop. (%)	73.7	72.3[g]
Employment in industrial sector (%)	24.0	24.7
Employment in agricultural sector (%)	36.0	29.5
Agricultural production index (1999-2001=100)	104	104[g]
Food production index (1999-2001=100)	105	104[g]
Primary energy production (000s Mt oil equiv.)	19752	17000[g]
Motor vehicles (per 1,000 inhabitants)	88.0	107.7[g]
Telephone lines (per 100 inhabitants)	27.0	25.9
Internet users, estimated (000s)	2500.0	11204.3

Total trade		Major trading partners			2005
	(million US$)	(% of exports)		(% of imports)	
Exports	73476.4	Germany	13	Germany	12
Imports	116774.2	UK	8	Russian Fed.	11
		Italy	8	Italy	6

Social indicators	2000-2006
Population growth rate 2000-2005 (% per annum)	1.4
Population aged 0-14 years (%)	29.0
Population aged 60+ years (women and men, % of total)	9.0/7.0
Sex ratio (women per 100 men)	99
Life expectancy at birth 2000-2005 (women and men, years)	71/66
Infant mortality rate 2000-2005 (per 1,000 births)	42
Total fertility rate 2000-2005 (births per woman)	2.5
Urban population (%)	67
Urban population growth rate 2000-2005 (% per annum)	2.2
Rural population growth rate 2000-2005 (% per annum)	−0.1
Foreign born (%)	1.9
Refugees and others of concern to UNHCR[h]	8692
Government education expenditure (% of GNP)	3.8
Primary-secondary gross enrolment ratio (w and m per 100)	80/94
Third-level students (women and men, % of total)	41/59
Television receivers (per 1,000 inhabitants)	537
Parliamentary seats (women and men, % of total)	4/96

Environment	2000-2006
Threatened species	123
Forested area (% of land area)	13
CO2 emissions (000s Mt of carbon dioxide/per capita)	220409/3.1
Energy consumption per capita (kilograms oil equiv.)	1031
Precipitation (mm)	678[i]
Average minimum and maximum temperatures (centigrade)[i]	10.3/18.4

a Redenominated the Lira. The ISO code of the new Turkish Lira is TRY and 1 TRY is equivalent to 1,000,000 old Turkish Lira (TRL). Use of the operational rate system began in 1 Jan 2005. b October 2006. c May 2006. d August 2006. e Persons aged 15 years and over. f Estimates based on the 2000 Population Census results. g 2004. h Provisional. i Istanbul.

Turkmenistan

Region	South-central Asia
Largest urban agglom. (pop., 000s)	Ashgabad (711)
Currency	manat
Population in 2006 (proj., 000s)	4899
Surface area (square kms)	488100
Population density (per square km)	10
United Nations membership date	2 March 1992

Economic indicators	2000	2005
Exchange rate (national currency per US$)[a]	5200.00	5200.00[b]
Tourist arrivals (000s)	300[c]	...
GDP (million current US$)	4157	5826
GDP (per capita current US$)	923	1205
Gross fixed capital formation (% of GDP)	35.0	30.0
Agricultural production index (1999-2001=100)	102	122[d]
Food production index (1999-2001=100)	101	131[d]
Primary energy production (000s Mt oil equiv.)	50232	63513[d]
Telephone lines (per 100 inhabitants)	8.2	7.7
Internet users, estimated (000s)	6.0	36.0[d]

Total trade		Major trading partners			2005
	(million US$)[e]	(% of exports)[e]		(% of imports)[e]	
Exports	2505.5	Russian Fed.	41	Russian Fed.	14
Imports	1785.5	Italy	16	Turkey	14
		Iran	10	Ukraine	12

Social indicators	2000-2006
Population growth rate 2000-2005 (% per annum)	1.4
Population aged 0-14 years (%)	32.0
Population aged 60+ years (women and men, % of total)	7.0/5.0
Sex ratio (women per 100 men)	103
Life expectancy at birth 2000-2005 (women and men, years)	67/58
Infant mortality rate 2000-2005 (per 1,000 births)	78
Total fertility rate 2000-2005 (births per woman)	2.8
Contraceptive use (% of currently married women)	62[f]
Urban population (%)	46
Urban population growth rate 2000-2005 (% per annum)	1.9
Rural population growth rate 2000-2005 (% per annum)	1.0
Refugees and others of concern to UNHCR[g]	11965
Newspaper circulation (per 1,000 inhabitants)	7
Television receivers (per 1,000 inhabitants)	183
Parliamentary seats (women and men, % of total)	16/84

Environment	2000-2006
Threatened species	45
Forested area (% of land area)	8
CO2 emissions (000s Mt of carbon dioxide/per capita)	43413/9.2
Energy consumption per capita (kilograms oil equiv.)	3536
Precipitation (mm)	227[h]
Average minimum and maximum temperatures (centigrade)[h]	10.4/23.2

a Data refer to non-commercial rates derived from the Operational Rates of Exchange for United Nations Programmes. b October 2006. c 1998. d 2004. e 2000. f Including the lactational amenorrhoea method and/or breastfeeding if reported as the current contraceptive method. g Provisional. h Ashgabat.

Tuvalu

Region	Oceania-Polynesia
Largest urban agglom. (pop., 000s)	Funafuti (5)
Currency	Australian dollar
Population in 2006 (proj., 000s)	10[a]
Surface area (square kms)	26
Population density (per square km)	368
United Nations membership date	5 September 2000

Economic indicators	2000	2005
Exchange rate (national currency per US$)	1.91	1.33[b]
Consumer price index (2000=100)	100	120[c]
Tourist arrivals (000s)	1	1[d]
GDP (million current US$)	12	26
GDP (per capita current US$)	1204	2516
Gross fixed capital formation (% of GDP)	55.0	56.0
Agricultural production index (1999-2001=100)	100	100[e]
Food production index (1999-2001=100)	100	100[e]
Telephone lines (per 100 inhabitants)	...	6.8[f]
Internet users, estimated (000s)	0.5	1.6[d]

Social indicators	2000-2006
Population growth rate 2000-2005 (% per annum)	0.5
Population aged 0-14 years (%)	41.0
Sex ratio (women per 100 men)	102[gh]
Life expectancy at birth 2000-2005 (women and men, years)	65/34[hi]
Total fertility rate 2000-2005 (births per woman)	3.7[ij]
Urban population (%)	48
Urban population growth rate 2000-2005 (% per annum)	1.4
Rural population growth rate 2000-2005 (% per annum)	−0.3
Primary-secondary gross enrolment ratio (w and m per 100)	97/92
Parliamentary seats (women and men, % of total)	—/100

Environment	2000-2006
Threatened species	11
CO2 emissions (000s Mt of carbon dioxide/per capita)[k]	5/0.5[l]

a 2005. b October 2006. c 1st quarter 2006. d 2004. e 2003. f 2001. g Refers to de facto population count. h 2002. i Published by the Secretariat of the Pacific Community. j 2000-2003. k Source: UNFCCC. l 1994.

Uganda

Region	Eastern Africa
Largest urban agglom. (pop., 000s)	Kampala (1319)
Currency	shilling
Population in 2006 (proj., 000s)	29857
Surface area (square kms)	241038
Population density (per square km)	124
United Nations membership date	25 October 1962

Economic indicators	2000	2005
Exchange rate (national currency per US$)[a]	1766.68	1819.77[b]
Consumer price index (2000=100)	100	134[c]
Unemployment (percentage of labour force)[d]	...	3.2[e]
Balance of payments, current account (million US$)	−825	−267
Tourist arrivals (000s)	193	512[f]
GDP (million current US$)	5734	9115
GDP (per capita current US$)	236	316
Gross fixed capital formation (% of GDP)	19.0	22.0
Employment in industrial sector (%)	6.3[g]	7.6[e]
Employment in agricultural sector (%)	89.6[g]	69.1[e]
Agricultural production index (1999-2001=100)	99	109[f]
Food production index (1999-2001=100)	99	109[f]
Primary energy production (000s Mt oil equiv.)	135	163[f]
Motor vehicles (per 1,000 inhabitants)	5.1	5.6[f]
Telephone lines (per 100 inhabitants)	0.3	0.4
Internet users, estimated (000s)	40.0	500.0

Total trade		Major trading partners			2005
	(million US$)	(% of exports)		(% of imports)	
Exports	812.8	Netherlands	11	Kenya	25
Imports	2054.1	Untd Arab Em	10	Japan	7
		Switzerland	9	South Africa	7

Social indicators	2000-2006
Population growth rate 2000-2005 (% per annum)	3.4
Population aged 0-14 years (%)	50.0
Population aged 60+ years (women and men, % of total)	4.0/4.0
Sex ratio (women per 100 men)	100
Life expectancy at birth 2000-2005 (women and men, years)	47/46
Infant mortality rate 2000-2005 (per 1,000 births)	81
Total fertility rate 2000-2005 (births per woman)	7.1
Contraceptive use (% of currently married women)	23[h]
Urban population (%)	13
Urban population growth rate 2000-2005 (% per annum)	4.2
Rural population growth rate 2000-2005 (% per annum)	3.3
Refugees and others of concern to UNHCR[i]	259089
Government education expenditure (% of GNP)	5.3
Primary-secondary gross enrolment ratio (w and m per 100)	76/78
Third-level students (women and men, % of total)	38/62
Newspaper circulation (per 1,000 inhabitants)	3
Television receivers (per 1,000 inhabitants)	17
Parliamentary seats (women and men, % of total)	30/70

Environment	2000-2006
Threatened species	170
Forested area (% of land area)	21
CO2 emissions (000s Mt of carbon dioxide/per capita)	1713/0.1
Energy consumption per capita (kilograms oil equiv.)	24
Precipitation (mm)[j]	1552[k]
Average minimum and maximum temperatures (centigrade)[k]	20.6/23.9

a Principal rate. b October 2006. c September 2006. d Persons aged 10 years and over. e 2003. f 2004. g 1994. h 2000/01. i Provisional. j January to November only. k Entebbe.

Ukraine

Region	Eastern Europe
Largest urban agglom. (pop., 000s)	Kiev (2672)
Currency	hryvnia
Population in 2006 (proj., 000s)	45986
Surface area (square kms)	603700
Population density (per square km)	76
United Nations membership date	24 October 1945

Economic indicators	2000	2005[b]
Exchange rate (national currency per US$)[a]	5.43	5.05[b]
Consumer price index (2000=100)	100	157[c]
Industrial production index (1995=100)	109	166[d]
Unemployment (percentage of labour force)[e]	11.6	7.2
Balance of payments, current account (million US$)	1481	2531
Tourist arrivals (000s)	6431	15629[f]
GDP (million current US$)	31262	81669
GDP (per capita current US$)	636	1757
Gross fixed capital formation (% of GDP)	20.0	21.0
Labour force participation, adult female pop. (%)	50.7	57.9[g]
Labour force participation, adult male pop. (%)	64.4	67.3[g]
Employment in industrial sector (%)	31.4	24.2
Employment in agricultural sector (%)	20.5	19.4
Agricultural production index (1999-2001=100)	101	115[f]
Food production index (1999-2001=100)	101	115[f]
Primary energy production (000s Mt oil equiv.)[h]	60327	61447[f]
Motor vehicles (per 1,000 inhabitants)	124.0	138.3
Telephone lines (per 100 inhabitants)	21.2	25.8
Internet users, estimated (000s)	350.0	4560.0

Total trade		Major trading partners			2005
	(million US$)	(% of exports)		(% of imports)	
Exports	34228.0	Russian Fed.	22	Russian Fed.	36
Imports	36122.0	Turkey	6	Germany	9
		Italy	6	Turkmenistan	7

Social indicators	2000-2006
Population growth rate 2000-2005 (% per annum)	−1.1
Population aged 0-14 years (%)	15.0
Population aged 60+ years (women and men, % of total)	25.0/16.0
Sex ratio (women per 100 men)	118
Life expectancy at birth 2000-2005 (women and men, years)	72/60
Infant mortality rate 2000-2005 (per 1,000 births)	16
Total fertility rate 2000-2005 (births per woman)	1.1
Urban population (%)	68
Urban population growth rate 2000-2005 (% per annum)	−0.9
Rural population growth rate 2000-2005 (% per annum)	−1.5
Refugees and others of concern to UNHCR[i]	76851
Government education expenditure (% of GNP)	4.6
Primary-secondary gross enrolment ratio (w and m per 100)	93/94
Third-level students (women and men, % of total)	54/46
Newspaper circulation (per 1,000 inhabitants)	177
Television receivers (per 1,000 inhabitants)	357
Intentional homicides (per 100,000 inhabitants)	12
Parliamentary seats (women and men, % of total)	9/81

Environment	2000-2006
Threatened species	61
Forested area (% of land area)	17
CO2 emissions (000s Mt of carbon dioxide/per capita)[j]	313140/6.6
Energy consumption per capita (kilograms oil equiv.)	2698
Precipitation (mm)	648[k]
Average minimum and maximum temperatures (centigrade)[k]	4.3/11.9

a Official rate. b October 2006. c May 2006. d June 2006. e Persons aged 15-70 years.
f 2004. g 2003. h On 1 of August. i Provisional. j Source: UNFCCC. k Kyiv.

United Arab Emirates

Region	Western Asia
Largest urban agglom. (pop., 000s)	Dubai (1330)
Currency	dirham
Population in 2006 (proj., 000s)	4657
Surface area (square kms)	83600
Population density (per square km)	56
United Nations membership date	9 December 1971

Economic indicators	2000	2005
Exchange rate (national currency per US$)[a]	3.67	3.67[b]
Unemployment (percentage of labour force)	2.3	
Tourist arrivals (000s)[c]	3907	5871[d]
GDP (million current US$)	70522	133757
GDP (per capita current US$)	21719	29751
Gross fixed capital formation (% of GDP)	22.0	19.0
Labour force participation, adult female pop. (%)	31.3[e]	...
Labour force participation, adult male pop. (%)	92.2[e]	...
Employment in industrial sector (%)	33.4	...
Employment in agricultural sector (%)	7.9	...
Agricultural production index (1999-2001=100)	141	64[f]
Food production index (1999-2001=100)	141	64[f]
Primary energy production (000s Mt oil equiv.)	166530	179471[f]
Motor vehicles (per 1,000 inhabitants)[g]	369.1	364.7[h]
Telephone lines (per 100 inhabitants)	31.4	27.5
Internet users, estimated (000s)	765.0	1397.2

Total trade		Major trading partners		2005
	(million US$)[f]	(% of exports)		(% of imports)[f]
Exports	82750.0	...	India	12
Imports	61587.5	...	China	9
		...	Japan	7

Social indicators	2000-2006
Population growth rate 2000-2005 (% per annum)	6.5
Population aged 0-14 years (%)	22.0
Population aged 60+ years (women and men, % of total)	2.0/1.0
Sex ratio (women per 100 men)	47
Life expectancy at birth 2000-2005 (women and men, years)	81/76
Infant mortality rate 2000-2005 (per 1,000 births)	9
Total fertility rate 2000-2005 (births per woman)	2.5
Urban population (%)	77
Urban population growth rate 2000-2005 (% per annum)	6.3
Rural population growth rate 2000-2005 (% per annum)	7.1
Foreign born (%)	68.2[i]
Refugees and others of concern to UNHCR[j]	183
Primary-secondary gross enrolment ratio (w and m per 100)	74/73
Third-level students (women and men, % of total)	66/34
Television receivers (per 1,000 inhabitants)	197

Environment	2000-2006
Threatened species	31
Forested area (% of land area)	4
CO2 emissions (000s Mt of carbon dioxide/per capita)	135285/33.6
Energy consumption per capita (kilograms oil equiv.)	13438
Precipitation (mm)	89[k]
Average minimum and maximum temperatures (centigrade)[k]	20.2/33.7

a Official rate. b October 2006. c Arrivals in hotels only. Including domestic tourism and nationals residing abroad. d 2003. e 1995. f 2004. g Source: World Automotive Market Report, Auto and Truck International (Illinois). h 2001. i Estimated data. j Provisional. k Abu Dhabi.

United Kingdom

Region	Northern Europe
Largest urban agglom. (pop., 000s)	London (8505)
Currency	pound sterling
Population in 2006 (proj., 000s)	59847
Surface area (square kms)	242900
Population density (per square km)	246
United Nations membership date	24 October 1945

Economic indicators	2000	2005
Exchange rate (national currency per US$)	0.67	0.53[a]
Consumer price index (2000=100)	100	115[b]
Industrial production index (1995=100)[c]	107	98[d]
Unemployment (percentage of labour force)[e]	5.5	5.0
Balance of payments, current account (million US$)	−37	−58
Tourist arrivals (000s)	25209	27755[f]
GDP (million current US$)	1442777	2198796
GDP (per capita current US$)	24592	36851
Gross fixed capital formation (% of GDP)	17.0	17.0
Labour force participation, adult female pop. (%)	54.9	55.7[f]
Labour force participation, adult male pop. (%)	71.1	70.1[f]
Employment in industrial sector (%)	25.3	22.0
Employment in agricultural sector (%)	1.5	1.4
Agricultural production index (1999-2001=100)	102	98[f]
Food production index (1999-2001=100)	102	98[f]
Primary energy production (000s Mt oil equiv.)	265038	217947[f]
Motor vehicles (per 1,000 inhabitants)	485.0	532.2[f]
Telephone lines (per 100 inhabitants)[g]	58.9	56.4
Internet users, estimated (000s)[g]	15800.0	28515.0

Total trade		Major trading partners			2005
	(million US$)	(% of exports)			(% of imports)
Exports	384365.0	USA	15	Germany	13
Imports	515782.2	Germany	11	USA	8
		France	9	France	7

Social indicators	2000-2006
Population growth rate 2000-2005 (% per annum)	0.3
Population aged 0-14 years (%)	18.0
Population aged 60+ years (women and men, % of total)	23.0/19.0
Sex ratio (women per 100 men)	105[h]
Life expectancy at birth 2000-2005 (women and men, years)	81/76
Infant mortality rate 2000-2005 (per 1,000 births)	5
Total fertility rate 2000-2005 (births per woman)	1.7
Contraceptive use (% of currently married women)	84
Urban population (%)	90
Urban population growth rate 2000-2005 (% per annum)	0.4
Rural population growth rate 2000-2005 (% per annum)	−0.4
Refugees and others of concern to UNHCR[i]	307064
Government education expenditure (% of GNP)	5.4
Primary-secondary gross enrolment ratio (w and m per 100)	106/105
Third-level students (women and men, % of total)	57/43
Newspaper circulation (per 1,000 inhabitants)	327
Television receivers (per 1,000 inhabitants)[g]	1101
Intentional homicides (per 100,000 inhabitants)	1
Parliamentary seats (women and men, % of total)	19/81

Environment	2000-2006
Threatened species	60
Forested area (% of land area)	12
CO2 emissions (000s Mt of carbon dioxide/per capita)[j]	557460/9.4
Energy consumption per capita (kilograms oil equiv.)	3711
Precipitation (mm)	611[k]
Average minimum and maximum temperatures (centigrade)[k]	7.5/14.4

a October 2006. b April 2006. c Monthly indices are adjusted for differences in the number of working days. d August 2006. e Persons aged 16 years and over, March - May of each year. f 2004. g Year beginning 1 April. h United Kingdom of Great Britain and Northern Ireland. i Provisional. j Source: UNFCCC. k London.

United Republic of Tanzania

Region	Eastern Africa
Largest urban agglom. (pop., 000s)	Dar es Salaam (2676)
Currency	shilling
Population in 2006 (proj., 000s)	39025
Surface area (square kms)	945087
Population density (per square km)	41
United Nations membership date	14 December 1961

Economic indicators	2000	2005
Exchange rate (national currency per US$) [a]	803.26	1287.14 [b]
Consumer price index (2000=100) [c]	...	612 [d]
Industrial production index (1995=100) [e]	137	208 [f]
Unemployment (percentage of labour force) [g]	...	5.1 [h]
Balance of payments, current account (million US$)	−499	−536
Tourist arrivals (000s)	459	566 [i]
GDP (million current US$)	9093	12586
GDP (per capita current US$)	269	337
Gross fixed capital formation (% of GDP)	17.0	22.0
Labour force participation, adult female pop. (%)	...	86.9 [j]
Labour force participation, adult male pop. (%)	...	89.8 [h]
Employment in industrial sector (%)	2.6 [k]	2.6 [h]
Employment in agricultural sector (%)	84.2 [k]	82.1 [h]
Agricultural production index (1999-2001=100)	100	108 [i]
Food production index (1999-2001=100)	101	106 [i]
Primary energy production (000s Mt oil equiv.)	240	367 [i]
Motor vehicles (per 1,000 inhabitants) [l]	3.9	3.7 [m]
Telephone lines (per 100 inhabitants)	0.5	0.4
Internet users, estimated (000s)	40.0	333.0 [i]

Total trade		Major trading partners			2005
	(million US$)	(% of exports)		(% of imports)	
Exports	1415.0	South Africa	19	South Africa	14
Imports	2757.2	UK	8	China	8
		Switzerland	8	Japan	7

Social indicators	2000-2006
Population growth rate 2000-2005 (% per annum)	2.0
Population aged 0-14 years (%)	43.0
Population aged 60+ years (women and men, % of total)	6.0/5.0
Sex ratio (women per 100 men)	101
Life expectancy at birth 2000-2005 (women and men, years)	46/46
Infant mortality rate 2000-2005 (per 1,000 births)	104
Total fertility rate 2000-2005 (births per woman)	5.0
Urban population (%)	24
Urban population growth rate 2000-2005 (% per annum)	3.6
Rural population growth rate 2000-2005 (% per annum)	1.5
Foreign born (%)	2.6 [n]
Refugees and others of concern to UNHCR [o]	549131
Government education expenditure (% of GNP)	2.2 [p]
Third-level students (women and men, % of total)	29/71
Television receivers (per 1,000 inhabitants)	41
Parliamentary seats (women and men, % of total)	30/70

Environment	2000-2006
Threatened species	533
Forested area (% of land area)	44
CO_2 emissions (000s Mt of carbon dioxide/per capita)	3811/0.1
Energy consumption per capita (kilograms oil equiv.)	39
Precipitation (mm)	1148 [q]

a Official rate. b October 2006. c Base: 1990=100, Zanzibar. d 2005. e Manufacturing. f 2nd quarter 2006. g Persons aged 10 years and over. h 2001. i 2004. j 2003. k 1991. l Source: World Automotive Market Report, Auto and Truck International (Illinois). m 2002. n Estimated data. o Provisional. p 1999. q Dar Es Salaam.

United States of America

Region	Northern America
Largest urban agglom. (pop., 000s)	New York-Newark (18718)
Currency	dollar
Population in 2006 (proj., 000s)	301029
Surface area (square kms)	9629091
Population density (per square km)	31
United Nations membership date	24 October 1945

Economic indicators	2000	2005
Exchange rate (national currency per US$)	1.00	1.00
Consumer price index (2000=100)[a]	100	118[b]
Industrial production index (1995=100)[c]	129	144[d]
Unemployment (percentage of labour force)[e]	4.0	5.1
Balance of payments, current account (million US$)	−415	−792
Tourist arrivals (000s)[f]	51237	46085[g]
GDP (million current US$)	9764800	11713000[g]
GDP (per capita current US$)	34364	39650[g]
Gross fixed capital formation (% of GDP)	20.0	16.0
Labour force participation, adult female pop. (%)	59.9	59.2[g]
Labour force participation, adult male pop. (%)	74.8	73.3[g]
Employment in industrial sector (%)	23.2	20.6
Employment in agricultural sector (%)	2.6	1.6
Agricultural production index (1999-2001=100)	101	108[g]
Food production index (1999-2001=100)	102	108[g]
Primary energy production (000s Mt oil equiv.)	1470444	1424929[g]
Motor vehicles (per 1,000 inhabitants)	779.4	790.8[h]
Telephone lines (per 100 inhabitants)	68.4	60.6
Internet users, estimated (000s)	124000.0	197800.0

Total trade		Major trading partners		2005	
	(million US$)[i]	(% of exports)[i]		(% of imports)[i]	
Exports	904339.5	Canada	23	Canada	17
Imports	1732321.0	Mexico	13	China	15
		Japan	6	Mexico	10

Social indicators	2000-2006
Population growth rate 2000-2005 (% per annum)	1.0
Population aged 0-14 years (%)	21.0
Population aged 60+ years (women and men, % of total)	19.0/15.0
Sex ratio (women per 100 men)	103
Life expectancy at birth 2000-2005 (women and men, years)	80/75
Infant mortality rate 2000-2005 (per 1,000 births)	7
Total fertility rate 2000-2005 (births per woman)	2.0
Urban population (%)	81
Urban population growth rate 2000-2005 (% per annum)	1.4
Rural population growth rate 2000-2005 (% per annum)	−0.7
Foreign born (%)	10.4
Refugees and others of concern to UNHCR[j]	549083
Government education expenditure (% of GNP)	5.8
Primary-secondary gross enrolment ratio (w and m per 100)	96/97
Third-level students (women and men, % of total)	57/43
Newspaper circulation (per 1,000 inhabitants)	197
Television receivers (per 1,000 inhabitants)	882
Intentional homicides (per 100,000 inhabitants)	6
Parliamentary seats (women and men, % of total)	15/85

Environment	2000-2006
Threatened species	1178
Forested area (% of land area)	25
CO2 emissions (000s Mt of carbon dioxide/per capita)[k]	5841500/20.0[l]
Energy consumption per capita (kilograms oil equiv.)	6980
Precipitation (mm)	981[m]
Average minimum and maximum temperatures (centigrade)[m]	9.6/19.4

a All urban consumers. b September 2006. c Monthly indices are adjusted for differences in the number of working days. d August 2006. e Persons aged 16 years and over. f Including Mexicans staying one or more nights in the US. g 2004. h 2003. i Including Puerto Rico and US Virgin Islands. j Provisional. k Source: UNFCCC. l Including territories. m Washington, DC.

Uruguay

Region	South America
Largest urban agglom. (pop., 000s)	Montevideo (1264)
Currency	peso
Population in 2006 (proj., 000s)	3487
Surface area (square kms)	175016
Population density (per square km)	20
United Nations membership date	18 December 1945

Economic indicators	2000	2005
Exchange rate (national currency per US$)	12.52	23.80[a]
Consumer price index (2000=100)[b]	100	175[c]
Industrial production index (1995=100)[d]	128	219[c]
Unemployment (percentage of labour force)	13.6	12.2
Balance of payments, current account (million US$)	−566	−2
Tourist arrivals (000s)	1968	1756[e]
GDP (million current US$)	20086	13215[e]
GDP (per capita current US$)	6011	3842[e]
Gross fixed capital formation (% of GDP)	13.0	13.0
Labour force participation, adult female pop. (%)	49.3	49.0[f]
Labour force participation, adult male pop. (%)	71.9	69.0[f]
Employment in industrial sector (%)	24.7	21.9
Employment in agricultural sector (%)	4.1	4.6
Agricultural production index (1999-2001=100)	103	113[e]
Food production index (1999-2001=100)	103	116[e]
Primary energy production (000s Mt oil equiv.)	606	411[e]
Motor vehicles (per 1,000 inhabitants)	217.8	153.6
Telephone lines (per 100 inhabitants)	6.7	30.9
Internet users, estimated (000s)	350.0	668.0

Total trade		Major trading partners			2005
	(million US$)	(% of exports)		(% of imports)	
Exports	3404.5	USA	23	Brazil	21
Imports	3878.9	Brazil	13	Argentina	20
		Argentina	8	Russian Fed.	8

Social indicators	2000-2006
Population growth rate 2000-2005 (% per annum)	0.7
Population aged 0-14 years (%)	24.0
Population aged 60+ years (women and men, % of total)	20.0/15.0
Sex ratio (women per 100 men)	106
Life expectancy at birth 2000-2005 (women and men, years)	79/72
Infant mortality rate 2000-2005 (per 1,000 births)	13
Total fertility rate 2000-2005 (births per woman)	2.3
Urban population (%)	92
Urban population growth rate 2000-2005 (% per annum)	0.9
Rural population growth rate 2000-2005 (% per annum)	−0.8
Refugees and others of concern to UNHCR[g]	130
Government education expenditure (% of GNP)	2.3
Primary-secondary gross enrolment ratio (w and m per 100)	112/105
Third-level students (women and men, % of total)	66/34
Television receivers (per 1,000 inhabitants)	259
Intentional homicides (per 100,000 inhabitants)	6
Parliamentary seats (women and men, % of total)	11/89

Environment	2000-2006
Threatened species	64
Forested area (% of land area)	7
CO2 emissions (000s Mt of carbon dioxide/per capita)	4380/1.3
Energy consumption per capita (kilograms oil equiv.)	663
Precipitation (mm)	1101[b]
Average minimum and maximum temperatures (centigrade)[b]	12.4/21.4

a October 2006. b Montevideo. c August 2006. d Manufacturing. e 2004. f 2003.
g Provisional.

US Virgin Islands

Region	Caribbean
Largest urban agglom. (pop., 000s)	Charlotte Amalie (52)
Currency	US dollar
Population in 2006 (proj., 000s)	112
Surface area (square kms)	347
Population density (per square km)	322

Economic indicators	2000	2005
Exchange rate (national currency per US$)	1.00	1.00
Unemployment (percentage of labour force)[a]	5.9[b]	...
Tourist arrivals (000s)	546	544[c]
Agricultural production index (1999-2001=100)	100	105[c]
Food production index (1999-2001=100)	100	99[d]
Telephone lines (per 100 inhabitants)	29.0	63.9
Internet users, estimated (000s)	15.0	30.0[e]

Social indicators	2000-2006
Population growth rate 2000-2005 (% per annum)	0.2
Population aged 0-14 years (%)	24.0
Population aged 60+ years (women and men, % of total)	17.0/16.0
Sex ratio (women per 100 men)	110
Life expectancy at birth 2000-2005 (women and men, years)	83/75
Infant mortality rate 2000-2005 (per 1,000 births)	9
Total fertility rate 2000-2005 (births per woman)	2.2
Urban population (%)	94
Urban population growth rate 2000-2005 (% per annum)	0.5
Rural population growth rate 2000-2005 (% per annum)	−4.7
Foreign born (%)	32.0[f]
Television receivers (per 1,000 inhabitants)	663
Intentional homicides (per 100,000 inhabitants)	21

Environment	2000-2006
Threatened species	35
Forested area (% of land area)	41
CO2 emissions (000s Mt of carbon dioxide/per capita)	13548/121.3
Energy consumption per capita (kilograms oil equiv.)	16622[e]

a Persons aged 16 to 65 years. b 1997. c 2004. d 2003. e 2002. f Estimated data.

Uzbekistan

Region	South-central Asia
Largest urban agglom. (pop., 000s)	Tashkent (2181)
Currency	som
Population in 2006 (proj., 000s)	26980
Surface area (square kms)	447400
Population density (per square km)	60
United Nations membership date	2 March 1992

Economic indicators	2000	2005
Exchange rate (national currency per US$)[a]	974.22[b]	1230.00[c]
Unemployment (percentage of labour force)[d]	0.4[e]	...
Tourist arrivals (000s)	302	262[f]
GDP (million current US$)	13759	11788[f]
GDP (per capita current US$)	557	450[f]
Gross fixed capital formation (% of GDP)	24.0	23.0
Employment in industrial sector (%)	19.4[g]	...
Employment in agricultural sector (%)	38.5[g]	...
Agricultural production index (1999-2001=100)	101	107[f]
Food production index (1999-2001=100)	102	105[f]
Primary energy production (000s Mt oil equiv.)	60062	62217[f]
Telephone lines (per 100 inhabitants)	3.5	6.7
Internet users, estimated (000s)	120.0	880.0[f]

Social indicators	2000-2006
Population growth rate 2000-2005 (% per annum)	1.5
Population aged 0-14 years (%)	33.0
Population aged 60+ years (women and men, % of total)	7.0/5.0
Sex ratio (women per 100 men)	101
Life expectancy at birth 2000-2005 (women and men, years)	70/63
Infant mortality rate 2000-2005 (per 1,000 births)	58
Total fertility rate 2000-2005 (births per woman)	2.7
Contraceptive use (% of currently married women)	68
Urban population (%)	37
Urban population growth rate 2000-2005 (% per annum)	1.1
Rural population growth rate 2000-2005 (% per annum)	1.6
Refugees and others of concern to UNHCR[h]	44537
Primary-secondary gross enrolment ratio (w and m per 100)	95/97
Third-level students (women and men, % of total)	44/56
Television receivers (per 1,000 inhabitants)	290
Intentional homicides (per 100,000 inhabitants)	3
Parliamentary seats (women and men, % of total)	16/84

Environment	2000-2006
Threatened species	35
Forested area (% of land area)	5
CO2 emissions (000s Mt of carbon dioxide/per capita)	123840/4.8
Energy consumption per capita (kilograms oil equiv.)	2332
Precipitation (mm)	419[i]
Average minimum and maximum temperatures (centigrade)[i]	8.3/21.0

a Data refer to non-commercial rates derived from the Operational Rates of Exchange for United Nations Programmes. b October 2003. c October 2006. d Employment office records. e 1995. f 2004. g 1999. h Provisional. i Tashkent.

Vanuatu

Region	Oceania-Melanesia
Largest urban agglom. (pop., 000s)	Vila (36)
Currency	vatu
Population in 2006 (proj., 000s)	215
Surface area (square kms)	12189
Population density (per square km)	18
United Nations membership date	15 September 1981

Economic indicators	2000	2005
Exchange rate (national currency per US$)[a]	142.81	109.65[b]
Consumer price index (2000=100)	100	112
Balance of payments, current account (million US$)	−14	−64
Tourist arrivals (000s)	58	61[c]
GDP (million current US$)	223	291[c]
GDP (per capita current US$)	1164	1405[c]
Gross fixed capital formation (% of GDP)	21.0	21.0
Agricultural production index (1999-2001=100)	98	97[c]
Food production index (1999-2001=100)	98	97[c]
Motor vehicles (per 1,000 inhabitants)[d]	36.6	53.2[e]
Telephone lines (per 100 inhabitants)	10.5	3.2
Internet users, estimated (000s)	4.0	7.5

Total trade		Major trading partners			2005
	(million US$)[f]	(% of exports)[f]		(% of imports)[f]	
Exports	23.2	Bangladesh	22	Australia	44
Imports	86.7	Japan	12	New Zealand	12
		UK	10	Fiji	9

Social indicators	2000-2006
Population growth rate 2000-2005 (% per annum)	2.0
Population aged 0-14 years (%)	40.0
Population aged 60+ years (women and men, % of total)	5.0/5.0
Sex ratio (women per 100 men)	96
Life expectancy at birth 2000-2005 (women and men, years)	70/67
Infant mortality rate 2000-2005 (per 1,000 births)	34
Total fertility rate 2000-2005 (births per woman)	4.2
Urban population (%)	23
Urban population growth rate 2000-2005 (% per annum)	3.5
Rural population growth rate 2000-2005 (% per annum)	1.5
Foreign born (%)	1.6[g]
Government education expenditure (% of GNP)	10.0
Primary-secondary gross enrolment ratio (w and m per 100)	77/82
Third-level students (women and men, % of total)	36/64
Television receivers (per 1,000 inhabitants)	13
Parliamentary seats (women and men, % of total)	4/96

Environment	2000-2006
Threatened species	34
Forested area (% of land area)	37
CO2 emissions (000s Mt of carbon dioxide/per capita)	89/0.4
Energy consumption per capita (kilograms oil equiv.)	139[h]
Precipitation (mm)	2222[i]
Average minimum and maximum temperatures (centigrade)[i]	21.5/28.2

a Official rate. b August 2006. c 2004. d Source: World Automotive Market Report, Auto and Truck International (Illinois). e 2002. f 2000. g 1989. h Estimated data. i Port Vila.

Venezuela (Bolivarian Rep. of)

Region	South America
Largest urban agglom. (pop., 000s)	Caracas (2913)
Currency	bolívar
Population in 2006 (proj., 000s)	27216
Surface area (square kms)	912050
Population density (per square km)	30
United Nations membership date	15 November 1945

Economic indicators	2000	2005
Exchange rate (national currency per US$)[a]	699.75	2147.00[b]
Consumer price index (2000=100)[c]	100	255
Industrial production index (1995=100)[d]	92	116[e]
Unemployment (percentage of labour force)[f]	13.9	15.8[g]
Balance of payments, current account (million US$)	11853	25359
Tourist arrivals (000s)	469	492[h]
GDP (million current US$)	121258	111958[h]
GDP (per capita current US$)	4966	4260[h]
Gross fixed capital formation (% of GDP)	21.0	18.0
Labour force participation, adult female pop. (%)	47.3	54.7[g]
Labour force participation, adult male pop. (%)	82.0	83.6[g]
Employment in industrial sector (%)	22.3	19.8[i]
Employment in agricultural sector (%)	10.2	10.7[i]
Agricultural production index (1999-2001=100)	101	98[h]
Food production index (1999-2001=100)	101	98[h]
Primary energy production (000s Mt oil equiv.)	214935	197409[h]
Motor vehicles (per 1,000 inhabitants)	98.5	119.6[h]
Telephone lines (per 100 inhabitants)	3.2	13.5
Internet users, estimated (000s)	820.0	3308.4

Total trade		Major trading partners			2005
	(million US$)		(% of exports)		(% of imports)
Exports	55487.0	USA	43	USA	31
Imports	21848.1	Colombia	2	Colombia	11
		Canada	1	Brazil	9

Social indicators	2000-2006
Population growth rate 2000-2005 (% per annum)	1.8
Population aged 0-14 years (%)	31.0
Population aged 60+ years (women and men, % of total)	8.0/7.0
Sex ratio (women per 100 men)	99
Life expectancy at birth 2000-2005 (women and men, years)	76/70
Infant mortality rate 2000-2005 (per 1,000 births)	18
Total fertility rate 2000-2005 (births per woman)	2.7
Urban population (%)	93
Urban population growth rate 2000-2005 (% per annum)	2.3
Rural population growth rate 2000-2005 (% per annum)	−4.2
Foreign born (%)	4.4
Refugees and others of concern to UNHCR[j]	206320
Primary-secondary gross enrolment ratio (w and m per 100)	92/89
Third-level students (women and men, % of total)	51/49
Television receivers (per 1,000 inhabitants)	191
Intentional homicides (per 100,000 inhabitants)	26
Parliamentary seats (women and men, % of total)	18/82

Environment	2000-2006
Threatened species	233
Forested area (% of land area)	56
CO2 emissions (000s Mt of carbon dioxide/per capita)	144227/5.6
Energy consumption per capita (kilograms oil equiv.)	2329
Precipitation (mm)	913[k]
Average minimum and maximum temperatures (centigrade)[k]	16.0/30.3

a Official rate. b October 2006. c Metropolitan area, Caracas. d Manufacturing, excluding petroleum refineries. e June 2006. f Persons aged 15 years and over. g 2002. h 2004. i 2003. j Provisional. k Caracas.

Viet Nam

Region	South-eastern Asia
Largest urban agglom. (pop., 000s)	Ho Chi Minh City (5065)
Currency	dong
Population in 2006 (proj., 000s)	85344
Surface area (square kms)	331689
Population density (per square km)	257
United Nations membership date	20 September 1977

Economic indicators	2000	2005
Exchange rate (national currency per US$)	14514.00	15927.00[a]
Consumer price index (2000=100)	100	133[b]
Unemployment (percentage of labour force)[c]	2.3	2.1[d]
Balance of payments, current account (million US$)	1106	−926[d]
Tourist arrivals (000s)	1383[e]	2928[df]
GDP (million current US$)	31173	45819[d]
GDP (per capita current US$)	396	551[d]
Gross fixed capital formation (% of GDP)	28.0	33.0
Labour force participation, adult female pop. (%)	67.3	66.7[g]
Labour force participation, adult male pop. (%)	74.3	74.4[g]
Employment in industrial sector (%)	12.4	17.4[d]
Employment in agricultural sector (%)	65.3	57.9[d]
Agricultural production index (1999-2001=100)	101	124[d]
Food production index (1999-2001=100)	100	124[d]
Primary energy production (000s Mt oil equiv.)	27287	45408[d]
Motor vehicles (per 1,000 inhabitants)[h]	0.9	2.5[d]
Telephone lines (per 100 inhabitants)	62.9	18.8
Internet users, estimated (000s)	200.0	10711.0

Total trade		Major trading partners			2005
	(million US$)[d]	(% of exports)[d]		(% of imports)[d]	
Exports	26316.2	USA	19	China	14
Imports	31079.4	Japan	13	Singapore	12
		China	11	Japan	11

Social indicators	2000-2006
Population growth rate 2000-2005 (% per annum)	1.4
Population aged 0-14 years (%)	30.0
Population aged 60+ years (women and men, % of total)	8.0/7.0
Sex ratio (women per 100 men)	100
Life expectancy at birth 2000-2005 (women and men, years)	72/68
Infant mortality rate 2000-2005 (per 1,000 births)	30
Total fertility rate 2000-2005 (births per woman)	2.3
Contraceptive use (% of currently married women)	79
Urban population (%)	26
Urban population growth rate 2000-2005 (% per annum)	3.0
Rural population growth rate 2000-2005 (% per annum)	0.8
Foreign born (%)	<
Refugees and others of concern to UNHCR[i]	17536
Primary-secondary gross enrolment ratio (w and m per 100)	81/86
Third-level students (women and men, % of total)	43/57
Television receivers (per 1,000 inhabitants)	208
Parliamentary seats (women and men, % of total)	27/73

Environment	2000-2006
Threatened species	310
Forested area (% of land area)	30
CO2 emissions (000s Mt of carbon dioxide/per capita)	76242/0.9
Energy consumption per capita (kilograms oil equiv.)	352
Precipitation (mm)	1931[j]
Average minimum and maximum temperatures (centigrade)[j]	23.7/32.3

a 1st quarter 2006. b March 2006. c Persons aged 15 years and over, July of each year. d 2004. e Arrivals of non-resident tourists at national borders (excluding same-day visitors). f Arrivals of non-resident visitors at national borders (including tourists and same-day visitors). g 2003. h Commercial vehicles only. i Provisional. j Ho Chi Minh City.

Western Sahara

Region	Northern Africa
Largest urban agglom. (pop., 000s)	El Aaiún (185)
Population in 2006 (proj., 000s)	356
Surface area (square kms)	266000
Population density (per square km)	1

Social indicators	2000-2006
Population growth rate 2000-2005 (% per annum)	2.6
Population aged 0-14 years (%)	34.0
Population aged 60+ years (women and men, % of total)	6.0/7.0
Sex ratio (women per 100 men)	93
Life expectancy at birth 2000-2005 (women and men, years)	66/62
Infant mortality rate 2000-2005 (per 1,000 births)	53
Total fertility rate 2000-2005 (births per woman)	3.9
Urban population (%)	92
Urban population growth rate 2000-2005 (% per annum)	2.7
Rural population growth rate 2000-2005 (% per annum)	1.7

Environment	2000-2006
Threatened species	30
CO_2 emissions (000s Mt of carbon dioxide/per capita)	240/0.7
Energy consumption per capita (kilograms oil equiv.)	294[a]

a Estimated data.

Yemen

Region	Western Asia
Largest urban agglom. (pop., 000s)	Sana'a (1801)
Currency	rial
Population in 2006 (proj., 000s)	21639
Surface area (square kms)	527968
Population density (per square km)	41
United Nations membership date	30 September 1947

Economic indicators	2000	2005
Exchange rate (national currency per US$)	165.59	197.74[a]
Consumer price index (2000=100)	100	201[a]
Unemployment (percentage of labour force)[b]	11.5[c]	...
Balance of payments, current account (million US$)	1337	1215
Tourist arrivals (000s)	73	274[d]
GDP (million current US$)	9520	13080[d]
GDP (per capita current US$)	531	643[d]
Gross fixed capital formation (% of GDP)	17.0	17.0
Labour force participation, adult female pop. (%)	29.2[e]	...
Employment in industrial sector (%)	11.1[c]	...
Employment in agricultural sector (%)	54.1[c]	...
Agricultural production index (1999-2001=100)	99	111[d]
Food production index (1999-2001=100)	99	111[d]
Primary energy production (000s Mt oil equiv.)	21413	20050[d]
Motor vehicles (per 1,000 inhabitants)[f]	49.3	50.5[g]
Telephone lines (per 100 inhabitants)	1.9	3.9
Internet users, estimated (000s)	15.0	180.0[d]

Total trade		Major trading partners			2005
	(million US$)	(% of exports)		(% of imports)	
Exports	5608.9	China	35	Untd Arab Em	19
Imports	4862.7	India	16	Saudi Arabia	9
		Thailand	12	Switzerland	8

Social indicators	2000-2006
Population growth rate 2000-2005 (% per annum)	3.1
Population aged 0-14 years (%)	46.0
Population aged 60+ years (women and men, % of total)	4.0/3.0
Sex ratio (women per 100 men)	97
Life expectancy at birth 2000-2005 (women and men, years)	62/59
Infant mortality rate 2000-2005 (per 1,000 births)	69
Total fertility rate 2000-2005 (births per woman)	6.2
Urban population (%)	27
Urban population growth rate 2000-2005 (% per annum)	4.5
Rural population growth rate 2000-2005 (% per annum)	2.6
Foreign born (%)	1.4[h]
Refugees and others of concern to UNHCR[i]	82741
Government education expenditure (% of GNP)	10.3
Primary-secondary gross enrolment ratio (w and m per 100)	53/85
Third-level students (women and men, % of total)	26/74
Television receivers (per 1,000 inhabitants)	337
Parliamentary seats (women and men, % of total)	1/99

Environment	2000-2006
Threatened species	204
Forested area (% of land area)	1
CO2 emissions (000s Mt of carbon dioxide/per capita)	17082/0.9
Energy consumption per capita (kilograms oil equiv.)	261

a July 2006. b Persons aged 15 years and over. c 1999. d 2004. e 1995. f Source: United Nations Economic and Social Commission for Western Asia (ESCWA). g 2001. h Estimated data. i Provisional.

Zambia

Region	Eastern Africa
Largest urban agglom. (pop., 000s)	Lusaka (1260)
Currency	kwacha
Population in 2006 (proj., 000s)	11861
Surface area (square kms)	752618
Population density (per square km)	16
United Nations membership date	1 December 1964

Economic indicators	2000	2005
Exchange rate (national currency per US$)[a]	4157.83	3851.48[b]
Consumer price index (2000=100)	100	269[c]
Industrial production index (1995=100)	93	139[d]
Unemployment (percentage of labour force)[e]	12.0[f]	...
Balance of payments, current account (million US$)	−584	...
Tourist arrivals (000s)	457	515[g]
GDP (million current US$)	3239	5315[g]
GDP (per capita current US$)	303	463[g]
Gross fixed capital formation (% of GDP)	17.0	25.0
Labour force participation, adult female pop. (%)	56.0[f]	...
Labour force participation, adult male pop. (%)	68.0[f]	...
Employment in industrial sector (%)	7.0[f]	...
Agricultural production index (1999-2001=100)	98	105[g]
Food production index (1999-2001=100)	101	108[g]
Primary energy production (000s Mt oil equiv.)	781	865[g]
Motor vehicles (per 1,000 inhabitants)	0.8[h]	...
Telephone lines (per 100 inhabitants)[i]	0.8	0.8
Internet users, estimated (000s)[i]	20.0	231.0[g]

Total trade		Major trading partners			2005
	(million US$)	(% of exports)		(% of imports)	
Exports	1851.6	Switzerland	29	South Africa	47
Imports	2574.7	South Africa	19	UK	12
		UK	14	Zimbabwe	5

Social indicators	2000-2006
Population growth rate 2000-2005 (% per annum)	1.7
Population aged 0-14 years (%)	46.0
Population aged 60+ years (women and men, % of total)	5.0/4.0
Sex ratio (women per 100 men)	100
Life expectancy at birth 2000-2005 (women and men, years)	37/38
Infant mortality rate 2000-2005 (per 1,000 births)	95
Total fertility rate 2000-2005 (births per woman)	5.7
Contraceptive use (% of currently married women)	34[j]
Urban population (%)	35
Urban population growth rate 2000-2005 (% per annum)	1.8
Rural population growth rate 2000-2005 (% per annum)	1.7
Foreign born (%)	3.6[k]
Refugees and others of concern to UNHCR[l]	155864
Government education expenditure (% of GNP)	2.9
Primary-secondary gross enrolment ratio (w and m per 100)	68/73
Third-level students (women and men, % of total)	32/68
Newspaper circulation (per 1,000 inhabitants)	21
Television receivers (per 1,000 inhabitants)[i]	64
Parliamentary seats (women and men, % of total)	15/85

Environment	2000-2006
Threatened species	46
Forested area (% of land area)	42
CO2 emissions (000s Mt of carbon dioxide/per capita)	2200/0.2
Energy consumption per capita (kilograms oil equiv.)	117
Precipitation (mm)	843[m]
Average minimum and maximum temperatures (centigrade)[m]	14.9/26.4

a Official rate. b October 2006. c May 2006. d 1st quarter 2006. e Persons aged 12 years and over. f 1998. g 2004. h 1996. i Year beginning 1 April. j 2001/02. k Estimated data. l Provisional. m Lusaka.

Zimbabwe

Region	Eastern Africa
Largest urban agglom. (pop., 000s)	Harare (1515)
Currency	dollar
Population in 2006 (proj., 000s)	13085
Surface area (square kms)	390757
Population density (per square km)	33
United Nations membership date	25 August 1980

Economic indicators	2000	2005
Exchange rate (national currency per US$) [ab]	55.07	250.00[c]
Consumer price index (2000=100) [d]	100	98321[e]
Industrial production index (1995=100) [f]	97	73[g]
Unemployment (percentage of labour force) [h]	6.0[i]	...
Tourist arrivals (000s)	1868[j]	1854[kl]
GDP (million current US$)	5628	4546[l]
GDP (per capita current US$)	447	351[l]
Gross fixed capital formation (% of GDP)	11.0	3.0
Labour force participation, adult female pop. (%)	64.5[i]	63.0[m]
Labour force participation, adult male pop. (%)	78.8	78.6[i]
Employment in agricultural sector (%)	70.0[n]	
Agricultural production index (1999-2001=100)	107	76[l]
Food production index (1999-2001=100)	105	86[l]
Primary energy production (000s Mt oil equiv.)	3360	2853[l]
Motor vehicles (per 1,000 inhabitants)	48.6	54.2[l]
Telephone lines (per 100 inhabitants) [o]	2.2	2.8
Internet users, estimated (000s) [o]	50.0	1200.0

Total trade		Major trading partners			2005
	(million US$)[l]	(% of exports)[l]		(% of imports)[l]	
Exports	1926.1	South Africa	29	South Africa	53
Imports	2203.8	Switzerland	7	Botswana	4
		UK	7	UK	4

Social indicators	2000-2006
Population growth rate 2000-2005 (% per annum)	0.6
Population aged 0-14 years (%)	40.0
Population aged 60+ years (women and men, % of total)	6.0/5.0
Sex ratio (women per 100 men)	101
Life expectancy at birth 2000-2005 (women and men, years)	37/38
Infant mortality rate 2000-2005 (per 1,000 births)	62
Total fertility rate 2000-2005 (births per woman)	3.6
Urban population (%)	36
Urban population growth rate 2000-2005 (% per annum)	1.9
Rural population growth rate 2000-2005 (% per annum)	-<
Foreign born (%)	5.2[p]
Refugees and others of concern to UNHCR [q]	13968
Government education expenditure (% of GNP)	4.9
Primary-secondary gross enrolment ratio (w and m per 100)	67/70
Third-level students (women and men, % of total)	39/61
Television receivers (per 1,000 inhabitants) [r]	61
Parliamentary seats (women and men, % of total)	17/83

Environment	2000-2006
Threatened species	50
Forested area (% of land area)	49
CO$_2$ emissions (000s Mt of carbon dioxide/per capita)	11487/0.9
Energy consumption per capita (kilograms oil equiv.)	272
Precipitation (mm)	841[s]
Average minimum and maximum temperatures (centigrade) [s]	12.3/25.5

a The Zimbabwe dollar was redenominated by removing three zeros effective 2 Aug. 2006. b Official rate. c August 2006. d Annual average is calculated as geometric mean of monthly indices. e January 2006. f Calculated by the Statistics Division of the United Nations from component national indices. g 2003. h Persons aged 15 years and over. i 1999. j Arrivals of non-resident tourists at national borders (excluding same-day visitors). k Arrivals of non-resident visitors at national borders (including tourists and same-day visitors). l 2004. m 2002. n 1998. o Year beginning 30 June. p Estimated data. q Provisional. r Year ending 30 June. s Harare.

Technical notes

Geographical coverage

The geographical designations, units employed and presentation of the material in this publication has been adopted solely for the purpose of providing a convenient geographical basis for the statistical series.

Because of space limitations, the country and area names used in the tables are generally the commonly employed short titles in use in the United Nations, the full titles being used only when a short form is not available. Countries or areas are listed in English alphabetical order.

Notes on the indicators

Terms given below in italic are defined in the "Data dictionary", which begins on p. 227.

General indicators

Region is given according to regional groupings of countries and areas based mainly on continents. This information is from *Standard Country or Area Codes and Geographical Regions for Statistical Use*, Revision 4 (United Nations publication, excerpted at the Statistics Division Internet site, <http://unstats.un.org/unsd/methods/m49/m49.htm>).

Currency shows the national monetary unit and is from table 44 in the United Nations *Monthly Bulletin of Statistics*.[13]

Population projections for 2006 were prepared by the United Nations Population Division and published in *World Population Prospects: The 2004 Revision*. They are available also at < http://esa.un.org/unpp/>.[19]

Surface area (excluding polar regions and uninhabited islands) is from table 3 in the United Nations *Demographic Yearbook*.[10]

Population density refers to population per square kilometre of surface area. This series is from table 3 in the United Nations *Demographic Yearbook*.[10] and *World Population Prospects: The 2004 Revision*.

Largest urban agglomeration shows the population of the largest urban agglomeration (city plus contiguous built-up areas) or, if unavailable, largest city according to its administrative boundaries for each country or area. This series is from Tables A.12 and 13 of the *World Urbanization Prospects: The 2005 Revision.*[18]

< http:// esa.un.org/unup/>. See also <http://www.un.org/esa/population/publications/ WUP2005/2005wup.htm>.

United Nations membership date is from the United Nations *Terminology Bulletin, <http://unterm.un.org/ >* and List of Member States, <http://www.un.org/members/list.shtml>.

Economic indicators

Exchange rates are shown in units of national currency per US dollar and refer to end-of-period quotations. Unless otherwise stated, the table refers to the midpoint market rates (average of buying and selling rates). This series is compiled by the International Monetary Fund and is published as table 44 in the United Nations *Monthly Bulletin of Statistics.*[13] For currencies for which IMF does not publish exchange rates, non-commercial rates derived from the operational rates of exchange for United Nations programmes are shown.

Consumer price index numbers published in table 6 in the United Nations *Monthly Bulletin of Statistics* [13] are designed to show changes over time in the cost of selected goods and services that are considered as representative of the consumption habits of the population concerned. The indices here generally refer to "all items" and to the country as a whole.

The *industrial production index* shown here generally covers mining, manufacturing and electricity, gas and water. It does not include construction unless otherwise indicated. This series is from table 5 in the United Nations *Monthly Bulletin of Statistics.* [13]

Unemployment is defined to include persons above a certain age who during a specified period of time were without work, currently available for work and seeking work. National definitions of unemployment often differ from the recommended international standard definitions and thereby limit international comparability. Inter-country comparisons are also complicated by the different types of data collection systems used to obtain information on unemployed persons. Unless otherwise noted, these data are national employment office statistics compiled by the International Labour Office and published in table 11 in the United Nations *Monthly Bulletin of Statistics*.[13] Supplementary data were obtained from the International Labour Office, *Yearbook of Labour Statistics*. [5]. See also <http://laborsta.ilo.org/>

Balance of payments, current account: This series refers to the current account balance and is from the International Monetary Fund's *International Financial Statistics*.[6]

Tourist arrivals data are those compiled by the World Tourism Organization. They are published in the *Yearbook of Tourism Statistics* [25], and in the United Nations *Statistical Yearbook*.[16]

Gross domestic product total in current United States dollars are estimates of the total production of goods and services of the countries represented in economic terms, not as measures of the standard of living of their inhabitants. In order to have comparable coverage for as many countries as possible, these US dollar estimates are based on official GDP national currency data, supplemented by national currency estimates prepared by the Statistics Division using additional data from national and international sources.

The estimates given here are in most cases those accepted by the United Nations General Assembly Committee on Contributions for determining United Nations members' contributions to the United Nations regular budget.

The exchange rates for the conversion of GDP national currency data into United States dollars are the average market rates published by the International Monetary Fund in its monthly publication *International Financial Statistics*.[6] Official exchange rates are used only when free market rates are not available. For non-members of the Fund, the conversion rates used are the average of United Nations operational rates of exchange. It should be noted that the conversion from local currency into US dollars introduces deficiencies in comparability over time and among countries which should be considered when using the data. For example, comparability over time is distorted when exchange rate fluctuations differ substantially from domestic inflation rates. These series are published in the *National Accounts Statistics: Analysis of Main Aggregates* [15], and in the United Nations *Statistical Yearbook.* [16]

Gross domestic product per capita estimates are the value of all goods and services produced in the economy divided by population. These estimates are also published in the *National Accounts Statistics: Analysis of Main Aggregates* [15], and in the United Nations *Statistical Yearbook.* [16]

Gross fixed capital formation data are based on the percentage distribution of GDP in current prices. This series is from the National Accounts Database compiled from national data provided to the United Nations Statistics Division and is published in the United Nations *Statistical Yearbook.*[16] Data in national currency are published in *National Accounts Statistics: Main Aggregates and Detailed Tables.* [14]

Labour force participation rate for the adult population (15 years and over) refers to the total of employed persons (including employers, persons working on their own account, salaried employees and wage earners and, in so far as data are available, unpaid family workers) and of unemployed persons at the time of the census or survey which provided the data. In general, the economically active population does not include full-time students who are not working, persons occupied solely in household work, retired persons living entirely on their own means and persons wholly dependent upon others. These series are from the estimates and projections published in the *Key Indicators of the Labour Market* prepared by the International Labour Office.[4]

Employment in industrial and agricultural sectors refer to the population above a specified age who perform any work at all, in the reference period, for pay or profit in industry (mining, manufacturing and electricity, gas and water) and agriculture. These include persons who are temporarily absent from a job, for such reasons as illness, maternity or parental leave, holiday, training or industrial dispute. These percentages in which employment in the sector is the numerator and total employment the denominator are published in the *Key Indicators of the Labour Market* prepared by the International Labour Office. [4]

Agricultural production index covers all crops and livestock products. This series is from the Internet site of the Food and Agriculture Organization of the United Nations <apps.fao.org>. It is published in FAO *Yearbook: Production* [2], and in the United Nations *Statistical Yearbook*. [16]

Food production index covers commodities that are considered edible and contain nutrients. (Coffee and tea are therefore excluded because they have practically no nutritional value). The index numbers shown may differ from those produced by countries themselves because of differences in concepts of production, coverage, weights, time reference of data, and methods of evaluation. The series include estimates made by FAO in cases where no official or semi-official figures are available from the countries. This series is from the Internet site of the Food and Agriculture Organization of the United Nations <apps.fao.org>. It is published in the FAO *Yearbook: Production* [2], and in the United Nations *Statistical Yearbook*. [16]

Primary energy production refers to the first stage of production of various forms of energy, converted into a common unit (metric ton of oil equivalent). This series is from the Energy Statistics Database of the United Nations Statistics Division. It is published in the *Energy Statistics Yearbook* [11], and in the United Nations *Statistical Yearbook*. [16]

Motor vehicles in use series is calculated from data compiled from national statistical sources and is published in the United Nations *Statistical Yearbook*. [16] It refers to passenger cars and commercial vehicles in use according to census on registration figures for year's census or annual registration took place.

Telephones lines series is calculated from the number of main telephone lines in operation. The source of data is the International Telecommunications Union's publications, World Telecommunication Report [8], and *Yearbook of Statistics*. [9] It is published in the United Nations *Statistical Yearbook*. [16]

Internet users are mainly based on reported estimates, or derivations based on reported Internet access pro-

vider subscriber counts and in a few cases, calculated by multiplying the number of hosts by an estimated multiplier. This series is from the *International Telecommunication Union Yearbook of Statistics* [8]. It is also published in the United Nations
Statistical Yearbook. [16] See also
< http://www.itu.int/ITU-D/ict/statistics/>

Total Trade: exports and imports show the movement of
goods out of and into a country as shown in Tables
1 and 2 of the United Nations *International Trade
Statistics Yearbook.*[12]. Exports are generally valued at the frontier of the importing country (f.o.b.
valuation). Imports are valued at the frontier of the
importing country (c.i.f. valuation). Both imports
and exports are shown in United States dollars.
Conversion from national currencies is made by
means of currency conversion factors based on official exchange rates (par values or weighted average exchange rates.

Major export and import trading partners are expressed
as percentages of total exports and imports of the
country or area, as estimated by the United Nations
Statistics Division from its Commodity Trade Statistics Database (COMTRADE). These series are
published in the United Nations *International Trade
Statistics Yearbook.* [12]

Social indicators

The *population annual growth rate* is the average annual
percentage change in total population size in the
period 2000-2005. This series is from table A.11 in
World Population Prospects: The 2004 Revision. [19]

Population age group 0-14 years refers to the population
aged 0-14 years of both sexes as a percentage of
total population. Age group 60 years and over refers to elderly men as a percentage of all males
and elderly women as a percentage of all females.
These series are from the United Nations publication *World Population Prospects: The 2004 Revision.* [19}

Sex ratio is calculated from data prepared by the United
Nations Population Division and is published in
World Population Prospects: The 2004 Revision.[19]

Life expectancy at birth and infant mortality rate are five-
year averages for the period 2000-2005 and are

from tables A.26 and A.27 respectively in *World Population Prospects: The 2004 Revision.* [19]

Total fertility rate is a five-year average for the period 2000-2005 and is from table A.20 in the United Nations publication *World Population Prospects: The 2004 Revision.*[19] Supplementary data are from the United Nations publication *Demographic Yearbook.*[10]

Contraceptive use refers to use by currently married women of child-bearing age, of any method and is expressed as a percentage. The source of data is the contraceptive use database compiled by the United Nations Population Division and is published in the *World Contraceptive Use 2005*, CD-ROM Edition – Data in digital form (POP/DB/CP/Rev. 2005). See also < http://www.un.org/esa/population/publications/contraceptive2005/WCU2005.htm>.

Urban population, urban population growth rate and rural population growth rate series are based on the number of persons defined as urban or rural according to national definitions of this concept. In most cases these definitions are those used in the most recent population census. These series are from *World Urbanization Prospects: The 2005 Revision.*[18]

Foreign-born population refers to persons born outside the country or area in which they are enumerated. The country or area of birth is based on the national boundaries existing at the time of census. This series is from Demographic Statistics Database of the United Nations Statistics Division; and the databases on world migrant populations, and on the foreign-born maintained by the United Nations Population Division.

The term *refugee* in this series refers to persons granted a humanitarian status and/or those granted temporary protection. It includes persons, who have been granted temporary protection on a group basis. The series also includes returned refugees, asylum seekers and persons displaced internally within their own country for reasons that would make them of concern to the Office of the United Nations High Commissioner for Refugees (UNHCR) if they were outside their country of origin. This series is from *2005 Global Refugee Trends, Table 1. Asylum*

Seekers, refugees and others of concern to UNHCR.
[22] See also
<http://www.unhcr.org/statistics/STATISTICS/
4486ceb12.pdf>.

Government educational expenditures is from the World
Education Indicators database calculated from
Global Education Digest, UNESCO Institute for Statistics. It shows the general trends in public expenditure on public and private education expressed as
a percentage of the gross national product. The
data shown should be considered as approximate
indications of the public resources allocated to
education. [20] See also
<http://stats.uis.unesco.org/ReportFolders/ >

Primary-secondary gross enrolment ratio and percentage
of *third level students* are from World Education
Indicators database calculated from Global Education Digest, UNESCO Institute for Statistics. [20]
See also
<http://stats.uis.unesco.org/ >

For the first and second levels, the enrolment ratio generally is the total enrolment of all ages in first- and
second-level education, divided by the total population in the official ages of enrolment in the country
times 100. The gross enrolment ratio at the first
and second level should include all pupils whatever
their ages, whereas the population is limited to the
range of official school ages. Therefore, for countries with almost universal education among the
school-age population, the gross enrolment ratio
will exceed 100 if the actual age distribution of pupils extends beyond the official school ages.

Newspaper circulation data are compiled by UNESCO [20]
and are published in the United Nations *Statistical
Yearbook.* [16]

Television receivers in use refer to television receivers in
use and/or licenses issued per thousand inhabitants. This series is from the *World Telecommunication Indicators database*, International Telecommunication Union (ITU) [8].

Intentional homicides (homicide purposely inflicted) refers to death purposely inflicted by another person
per 100,000 population. Data are from the United
Nations *Demographic Yearbook* [10], where homicide and injury purposely inflicted by other persons
is reported as a cause of death.

Seats in parliament refers to the number of women and men in the lower chamber of parliament expressed as a percentage. These data are published in the *Women in National Parliaments* and the Internet site of the Inter-Parliamentary Union, <http://www.ipu.org/wmn-e/classif.htm> [7].

Environmental indicators

Data on the number of *threatened species* include plants and animals and are compiled by the World Conservation Union IUCN/Species Survival Commission (SSC), published in the IUCN Red List of Threatened Species.[23] See <http://www.redlist.org/info/tables/table5.html>.

Forested area data are from *State of the World's Forests*, published by the Food and Agriculture Organization of the United Nations [3] and are also published in the United Nations *Statistical Yearbook*.[16] See also the Internet site of the *State of the World's Forests* at <http://www.fao.org/docrep/003/y0900e/y0900e00.htm>

CO₂ emission estimates represent the mass of CO2 produced during the combustion of solid, liquid, and gaseous fuels, from gas flaring and the manufacture of cement. These estimates do not include bunker fuels used in international transportation due to the difficulty of apportioning these fuels among the countries benefiting from that transport. These estimates are from the Carbon Dioxide Information Analysis Center located
at Oak Ridge National Laboratory, United States of America [1]
<http://cdiac.esd.ornl.gov/trends/emis/tre_coun.htm>, and from the Secretariat of the *United Nations Framework Convention on Climate Change (UNFCCC), Greenhouse Gas and Inventory Database* [21], <http://ghg.unfccc.int/>. Relative to other industrial sources for which CO_2 emissions are calculated, statistics on gas flaring activities are sparse and sporadic and in countries where gas flaring activities account for a considerable proportion of the total CO_2 emission, the sporadic nature of gas flaring statistics may produce spurious or misleading trends in national CO_2 emissions. This series is also published in the United Nations *Statistical Yearbook*.[16]

CO2 Per capita emissions figures are obtained by dividing total emissions of carbon dioxide by the population for a particular country and year.

Commercial energy consumption refers to "apparent consumption" and is derived from the formula "production + imports - exports - bunkers +/- stock changes". Accordingly the series may in some cases represent only an indication of the magnitude of actual inland availability. This series was obtained from the Energy Statistics Database of the United Nations Statistics Division. It is published in the *Energy Statistics Yearbook* [11], and in the United Nations *Statistical Yearbook*.[16]

Total *amount of precipitation and average minimum and maximum temperatures* are measurements from the weather stations closest to the largest urban agglomeration or city. These series are from the World Meteorological Organization. [24],

Data dictionary

A

age group: The age distribution of a population is given either by individual years of age or by age groups, which may be quinquennial age groups or quinary age groups, or broad age groups, such as 0-19 years, 20-59 years, 60 years and over. Age is generally expressed in years, or years and months. Statisticians often round off the age to the number of complete years lived, and this is called age at last birthday. (United Nations, 1958, para. 322)*

agriculture (agriculture, hunting and related service activities): Comprises the following divisions of the International Standard Industrial Classification of All Economic Activities (ISIC), Revision 3.1: growing of crops, market gardening, horticulture; farming of animals; growing of crops combined with farming of animals (mixed farming); agricultural and animal husbandry service activities, except veterinary activities; hunting, trapping and game propagation, including related service activities; forestry, logging and related service activities; fishing, operation of fish hatcheries and fish farms; service activities incidental to fishing. (United Nations, 2002)

agricultural production and food production indices: The indices of agricultural production of the Food and Agriculture Organization of the United Nations (FAO) are based on the sum of price-weighted quantities of different agricultural commodities produced after deductions of quantities used as seed and feed weighted in a similar manner. All the indices at the country, regional and world levels are calculated by the Laspeyres formula. Production quantities of each commodity are weighted by average international commodity prices in the base period and summed for each year. To obtain the index, the aggregate for a given year is divided by the average aggregate for the base period. The commodities covered in the computation of indices of agricultural production are all crops and live-

---

*   References for the data dictionary are given in parenthesis; refer to the list of references beginning on page 239.

stock products originating in each country. Practically all products are covered, with the main exception of fodder crops. The category of food production includes commodities that are considered edible and that contain nutrients. Accordingly, coffee and tea are excluded because they have practically no nutritive value. (FAO, 1995, p. ix)

annual growth: See rate of change

area: See land and water area

B

balance of payments: A statistical statement that systematically summarizes, for a specific time period, the economic transactions of an economy with the rest of the world. Transactions, for the most part between residents and non-residents, consist of transactions involving goods, services and income; transactions involving financial claims on, and liabilities to, the rest of the world; and transactions (such as gifts) classified as transfers, which involve offsetting entries to balance—in an accounting sense— one-sided transactions. (IMF, 1993, para. 13) See also current account.

base period: The period of time for which data used as the base of an index number, or other ratio, have been collected. This period is frequently one year but it may be as short as one day or as long as the average of a group of years. (Kendall Buckland, 1982).

base year: See base period

C

CO_2 emissions: Carbon dioxide (CO_2) is a colourless, odourless and non-poisonous gas formed by combustion of carbon and in the respiration of living organisms and is considered a greenhouse gas. Emissions means the release of greenhouse gases and/or their precursors into the atmosphere over a specified area and period of time. (United Nations, 1992 and 1996)

commercial energy: Energy sold in the market. (United Nations, 1982, para. 55)

consumer price index: Measures changes over time in the general level of prices of goods and services that a reference population acquires, uses or pays for consumption. A consumer price index is estimated as a series of summary measures of the period-to-period proportional change in the prices of a fixed set of consumer goods and services of constant quantity and characteristics, acquired, used or paid for by the reference population. Each summary measure is constructed as a weighted average of a large number of elementary aggregate indices. Each of the elementary aggregate indices is esti-mated using a sample of prices for a defined set of goods and services obtained in, or by residents of, a specific region from a given set of outlets or other sources of consumption goods and services. (ILO, 1988)

contraception: In its narrow usage, measures excluding sterilization (and, in some discussions, permanent and periodic abstinence) which are taken in order to prevent sexual intercourse from resulting in con-ception. In broader usage, a contraceptive method is sometimes called a birth control method, which includes intentional abortion, sterilization and complete abstinence from coitus. (United Nations, 1958, para. 624)

currency: Those notes and coins in circulation that are commonly used to make payments. Commemorative coins that are not actually in circulation should be excluded. (United Nations and others, 1994, para. 11.70)

current account: All balance of payments transactions (other than those in financial items) that involve economic values and occur between resident and non-resident entities. Also covered are offsets to current economic values provided or acquired with-out a quid pro quo. The major classifications of transaction flows cover goods and services, income and current transfers. (IMF, 1993, para. 152)

D

daily newspaper circulation: Daily newspapers are peri-odic publications, issued at least four times a week, intended for the general public and mainly designed to be a primary source of written informa-tion on current events connected with public af-fairs, international questions, politics etc. Circula-tion comprises the average number of copies sold directly, by subscription, and mainly distributed

free of charge both in the country and abroad.
(UNESCO, 1985)

density of population: Number of population per unit of
total land area of a country. (United Nations, n.d.)
See also land and water area.

E

economically active population: ("usually active" or "currently active") comprises all persons of either sex
above a specified age who furnish the supply of labour for the production of economic goods and services (employed and unemployed, including those
seeking work for the first time), as defined by the
System of National Accounts (*SNA*), during a specified time reference period. The economically active
population may be related to the total population
for the derivation of the crude participation rate,
or, more appropriately, to the population above the
age prescribed for the measurement of the economically active population. Production includes all
individual or collective goods or services that are
supplied to units other than their producers, or intended to be so supplied, including the production
of goods or services used up in the process of pro-
ducing such goods or services; the own-account
production of all goods that are retained by their
producers for their own final consumption or gross
capital formation; the own-account production of
housing services by owner-occupiers and of domestic and personal services produced by employing
paid domestic staff. Not economically active population comprises the balance of the population.
(United Nations and others, 1994)

education expenditure: See government education expenditure

employment: The "employed" comprise all persons above
a specified age who during a specified brief reference period not longer than one week, were in
"paid employment" or in "self-employment" as defined below. "Persons in paid employment" comprise all persons in the following categories: (a) "at
work": persons who during the reference period
performed some work for wages, salary or related
payments, in cash or in kind; (b) "with a job but not
at work": persons who, having already worked in

their present job, were absent during the reference period and continued to have a strong attachment to their job. "Persons in self-employment" comprise all persons (a) "at work": persons who during the reference period performed some work for profit or family gain, in cash or in kind; (b) "with work but not at work": persons who during the reference period had work to be performed at the workplace but were temporarily absent due to illness or injury, vacation, holiday or ceremonies, bad weather or other similar reasons. Employers, own-account workers [other than those who were paid directly for services performed], members of producers' co-operatives, and unpaid family workers, irrespective of the number of hours worked, should be considered in self-employment and should be classified as "at work" or "not at work" as the case may be. (International Labour Organisation (ILO). Current International Recommendations on Labour Statistics, 2000 Edition. Geneva, 2000. (pages 49-51)

energy: Comprises primary energy from sources that involve only extraction or capture, with or without separation from contiguous material, cleaning or grading, before the energy embodied in that source can be converted into heat or mechanical work, and secondary energy from all sources of energy that results from transformation of primary sources. (United Nations, 1982, para. 29)

energy consumption: Apparent consumption of energy comprises inland deliveries of energy commodities, which is equal to imports plus production minus changes in stocks minus exports. (United Nations, 1982, paras. 161-165) See also energy.

enrolment ratio: See primary-secondary gross enrolment ratio.

exchange rate: Price in a given currency at which bills drawn in another currency may be bought. (Oxford University Press, 1982).

exports (merchandise): Goods leaving the statistical territory of a country. In the "general trade system", the definition of the statistical territory of a country coincides with its economic territory. In the "special trade system", the definition of the statistical territory comprises only a particular part of the economic territory, mainly that part which coincides with the free circulation area for goods. "The free circulation area" is a part of the economic territory of a country within which goods "may be disposed of without Customs restrictions". In the case of exports, the transaction value is the value at which the goods were sold by the exporter, including the cost of transportation and insurance, to bring the goods onto the transporting vehicle at the frontier of the exporting country (a FOB type valuation). (United Nations, M/52/Rev.2 and Series G)

F

food production index: See agricultural production and food production indices.

foreign-born: Individuals not born in the territory in which they live. (United Nations, 1958)

forest and other wooded land: Land under natural or planted stands of trees, whether productive or not, including land from which forest has been cleared but which will be reforested in the foreseeable future, and including areas occupied by roads, small cleared tracts and other small open areas within the forest that constitute an integral part of the forest. (FAO/United Nations ECE, 1995)

G

government education expenditure: General government expenditures for educational affairs and services at pre-primary, primary, secondary and tertiary levels and subsidiary services to education. Expenditures comprise final consumption expenditures, gross capital formation, subsidies and loans. General government comprises all central, state and local government units and non-profit institutions controlled and mainly financed by government units. (UNESCO, 1978)

gross domestic product (GDP): An aggregate measure of production equal to the sum of the gross values added of all resident institutional units engaged in production (plus any taxes, and minus any subsidies, on products not included in the value of their outputs). The sum of the final uses of goods and services (all uses except intermediate consumption) measured in purchasers' prices, less the value of imports of goods and services, or the sum of primary incomes distributed by resident producer units. (42, paras. 1.128 and 2.173-2.174) (United Nations and others, 1993)

gross fixed capital formation: The total value of a producer's acquisitions, less disposals, of fixed assets during the accounting period plus certain additions to the value of non-produced assets realized by the productive activity of institutional units. Fixed assets are tangible or intangible assets produced as outputs from processes of production that are themselves used repeatedly or continuously in other processes of production for more than one year. (United Nations and others, 1994, para. 10.33)

growth rate: See rate of change and rate of increase

H

homicide purposely inflicted (assault): Deaths from homicide and injuries inflicted by another person with intent to injure or kill, by any means, excluding injuries due to legal intervention and operations of war. (WHO, 1992, X85-Y09)

I

imports (merchandise): Goods which add to the stock of material resources of a country by entering its economic territory. Goods simply being transported through a country (goods in transit) or temporarily admitted (except for goods for inward processing) do not add to the stock of material resources of a country and are not included in the international merchandise trade statistics. In many cases, a country's economic territory largely coincides with its customs territory, which is the territory in which the customs law of a country applies in full. In the case of imports, the transaction value is the value at which the goods were purchased by the importer plus the cost of transportation and insurance to the frontier of the importing country (a CIF-type valuation). (United Nations, M/52/Rev.2 and Series G). See also exports.

industrial production index: Laspeyres's index of total value-added in all industrial production, where value added is the value of output less the values of both intermediate consumption and consumption of fixed capital. (United Nations, Series P) See also industry and Laspeyres's index.

infant mortality rate: Generally computed as the ratio of infant deaths (the deaths of children under one year of age) in a given year to the total number of live births in the same year. (United Nations, 1958, para. 411)

inhabitants: Inhabitants of a State may be subjects, citizens or nationals of that State, who enjoy certain political rights, or they may be aliens or foreigners who are citizens of another State, or citizens of no State at all and are called stateless. (United Nations, 1958) See also population.

L

land and water area: Total land area comprises agricultural land, forest and other wooded land, built-up and related land (excluding scattered farm buildings), wet open land, dry open land with special vegetation cover and open land without, or with insignificant, vegetation cover. Water area comprises inland waters and tidal waters. (United Nations, n.d.)

Laspeyres's index: A form of index number where prices, quantities or other units of measure over time are weighted according to their values in a specified base period. (Kendall and Buckland, 1982)

life expectancy at birth: Average number of years of life at birth (age 0) according to the expected mortality rates by age estimated for the reference year and population. (United Nations, 1958, and Series R)

long-term rate of change: See rate of change

M

motor vehicles: Motor cars and other motor vehicles in operation, principally designed for the transport of persons and goods. (United Nations, 1994, groups 781-783)

N

newspaper circulation: See daily newspaper circulation.

O

oil equivalent: A single average figure for the energy content of a specified quantity of oil. (United Nations, 1982)

P

parliament: Legislative assembly of persons forming the supreme legislature of a country. (Oxford University Press, 1982)

partner countries: Countries of origin and purchase in international merchandise trade transactions. (United Nations, 1994, para. 127)

population: The total population of a country may comprise either all usual residents of the country (de jure population) or all persons present in the country (de facto population) at the time of the census. For purposes of international comparisons, the de facto definition is recommended. (United Nations, 1958, and Series R)

population density: See density of population

precipitation: Quantity of rain, snow etc. falling to ground. Average annual normals over a long (multi-year) period. (Oxford University Press, 1982; WMO, 1982)

primary-secondary gross enrolment ratio: The total enrolment, regardless of age, divided by the population of the total age group defined in the national regulations for the first and second levels of education. Education at the first level provides the basic elements of education (e.g. at elementary school, primary school). Education at the second level is provided at middle school, secondary school, high school, teacher-training school at this level and schools of a vocational or technical nature. Enrolment is at the beginning of the school or academic year. (UNESCO, 1978)

R

rate of change: The ratio of total change in a specified time reference period to the value at the beginning of the period or at a specified earlier time reference. When changes over a period of more than one calendar year are studied, the mean annual rate of change may be computed. (Adapted from United Nations, 1958) See also rate of increase.

rate of increase (crude, of population): The ratio of total growth in a given period to the mean population of that period is called the crude rate of increase. When population increase over a period of more than one calendar year is studied, the mean annual rate of increase may be computed. (United Nations, 1958) See also rate of change.

refugees: Any person who, owing to well-founded fear of
being persecuted for reasons of race, religion, nationality, membership of particular social group or
political opinion, is outside the country of his nationality and is unable to, or owing to such fear, is
unwilling to avail himself of the protection of that
country; or who, not having a nationality and being
outside the country of his former habitual residence, is unable or, owing to such fear, is unwilling
to return to it. (United Nations, 1951, and 1967).

region (geographical): Macro geographical regions ar-
ranged according to continents and component
geographical regions used for statistical purposes
by the Population Division and Statistics Division of
the United Nations Secretariat. (United Nations,
M/49/Rev.3)

rural population: Population which is not urban. See urban population

S

sex ratio: The ratio of the number of one sex to that of
the other. (United Nations, 1958)

surface area: See land and water area.

T

telephone lines (telephone main lines): A telephone main
line connects the subscriber's terminal equipment
to the public switched network and has a dedicated
port in the telephone exchange equipment. (ITU,
n.d., p. 9)

television receivers: Apparatus for displaying pictures
transmitted by radio transmission, usually with appropriate sound. ISIC2 code 303201. (United Nations, 1968)

temperature, average: Average annual normals over a
long (multi-year) period. (WMO, 1982)

third-level students: Education provided at university,
teachers' college, higher professional school,
which requires, as a minimum condition of admission, the successful completion of education at the
second level, or evidence of the attainment of an
equivalent level of knowledge. Enrolment is at the
beginning of the school or academic year.
(UNESCO, 1978)

threatened species (animals): Species that have been
assessed and found to meet one of the standard

World Conservation Union status categories indicating threatened status: endangered, vulnerable, rare, indeterminate (known to be endangered, vulnerable or rare but where there is not enough information to say which is appropriate), insufficiently known (suspected but not definitely known to belong to any of the above categories, because of lack of information). (World Conservation Union, 1994, p. 20)

total fertility rate: The number of children that would be born per woman, assuming no female mortality at child bearing ages and the age-specific fertility rates of a specified country and reference period. (United Nations, 1958, para. 634)

tourist (international): Any person who travels to a country other than that in which s/he has his/her usual residence but outside his/her usual environment for a period not exceeding 12 months and whose main purpose of visit is other than the exercise of an activity remunerated from with the country visited, and who stay at least one night in a collective or private accommodation in the country visited. (United Nations and World Tourism Organization, 1994)

trading partner: See partner countries.

U

unemployment: All persons who during a specified reference period were: "without work", that is, were not in paid employment as specified by the international definition of employment; "currently available for work", that is, were available for paid employment or self-employment during the reference period; and "seeking work", that is, had taken specific steps in a specified recent period to seek paid employment or self-employment. In circumstances where employment opportunities are particularly limited and where persons not working do not have easy access to formal channels for seeking employment or face social and cultural barriers when looking for a job, the "seeking work" criterion should be relaxed. (ILO, 1988)

United Nations membership: The original Members of the United Nations are the States which, having participated in the United Nations Conference on International Organization at San Francisco, or having previously signed the Declaration by United Nations of 1 January 1942, signed the Charter and ratified it in accordance with Article 110. Membership in the United Nations is open to all other peace-loving States which accept the obligations contained in the Charter and, in the judgement of the Organization, are able and willing to carry out these obligations. The admission of any such State to membership in the United Nations will be effected by a decision of the General Assembly upon the recommendation of the Security Council. (United Nations, 1945)

urban agglomeration: Comprises a city or town proper and also the suburban fringe or thickly settled territory lying outside, but adjacent to, its boundaries. A single large urban agglomeration may comprise several cities or towns and their suburban fringes. (United Nations, 1998, para. 2.51)

urban population: Because of national differences in the characteristics that distinguish urban from rural areas, the distinction between urban and rural population is not amenable to a single definition that would be applicable to all countries. National definitions are most commonly based on size of locality. (United Nations, 1998).

V

vehicles: See motor vehicles

References

Food and Agriculture Organization of the United Nations (FAO, 1994). *Definition and Classification of Commodities* (Draft) (Rome). <http://www.fao.org/WAICENT/faoinfo/economic/faodef/FAODEFE.HTM>

Food and Agriculture Organization of the United Nations (FAO, 2003). *FAO Production Yearbook, 2003* (Rome).

Food and Agriculture Organization of the United Nations and United Nations, Economic Commission for Europe (FAO/ECE, 1995). Joint Working Party on Forest Economics and Statistics, as contained in *Forest Resources Assessment 1990: Global Synthesis*, FAO Forestry Paper 124 (Rome).

International Labour Organization (ILO, 1988). Current International Recommendations on Labour Statistics, 1988 Edition (Geneva).

International Monetary Fund (IMF, 1993). *Balance of Payments Manual*, Fifth Edition (Washington, D.C.).

International Telecommunication Union (ITU, n.d.). *Telecommunication Indicator Handbook*, Version 1.0 (Geneva).

Kendall, Sir Maurice G., and William R. Buckland for the International Statistical Institute (1982). *A Dictionary of Statistical Terms, Fourth Edition* (London, Longman Group).

Oxford University Press (1982). *The Concise Oxford Dictionary of Current English*, Seventh Edition (London).

United Nations (n.d.). *Readings in International Environment Statistics, ECE Standard Statistical Classification of Land Use*. Economic Commission for Europe, Conference of European Statisticians (United Nations publication).

United Nations (1945). Charter of the United Nations and Statute of the International Court of Justice.

United Nations (1951 and 1967). Convention relating to the Status of Refugees of 1951 (United Nations, *Treaty Series*, vol. 189 (1954), No. 2545, p. 137), art. 1) and Protocol relating to the Status of Refugees of 1967 (United Nations, *Treaty Series*, vol. 606 (1967), No. 8791, p. 267.

United Nations (1958). *Multilingual Demographic Dictionary, English Section.* Department of Economic and Social Affairs, Population Studies, No. 29 (United Nations publication, Sales No. E.58.XIII.4).

United Nations (1982). *Concepts and Methods in Energy Statistics, with Special Reference to Energy Accounts and Balances: A Technical Report.* Statistical Office, Series F, No. 29 and Corr. 1 (United Nations publication, Sales No. E.82.XVII.13 and corrigendum).

United Nations (1992). United Nations Framework Convention on Climate Change (A/AC.237/18 (Part II)/Add.1 and Corr.1). Opened for signature at Rio de Janeiro on 4 June 1992.

United Nations (1994). *Commodity Indexes for the Standard International Trade Classification, Revision 3.* Statistical Division, Series M, No. 38, Rev. 2 (United Nations publication, Sales No. E.94.XVII.10).

United Nations (1996). *Glossary of Environment Statistics.* Statistics Division, Series F, No. 67 (United Nations publication, Sales No. E.96.XVII.12).

United Nations (1998). *Principles and Recommendations for Population and Housing Censuses Rev. 1.* Statistics Division, Series M, No. 67, Rev. 1 (United Nations publication, Sales No. E.98.XVII.8).

United Nations (M/49/Rev.4). *Standard Country or Area Codes for Statistical Use* (ST/ESA/STAT/SER.M/49/Rev.4).

United Nations (M/52/Rev.2). International Merchandise Trade Statistics: Concepts and Definitions (ST/ESA/STAT/SER.M/52/Rev.2).

United Nations (Series G). *International Trade Statistics Yearbook.* Statistics Division, Series G (United Nations publication, annual).

United Nations (Series P). *Industrial Commodity Statistics Yearbook.* Statistics Division, Series P (United Nations sales publication, annual).

United Nations, Commission of the European Communities, International Monetary Fund, Organisation for Economic Cooperation and Development and World Bank (United Nations and others, 1993). *System of National Accounts 1993* (SNA 1993) (United Nations publication Sales No. E.94.XVII.4).

United Nations and World Tourism Organization (1994). *Recommendations on Tourism Statistics.* Series M,

No. 83 (United Nations publication, Sales No. E.94.XVII.6).

United Nations Educational, Scientific and Cultural Organization (UNESCO, 1978). *Revised Recommendation concerning the International Standardization of Educational Statistics* (Paris), as contained in *UNESCO Statistical Yearbook*, chap. 2 (Paris, annual). United Nations Educational, Scientific and Cultural Organization (UNESCO, 1985). Revised Recommendation concerning the International Standardization of Statistics on the Production and Distribution of Books, Newspapers and Periodicals (Paris), as contained in *UNESCO Statistical Yearbook*, chap. 7 (Paris, annual).

World Conservation Union, 1994. "Threatened species categories", as contained in World Conservation Monitoring Centre, *Biodiversity Data Sourcebook*, WCMC Biodiversity Series No. 1 (Cambridge, World Conservation Press).

World Health Organization (WHO, 1992). *International Statistical Classification of Diseases and Related Health Problems*, Tenth Revision (ICD-10), vol. 1 (Geneva).

World Meteorological Organization (WMO, 1982). Climatological Normals (CLINO) for Climate and Climate Ship Stations for the Period 1931-1960, 1971 edition, WMO No. 117 (Geneva).

Statistical sources

[1]   Carbon Dioxide Information Analysis Center, Global, Regional, and National CO_2, Emission Estimates from Fossil Fuel Burning, Cement Production, and Gas Flaring: 1751-1996 (Oak Ridge, Tennessee, USA). See also <http://cdiac.esd.ornl.gov/ftp/ndp030/nation1751_2002.ems>

[2]   Food and Agriculture Organization of the United Nations, *FAO Yearbook: Production,* annual (FAO, Rome). See also <apps.fao.org>.

[3]   _____, *State of the World's Forests* (FAO, Rome). See also <http://www.fao.org/docrep/007/y5574e/y5574e00.htm>

[4]   International Labour Office, *Key Indicators of the Labour Market* (Geneva). See also all ILO statistical databases at: http://www.ilo.org/public/english/support/lib/dblist.htm#statistics

[5]   _____, Yearbook of Labour Statistics (Geneva).

[6]   International Monetary Fund, *International Financial Statistics* (monthly and annual) (Washington, DC).

[7]   Inter-Parliamentary Union, *Women in National Parliaments.* See also <www.ipu.org/wmn-e/classif.htm>.

[8]   International Telecommunication Union (ITU), Geneva, the ITU database. See also < http://www.itu.int/ITU-D/ict/statistics/>

[9]   _____, Yearbook of Statistics (Geneva). See also <http://www.itu.int/publications/docs/bdt/stat03.html>

[10] United Nations, *Demographic Yearbook* (Series R, United Nations publication, annual).

[11] _____, *Energy Statistics Yearbook* (Series J, United Nations publication, annual).

[12] _____, *International Trade Statistics Yearbook* (Series G, United Nations publication, annual).

[13] _____, *Monthly Bulletin of Statistics* (Series Q, United Nations publication, monthly).

[14] _____, *National Accounts Statistics: Main Aggregates and Detailed Tables*, (Series X, United Nations publication, annual).

[15] _____, National Accounts Statistics: Analysis of Main Aggregates (Series X, United Nations publication, annual).

[16] _____, *Statistical Yearbook* (Series S, United Nations publication, annual).

[17] _____, *Terminology Bulletin No. 347*, Rev. 1 (United Nations publication, 1997) and corrigendum 1. See also <http://www.un.org/Overview/unmember.html>

[18] _____, *World Urbanization Prospects: The 2005 Revision*, (ESA/P/WP.190, United Nations publication, biennial). See also < http://www.un.org/esa.un.org/unup/>

[19] _____, *World Population Prospects: The 2004 Revision . Volume I Comprehensive Tables* (ST/ESA/SER. A/222, United Nations publication, biennial). See also < http://esa.un.org/unpp/>

[20] United Nations Educational, Scientific and Cultural Organization (UNESCO) Institute for Statistics, Montreal, the UNESCO statistics database. See also <http://stats.uis.unesco.org/ >

[21] United Nations Framework Convention on Climate Change (UNFCCC), *Greenhouse Gas Inventory Database* <http://ghg.unfccc.int>

[22] United Nations High Commissioner for Refugees, *2004 UNHCR Population Statistics,* Table 1. Asylum-seekers, refugees and others of concern to UNHCR, 2004. See also < http://www.unhcr.ch/cgi-bin/texis/vtx/statistics >

[23] The World Conservation Union, *2004 IUCN Red List of Threatened Species*. See also < http://www.redlist.org/info/stats >.

[24] World Meteorological Organization, *Climatological Normals (CLINO) for the Period 1961-1990*, (Geneva). See also

[25] World Tourism Organization (WTO), Madrid, WTO statistics database and the "Yearbook of Tourism Statistics". See also < http://www.world-tourism.org/facts/menu.html>

Current United Nations statistical publications

For a complete catalogue of publications prepared by the UN Statistics Division, contact the Division at:

United Nations DC2-1620
New York NY 10017 USA
facsimile: +1 (212) 963-4569
e-mail:statistics@un.org
http://unstats.un.org/unsd

To order publications:

UN Publications DC2-853
New York NY 10017 USA
tel. +1 (212)963-3489
https://unp.un.org/